ADVANCES IN CHEMICAL PHYSICS

VOLUME 127

EDITORIAL BOARD

Advances in
CHEMICAL PHYSICS

Edited by

I. PRIGOGINE

Center for Studies in Statistical Mechanics and Complex Systems
The University of Texas
Austin, Texas
and
International Solvay Institutes
Université Libre de Bruxelles
Brussels, Belgium

and

STUART A. RICE

Department of Chemistry
and
The James Franck Institute
The University of Chicago
Chicago, Illinois

VOLUME 127

AN INTERSCIENCE PUBLICATION
JOHN WILEY & SONS, INC.

Published by John Wiley & Sons, Inc., Hoboken, New Jersey.
Published simultaneously in Canada.

Library of Congress Catalog Number: 58-9935

ISBN 0-471-23583-0

Printed in the United States of America

10 9 8 7 6 5 4 3 2 1

CONTRIBUTORS TO VOLUME 127

Alastair D. Bruce, Department of Physics and Astronomy, The University of Edinburgh, Mayfield Road, Edinburgh EH9 3JZ United Kingdom

Vladimir I. Gaiduk, Institute of Electronics and Radio Engineering, Russian Academy of Sciences, Fryazino 141190, Moscow Region, Russia

Boris M. Tseitlin, Institute of Electronics and Radio Engineering, Russian Academy of Sciences, Fryazino 141190, Moscow Region, Russia

Nigel B. Wilding, Department of Physics, University of Bath, Bath BA2 7AY, United Kingdom

INTRODUCTION

Few of us can any longer keep up with the flood of scientific literature, even in specialized subfields. Any attempt to do more and be broadly educated with respect to a large domain of science has the appearance of tilting at windmills. Yet the synthesis of ideas drawn from different subjects into new, powerful, general concepts is as valuable as ever, and the desire to remain educated persists in all scientists. This series, *Advances in Chemical Physics*, is devoted to helping the reader obtain general information about a wide variety of topics in chemical physics, a field that we interpret very broadly. Our intent is to have experts present comprehensive analyses of subjects of interest and to encourage the expression of individual points of view. We hope that this approach to the presentation of an overview of a subject will both stimulate new research and serve as a personalized learning text for beginners in a field.

I. Prigogine
Stuart A. Rice

CONTENTS

Computational Strategies for Mapping Equilibrium Phase Diagrams — 1
By Alastair D. Bruce and Nigel B. Wilding

Molecular Models for Calculation of Dielectric/far-Infrared Spectra of Liquid Water — 65
By Vladimir I. Gaiduk and Boris M. Tseitlin

Author Index — 333

Subject Index — 339

COMPUTATIONAL STRATEGIES FOR MAPPING EQUILIBRIUM PHASE DIAGRAMS

ALASTAIR D. BRUCE

Department of Physics and Astronomy, The University of Edinburgh, Edinburgh, EH9 3JZ, United Kingdom

NIGEL B. WILDING

Department of Physics, University of Bath, Bath, BA2 7AY, United Kingdom

CONTENTS

I. Introduction: Defining Our Problem
II. Basic Equipment
  A. Formulation: Statistical Mechanics and Thermodynamics
  B. Tools: Elements of Monte Carlo
III. Paths
  A. Meaning and Specification
  B. Generic Routes
  C. Generic Sampling Strategies
    1. Serial Sampling
    2. Parallel Sampling
    3. Extended Sampling
IV. Path-Based Techniques: A Guided Tour
  A. Keeping It Simple: Numerical Integration and Reference States
    1. Strategy
    2. Critique
  B. Parallel Tempering; Around the Critical Point
    1. Strategy
    2. Critique
  C. Extended Sampling: Traversing the Barrier
    1. Strategy
    2. Critique

Advances in Chemical Physics, Volume 127, Edited by I. Prigogine and Stuart A. Rice
ISBN 0-471-23583-0

D. From Paths to Wormholes: Phase Switch
1. Strategy
2. Examples
3. Critique
V. Doing It Without Paths: Alternative Strategies
A. The Gibbs Ensemble Monte Carlo Method
1. Strategy
2. Critique
B. The NPT and Test Particle Method
1. Strategy
2. Critique
C. Beyond Equilibrium Sampling: Fast Growth Methods
1. Strategy
2. Critique
VI. Determining the Phase Boundary: Extrapolation, Tracking, and the Thermodynamic Limit
A. Extrapolation to the Phase Boundary
B. Tracking the Phase Boundary
C. Finite-Size Effects
1. Noncritical Systems
2. Near-Critical Systems
VII. Dealing with Imperfection
A. Polydispersity
B. Crystalline Defects
VIII. Outlook
Appendix A: Building Extended Sampling Distributions
Appendix B: Histogram Reweighting
References

I. INTRODUCTION: DEFINING OUR PROBLEM

The phenomenon of *phase behavior* is generic to science, so we should begin by defining it in a general way: It is the organization of many-body systems into forms that reflect the interplay between constraints imposed macroscopically (through the prevailing external conditions) and microscopically (through the interactions between the elementary constituents). In addition to its most familiar manifestations in condensed matter science, the phenomenon (and its attendant challenges) features in areas as diverse as gauge theories of subnuclear structure, the folding of proteins, and the self-organization of neural networks.

We will of course be rather more focused here. We shall be concerned with the *generic computational strategies* needed to address the problems of phase behavior. The physical context we shall explore will not extend beyond the *structural* organization of the elementary phases (liquid, vapor, crystalline) of matter, although the *strategies* are much more widely applicable than this. We shall have nothing to say about a wide spectrum of techniques (density functional theory [1], integral equation theories [2], anharmonic perturbation theory [3], virial expansions [4]) which have done much to advance our understanding of subclasses (solid–liquid, solid–solid, or liquid–gas) of phase behavior, but which are less than "generic." For the most part we shall also largely restrict ourselves to systems comprising simple classical particles in defect-free structures, interacting through a prescribed potential function.

There is one other rather more significant respect in which our objectives are limited; we need to identify it now. In its fullest sense the "problem" of phase behavior entails questions of *kinetics*: *how* phase transformations occur. Within a computational framework, such questions are naturally addressed with the techniques of Molecular Dynamics (MD), which numerically integrate the equations of motion associated with the many-body potential and thereby attempt to replicate, authentically, the phase-transformation *process* itself. There is a long, rich, and distinguished history of activity of this kind, providing insights into many (perhaps the most) challenging and interesting issues associated with phase behavior. But the "authenticity" carries a price: Like their laboratory counterparts, such computer experiments display phenomena (hysteresis and metastability) associated with the long time scales required for the reorganization processes. These phenomena are useful qualitative signatures of a phase transformation and worthy of study in their own right. But they tend to obscure the *intrinsically sharp* characteristics of the transformation (Fig. 1).

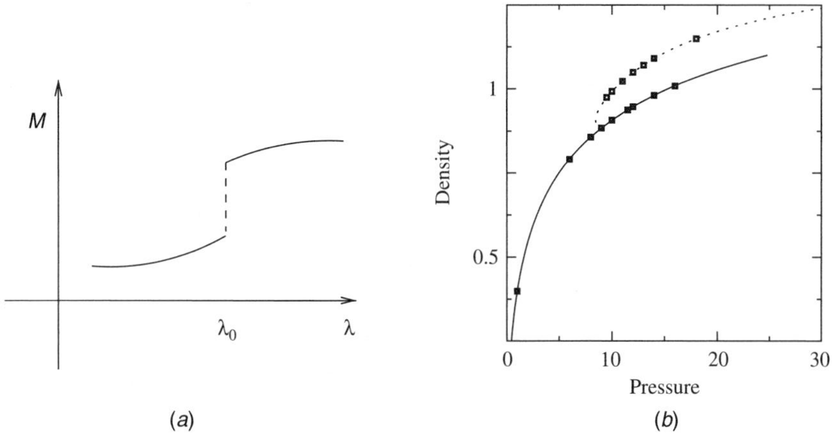

Figure 1. (a) Schematic representation of the results of an "ideal experiment" on phase behavior, which may take as much time as we need. The equilibrium value of some physical quantity M changes discontinuously at some sharply defined value, λ_0, of some field λ. (b) An example of the experimental reality. The results of a typical simulation study of solid–liquid phase behavior of hard spheres; the measured density continues to follow the branch (liquid or solid) on which the simulation is initiated, well beyond the coexistence pressure ($\simeq 11.3$ in these units) [5].

If our interest is restricted to those sharp characteristics—in particular *where* a phase transformation occurs in an "ideal experiment"—we may formulate (and in principle solve) the problem entirely within the framework of the equilibrium statistical mechanics of the competing phases (the framework which implies that "sharpness").

This is the stance we adopt here: We restrict our attention to the task of mapping *equilibrium phase boundaries*. This choice leads naturally to another. Freed of the need to capture the authentic dynamics, one finds that the natural computational tool becomes Monte Carlo (MC) rather than MD. The strategic advantage of this choice is the range of ways (not all of which have yet been dreamt of) in which MC algorithms may be engineered to move around configuration space. Indeed, understanding the distinctive features of that "space" which are implied by the *occurence* of the phase behavior, and engineering an algorithm (effectively a pseudo-dynamics) to match are the key themes of recent activities, which we shall highlight in this work.

We shall begin (Section II) by assembling the basic equipment. Section II.A formulates the problem in the complementary languages of thermodynamics and statistical mechanics. The shift in perspective—from "free energies" in the former to "probabilities" in the latter—helps to show what the *core problem* of phase behavior really is: a comparison of the *a priori* probabilities of two regions of configuration space. Section II.B outlines the standard portfolio of MC tools and explains why they are not equal to the challenge posed by this core problem.

Section III identifies and explores what proves to be *the* key concept here: the notion of a *path through configuration space*. Most approaches to the core problem utilize a "path" which, in some sense, "connects" the two regions of interest. They can be classified, helpfully we think, on the basis of the choices made in regard to that path: (a) the *route* it follows through configuration space and (b) the way in which one contrives to *sample* the configurations along that route. There are a limited number of generically distinct path-routing options; we summarize them in Section III.B. In Section III.C we identify the generically distinct path-sampling strategies. There are actually fewer than one might suppose; this is an area in which many wheels have been reinvented. Some of the new wheels at least run more quickly than their precursors; the advances here (the refinement of "extended sampling" techniques) account for much of the recent activity in this area. They are essential for would-be practitioners but not for the logic of the story. Accordingly, they are addressed separately, in an Appendix.

The relatively few generic choices in regard to routing and sampling have been deployed in combinations too numerous to mention, let alone discuss. We have, instead, elected to survey a small number of what we shall call *path-based techniques* (combinations of routing and sampling strategies) in some detail.

This survey (Section IV) extends from the standard technique of long-standing, numerical integration to reference macrostates, to one only recently introduced, in which the core problem is solved by a MC leap directly between the phases of interest.

We have chosen to organize our strategic survey around the concept of a path. There are some strategies which do not naturally fit into this framework, but which must certainly be included here; we discuss them in Section V.

Solving the core problem at some state point is the major part of the task of determining the phase boundary, but not quite all. To complete the job, one has to *track* a phase boundary (once one has found a point on it) and one has to *extrapolate* from the limited sizes of system that simulations can handle to the thermodynamic limit of interest. Section VI reviews the complementary tools needed.

We have chosen for the most part to focus on relatively idealized model systems; in Section VII we consider some of the issues associated with departures from the ideal.

Finally, Section VIII offers some thoughts on how we are likely to make further progress.

There are three further caveats about what the reader can hope to find in what follows. It is not easy (perhaps not possible) to be exhaustive and clear; we aim to be clear. One can adopt an organizational structure founded on either the history or the logic of the ideas; we have chosen the latter. We have surely devoted disproportionately much space to our own contributions; this is mainly because we can explain them best.

II. BASIC EQUIPMENT

A. Formulation: Statistical Mechanics and Thermodynamics

We will formulate the problem simply but generally. Consider a system of structureless, classical particles, characterized macroscopically by a set of thermodynamic coordinates (such as the temperature T) and microscopically by a set of model parameters that prescribe their interactions. The two sets of parameters play a strategically similar role; it is therefore convenient to denote them, collectively, by a single label, c (for "conditions" or "constraints" or "control parameters" in thermodynamic-and-model space).

We are fundamentally concerned with the phase behavior rooted in the *spatial* organization of the particles and reflected in the statistical behavior of their position coordinates $\vec{r}_i, i = 1, N$. The components of these coordinates are the principal members of a set of *generalized coordinates* $\{q\}$ locating the system in its *configuration space*. In some instances (dealing with fluid phases) it is advantageous to work with ensembles in which particle number N or system volume V is free to fluctuate; the coordinate set $\{q\}$ is then extended accordingly (to include N or V), and the control parameters c are extended to include the corresponding fields (chemical potential μ or pressure P).

The statistical behavior of interest is encapsulated in the equilibrium probability density function $P_0(\{q\}|c)$. This PDF is determined by an appropriate ensemble-dependent, dimensionless [6] configurational energy $\mathscr{E}(\{q\}, c)$. The relationship takes the generic form

$$P_0(\{q\}|c) = \frac{1}{Z(c)} e^{-\mathscr{E}(\{q\},c)} \tag{1}$$

where the normalizing prefactor (the partition function) is defined by

$$Z(c) \equiv \int \prod_i dq_i e^{-\mathscr{E}(\{q\},c)} \tag{2}$$

Some small print should appear here; we shall come back to it. The different phases that the system displays will in general be distinguished by the values of some macroscopic property loosely described as an *order parameter.* Thus, for example, the density serves to distinguish a liquid from a vapor; a structure factor distinguishes a liquid from a crystalline solid. A suitable order parameter, M, allows us to associate with each phase, α, a corresponding portion $\{q\}_\alpha$ of $\{q\}$-space. We write that statement concisely in the form

$$\{q\} \in \{q\}_\alpha \quad \text{iff} \quad M(\{q\}) \in [M]_\alpha \tag{3}$$

where $[M]_\alpha$ is the set of order parameter values consistent with phase α. The partitioning of $\{q\}$-space into distinct regions is the key feature of the core problem.

The equilibrium properties of a particular phase α follow from the conditional counterpart of Eq. (1):

$$P_0(\{q\}|\alpha, c) = \begin{cases} \frac{1}{Z_\alpha(c)} e^{-\mathscr{E}(\{q\},c)}, & \{q\} \in \{q\}_\alpha \\ 0, & \text{otherwise} \end{cases} \tag{4}$$

with

$$Z_\alpha(c) \equiv \int \prod_i dq_i \Delta_\alpha[M(\{q\})] e^{-\mathscr{E}(\{q\},c)} \equiv e^{-\mathscr{F}_\alpha(c)} \tag{5}$$

The last equation defines $\mathscr{F}_\alpha(c)$, the *free energy* of phase α, while

$$\Delta_\alpha[M] \equiv \begin{cases} 1, & M \in [M]_\alpha \\ 0, & \text{otherwise} \end{cases} \tag{6}$$

so that the integral is effectively confined to the set of configurations $\{q\}_\alpha$ associated with phase α.

Since the notation does not always make it apparent, we should note that α and c play operationally similar roles as *macrostate labels*: Together they identify the distinct sets of equilibrium macroscopic properties emerging from the equilibrium distribution, Eq. (1). For some purposes we will find it useful to concatenate the two labels into a single "grand" macrostate label

$$\alpha, c \rightarrow \mathscr{C} \tag{7}$$

For the time being we shall continue to display the two labels separately.

Now referring back to Eq. (1) we may write

$$Z_\alpha(c) = \int \prod_i dq_i \Delta_\alpha[M] e^{-\mathscr{E}(\{q\},c)} = Z(c) \int \prod_i dq_i \Delta_\alpha[M] P_0(\{q\}|c)$$

The *a priori* probability of phase α may thus be related to its free energy by

$$P_0(\alpha|c) \equiv \int \prod_i dq_i \Delta_\alpha[M] P_0(\{q\}|c) = \frac{Z_\alpha(c)}{Z(c)} = \frac{e^{-\mathscr{F}_\alpha(c)}}{Z(c)} \tag{8}$$

Alternatively,

$$P_0(\alpha|c) \equiv \int \prod_i dq_i \Delta_\alpha[M] P_0(\{q\}|c) = \int dM \Delta_\alpha[M] P_0(M|c) \tag{9}$$

where $P_0(M|c)$ is the equilibrium distribution of the chosen order parameter. Though seemingly nothing but a tautology, this representation proves remarkably fruitful.

For two phases, α and $\tilde{\alpha}$ say, it then follows that

$$\Delta\mathscr{F}_{\alpha\tilde{\alpha}}(c) \equiv \mathscr{F}_\alpha(c) - \mathscr{F}_{\tilde{\alpha}}(c) = \ln \frac{Z_{\tilde{\alpha}}(c)}{Z_\alpha(c)} = \ln \frac{P_0(\tilde{\alpha}|c)}{P_0(\alpha|c)} = \ln \frac{\int dM \Delta_{\tilde{\alpha}}[M] P_0(M|c)}{\int dM \Delta_\alpha[M] P_0(M|c)} \tag{10}$$

This is a key equation in several respects: It is conceptually helpful; it is cautionary; and it is suggestive, strategically.

At a conceptual level, Eq. (10) provides a helpful link between the languages of thermodynamics and statistical mechanics. According to the familiar mantra of thermodynamics, the favored phase will be that of *minimal free energy*; from a statistical mechanics perspective the favored phase is the one of *maximal probability*, *given* the probability partitioning implied by Eq. (1).

We also then see (the "cautionary" bit) that the thermodynamic mantra presupposes the validity of Eq. (1), and therefore that of its "small print," which we must now spell out. In general, Eq. (1) presupposes ergodicity on the space $\{q\}$. The framework can thus be trusted to tell us what we will "see" for some given c (the "favored phase") only to the extent that appropriate kinetic pathways exist to allow sampling (ultimately, *comparison*) of the distinct regions of configuration space associated with the different phases. In the context of laboratory experiments on real systems the relevant pathways typically entail the nucleation and growth of droplets of one phase embedded in the other; the associated time scales are long; and Eq (10) will be relevant only if the measurements extend over correspondingly long times. The fact that they frequently do not is signaled in the phenomena of metastability and hysteresis, which we have already touched on.

Finally, Eq. (10) helps to shape strategic thinking on how to broach the problem computationally. It reminds us that what is relevant here is the *difference* between free energies of two competing phases and that this free energy difference is a *ratio* of the *a priori* probabilities of the two phases. It implies that the phase boundary may be identified as the locus of points of equal *a priori* probability of the two phases, and that such points are in principle identifiable through the condition that the order parameter distribution will have equal integrated weights (areas) in the two phases. The discussion of the preceding section also suggests that the pathways by which our simulated system passes between the two regions of configuration space associated with the two phases will play a strategically crucial role. While, for the reasons just discussed, the details of those pathways are essential to the *physical applicability* of Eq. (10), they are *irrelevant* to the *values* of the quantities it defines; we are thus free to engineer whatever pathways we may wish.

These considerations lead one naturally to the Monte Carlo toolkit.

B. Tools: Elements of Monte Carlo

The Monte Carlo method probably ranks as the most versatile theoretical tool available for the exploration of many-body systems. It has been the subject of both general pedagogical texts [7] and applications-focused reviews [8]. Here we provide only its elements—enough to understand why, if implemented in its most familiar form, it does not deliver what we need, and to hint at the extended framework needed to make it do so.

The MC method generates a sequence (Markov chain) of configurations in $\{q\}$-space. The procedure can be constructed to ensure that, in the "long-enough-term," configurations will appear in that chain with *any* probability density, $P_{\mathcal{S}}(\{q\})$ (the $\mathcal{S}$ stands for "sampling"), we care to nominate. The key requirement (it is not strictly necessary [9]; and—as we shall see—it is not always sufficient) is that the transitions, from one configuration $\{q\}$ to another $\{q'\}$, should respect the detailed balance condition

$$P_{\mathscr{S}}(\{q\})P_{\mathscr{S}}(\{q\} \rightarrow \{q'\}) = P_{\mathscr{S}}(\{q'\})P_{\mathscr{S}}(\{q'\} \rightarrow \{q\}) \tag{11}$$

where $P_{\mathscr{S}}(\{q\} \rightarrow \{q'\})$ is the transition probability, the probability density of configuration $\{q'\}$ at Markov chain step $t+1$ given configuration $\{q\}$ at time t. [We have added a subscript to emphasize that its form is circumscribed by the choice of sampling density, through Eq. (11).] MC transitions satisfying this constraint are realized in a two-stage process. In the first stage, one generates a trial configuration $\{q'\} = \mathscr{T}\{q\}$, where $\mathscr{T}$ is some generally stochastic selection procedure; the probability density of a trial configuration $\{q'\}$ given $\{q\}$ is of the form

$$P_{\mathscr{T}}(\{q'\} \mid \{q\}) = \langle \delta(\{q'\} - \mathscr{T}\{q\}) \rangle_{\mathscr{T}} \tag{12}$$

where $\langle \cdot \rangle_{\mathscr{T}}$ represents an average with respect to the stochastic variables implicit in the procedure $\mathscr{T}$. In the second stage the "trial" configuration is accepted (the system "moves" from $\{q\}$ to $\{q'\}$ in configuration space) with probability $P_{\mathscr{A}}$ and is otherwise rejected (so the system "stays" at $\{q\}$); the form of the acceptance probability is prescribed by our choices for $P_{\mathscr{S}}$ and $P_{\mathscr{T}}$ since

$$P_{\mathscr{S}}(\{q\} \rightarrow \{q'\}) = P_{\mathscr{T}}(\{q'\} \mid \{q\})P_{\mathscr{A}}(\{q\} \rightarrow \{q'\})$$

It is then easy to verify that the detailed balance condition [Eq. (11)] is satisfied, if the acceptance probability is chosen as

$$P_{\mathscr{A}}(\{q\} \rightarrow \{q'\}) = \min\left\{1, \frac{P_{\mathscr{S}}(\{q'\})P_{\mathscr{T}}(\{q\} \mid \{q'\})}{P_{\mathscr{S}}(\{q\})P_{\mathscr{T}}(\{q'\} \mid \{q\})}\right\} \tag{13}$$

Suppose that, in this way, we build a Markov chain comprising a total of t_T steps; we set aside the first t_E configurations visited; we denote by $\{q\}^{(t)}(t = 1, \ldots, t_U)$ the configurations associated with the subsequent $t_U \equiv t_T - t_E$ steps. The promise on the MC package is that the expectation value $\langle Q \rangle_{\mathscr{S}}$ of some observable $Q = Q(\{q\})$ *defined* by

$$\langle Q \rangle_{\mathscr{S}} = \int \prod_i dq_i P_{\mathscr{S}}(\{q\})Q(\{q\}) \tag{14}$$

may be *estimated* by the sample average

$$\langle Q \rangle_{\mathscr{S}} \stackrel{eb}{=} \frac{1}{t_U} \sum_{t=1}^{t_U} Q(\{q\}^{(t)}) \tag{15}$$

Now we must consider the choices of $P_{\mathscr{S}}$ and $P_{\mathscr{T}}$. Tailoring those choices to whatever task one has in hand provides potentially limitless opportunity for ingenuity. But at this point we consider only the simplest possibilities. The sampling distribution $P_{\mathscr{S}}(\{q\})$ is chosen to be the appropriate *equilibrium* distribution $P_0(\{q\} \mid c)$ [Eq. (1)] so that the configurations visited are representative of a "real" system, even though their sequence is not an authentic representation of the "real" dynamics. We shall refer to this form of sampling distribution as *canonical*. The trial-coordinate selection procedure $\mathscr{T}$ is chosen to comprise some *small* change of *one* coordinate; the change is chosen to be small enough to guarantee a reasonable acceptance probability [Eq. (13)], but no smaller, or the Markov chain will wander unnecessarily slowly through the configuration space. We shall refer to this form of selection procedure as *local*. For such schemes (and sometimes for others) the selection probability density typically has the symmetry

$$P_{\mathscr{T}}(\{q\} \to \{q'\}) = P_{\mathscr{T}}(\{q'\} \to \{q\}) \tag{16}$$

With these choices, Eq. (13) becomes

$$P_{\mathscr{A}}(\{q\} \to \{q'\}) = \mathscr{A}(\Delta\mathscr{E}) \tag{17}$$

where

$$\Delta\mathscr{E} \equiv \mathscr{E}(\{q'\}, c) - \mathscr{E}(\{q\}, c) \tag{18}$$

and

$$\mathscr{A}(x) \equiv \min\{1, \exp[-x]\} \tag{19}$$

defines the Metropolis acceptance function [10].

These choices are not only the simplest, they are also the most frequent: The local–canonical strategy is the staple Monte Carlo method and has contributed enormously to our knowledge of many-body systems.

From what we have said, it would seem that this staple strategy would also deliver what we require here. If, as promised, the Markov chain visits configurations with the canonical probability [Eq. (1)], we should merely have to determine

$$P_0(\alpha|c) \equiv \langle \Delta_\alpha \rangle_0 \tag{20}$$

from its estimator [Eq. (15)], effectively the proportion of time the system is found in region $\{q\}_\alpha$. Equation (10) would then take care of the rest. But the local–canonical strategy fails us here. The Markov chain typically does not extend beyond the particular region of configuration space $\{q\}_\alpha$ in which it is initiated and the distribution of the (any) "order parameter" will capture only the contributions associated with that phase. The observations thus provide no basis for assigning a value to the relative probabilities of the two phases, and thus of estimating the location of the phase boundary.

This failure is a reflection of *both* of our "simple" choices. First, the *local* character of coordinate updating yields a dynamics that (though scarcely authentic) shares the essential problematic feature of the kinetic pathways supported by "real" dynamics: Evolution from one phase to another will typically require a traverse through intermediate regions in which the configurations have a two-phase character. Second, the choice of a *canonical* sampling distribution ensures that the probability of such intermediate configurations is extremely small; interphase traverses occur only on correspondingly long time scales. At this point, one remembers the small print ("in the long term") that accompanies the MC toolkit. As a result, the effective sampling distribution is not the nominal choice, $P(\{q\}|c)$ [Eq. (1)], but rather $P(\{q\}|\alpha, c)$ [Eq. (4)], with the phase α determined by our choice of initial configuration. Since the sampling distribution appears only in the acceptance probability [Eq. (13)], and then only as a *ratio*, the algorithm does not distinguish between $P(\{q\}|c)$ and $P(\{q\}|\alpha, c)$—as long as it is trapped in $\{q\}_\alpha$.

We conclude that the local–canonical MC strategy cannot directly deliver the simultaneous comparison between two phases which [Eq. (10) suggests] provides the most efficient resolution of the phase-boundary problem: In most circumstances, this strategy will simply explore a *single* phase. We must now ask whether we can get by with two *separate* (but still local–canonical) "single-phase" simulations, each determining the free-energy (or, equivalently, partition function) of one phase [Eq. (5)].

Let us first be clear about the circumstances in which a "single-phase" simulation makes sense. The brief and loose answer is: when the time (Markov chain length) $t_{e\alpha}$ typical of escape from phase α is long compared to the time $t_{s\alpha}$ required for effective sampling of the configuration space of that phase. More fully, and a little more formally: when there exists some $t_{s\alpha} < t_{e\alpha}$ such that the configuration set $\{q\}_\alpha$ defined by Eq. (3) is effectively equivalent to that defined by the condition

$$\{q\} \in \{q\}_\alpha \quad \text{iff} \quad \{q\} \text{ is reachable from } \{q\}_\alpha^0 \text{ within time } t_{s\alpha} \tag{21}$$

In this formulation, the configurations in $\{q\}_\alpha$ are identified as those that may be reached in a simulation of length $t_{s\alpha}$, initiated from some configuration $\{q\}_\alpha^0$ that is associated with phase α but is otherwise arbitrary. The equivalence of Eqs. (3) and (21) is assumed (usually tacitly) in all single-phase simulations.

Assuming these conditions are fulfilled, our single-phase local–canonical algorithm will allow us to estimate the single-phase canonical expectation value of any "observable" Q defined on the configurations $\{q\}$

$$\langle Q \rangle_{0,\alpha} = \int \prod_i dq_i P_0(\{q\}|\alpha, c) Q(\{q\}) \tag{22}$$

from the average [Eq. (15)] over a sample of the canonically distributed configurations. But the single-phase partition function Z_α *is not* [11] an "average over canonically distributed configurations"; rather, it measures the *total effective weight* of the configurations that contribute significantly to such averages. One can no more deduce it from a sample of single phase properties than one can infer the size of an electorate from a sample of their opinions. Thus local–canonical MC fails to deliver the absolute single-phase free energy also.

To determine the relative stability of two phases under conditions c thus requires a MC framework that, in some sense, does *more* than sample the equilibrium configurations appropriate to the (two) c-macrostates. We have seen where there is room for maneuver—in the choices we make in regard to $P_{\mathcal{S}}$ and $P_{\mathcal{T}}$. The possibilities inherent in the latter are intuitively obvious: It is better to find ways of bounding or leaping through configuration space than to be limited to the shuffle of local updating. The fact that we have flexibility in regard to the choice of sampling distribution is perhaps less obvious, so it is worth recording the simple result which shows us that we do.

Let $P_{\mathcal{S}}$ and $P_{\mathcal{S}'}$ be two arbitrary distributions of the coordinates $\{q\}$. Then the expectation values of some arbitrary observable Q with respect to the two distributions are formally related by the identity

$$\langle Q \rangle_{S'} = \int \prod_i dq_i P_{\mathcal{S}'}(\{q\}) Q(\{q\}) = \int \prod_i dq_i P_{\mathcal{S}}(\{q\}) Q(\{q\}) \frac{P_{\mathcal{S}'}(\{q\})}{P_{\mathcal{S}}(\{q\})} = \left\langle \frac{P_{\mathcal{S}'}}{P_{\mathcal{S}}} Q \right\rangle_S \tag{23}$$

Thus, in particular, we can—in principle—determine *canonical* expectation values from an ensemble defined by an *arbitrary* sampling distribution through the relationship

$$\langle Q \rangle_0 = \left\langle \frac{P_0}{P_{\mathcal{S}}} Q \right\rangle_S \tag{24}$$

We do not *have* to make the choice $P_{\mathcal{S}} = P_0$. The issue of what sampling distribution will be *optimal* was addressed in the earliest days of computer simulation [12]. The answer depends on the observable Q. In general the "obvious" choice $P_{\mathcal{S}} = P_0$, though not strictly optimal, is adequate. But for some observables the choice of a canonical sampling distribution is *so* "suboptimal" as to be useless [13]. The core problem we face here has a habit of presenting us with such quantities; we have already seen one example [Eq. (20)] and we shall see others.

III. PATHS

There are many ways of motivating, constructing, and describing the kind of MC sampling strategy we need; the core idea we shall appeal to here to structure our discussion is that of a *path*.

A. Meaning and Specification

For our purposes a *path* comprises a sequence of contiguous macrostates, $\mathscr{C}_1, \mathscr{C}_2 ... \mathscr{C}_\Omega \equiv \{\mathscr{C}\}$ [14]. By "contiguous" we mean that each adjacent pair in the sequence ($\mathscr{C}_j, \mathscr{C}_{j+1}$ say) has some configurations in common (or that a configuration of one lies arbitrarily close to a configuration of the other). A path thus comprises a quasi-continuous band through configuration space.

The physical quantities that distinguish the macrostates from one another will fall into one of two categories, which we shall loosely refer to as *fields* and *macrovariables*. In the former category we include thermodynamic fields, model parameters, and, indeed, the conjugate [15] of any "macrovariable." By "macrovariable" [16] we mean any collective property, aggregating contributions from all or large numbers of the constituent particles, free to fluctuate in the chosen ensemble, but in general sharply prescribed, in accordance with the Central Limit Theorem. Note that we do not restrict ourselves to quantities that appear on the map of thermodynamics, nor to the parameter space of the physical system itself: With simulation tools at our disposal, there are limitless varieties of parameters to vary and properties to observe.

B. Generic Routes

It may be evident (it should certainly not be surprising) that the extended MC framework needed to solve the phase-equilibrium problem entails exploration of a path that *links* the macrostates of the two competing phases, for the desired physical conditions $\mathscr{C}$. The generic choices here are distinguished by the way in which the path is *routed* in relation to the key landmark in the configuration space, namely, the two-phase region which separates the macrostates of the two phases and which confers on them their (at least meta-) stability. Figure 2 depicts four conceptually different possibilities.

First (Fig. 2a) the route may comprise two distinct sections, neither of which encroaches into the two-phase region and each of which terminates in a *reference* macrostate. By *reference* macrostate we mean one whose partition function (and thus free energy) is already known—on the basis of exact

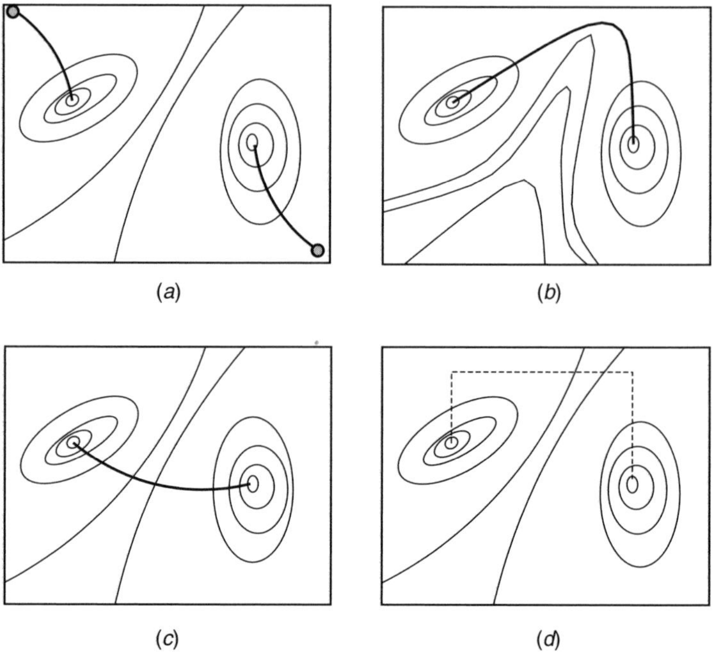

Figure 2. Schematic representation of the four conceptually different paths (the heavy lines) one may utilize to attack the phase-coexistence problem. Each figure depicts a configuration space spanned by two macroscopic properties (such as energy, density . . .); the contours link macrostates of equal probability, for some given conditions c (such as temperature, pressure . . .). The two mountaintops locate the equilibrium macrostates associated with the two competing phases, under these conditions. They are separated by a probability ravine (free-energy barrier). In case (a) the path comprises two disjoint sections confined to each of the two phases and terminating in appropriate reference macrostates. In (b) the path skirts the ravine. In (c) it passes through the ravine. In (d) it leaps the ravine.

calculation or previous measurement. The information accessible through MC study of the two sections of such a path has to be combined with the established properties of the reference macrostates to provide the desired link between the two equilibrium macrostates of interest. This is the traditional strategy for addressing the phase-coexistence problem.

In the case of a liquid–gas phase boundary, it is possible to choose a path (Fig. 2b) that links the macrostates of the two phases *without* passing through the two-phase region: One simply has to sail around the critical point at which the phase boundary terminates. The dependence on the existence of an adjacent critical point limits the applicability of this strategy.

The next option (it seems to be the only one left after Fig. 2a and Fig. 2b; but see Fig. 2d) is a route that negotiates the probability ravine (free-energy barrier) presented by the two-phase region (Fig. 2c). The extended sampling tools we discuss in the next section are essential here; with their refinement in recent years, this route has become increasingly attractive.

If either of the phases involved is a *solid*, there are *additional* reasons (we shall discuss them later) to avoid the ravine, over and above the low canonical probability of the macrostates that lie there. It is possible to do so. The necessary strategy (recently developed) is depicted in Fig. 2d. As in Fig. 2a, the path comprises two segments, each of which lies within a single-phase region. But in contrast to Fig. 2a the special macrostates to which these segments lead are not of the traditional reference form: The defining characteristic of *these* macrostates is that they should act as the ends of a "wormhole" along which the system may be transported by a collective Monte Carlo move, from one phase to the other. In this case, extended sampling methods are used to locate the wormhole ends.

C. Generic Sampling Strategies

The task of exploring (sampling) the macrostates that form a path can be accomplished in a number of conceptually different ways; we identify them here in a rudimentary way. We defer to subsequent sections the discussion of specific examples that will show how the information they provide is used to give the result we seek, along with their strengths and weaknesses.

1. *Serial Sampling*

The most obvious way of gathering information about the set of macrostates $\{\mathscr{C}\}$ is to use canonical Boltzmann sampling to explore them one at a time, with a sampling distribution set to

$$P_{\mathscr{S}}(\{q\}) = P_0(\{q\}|\mathscr{C}_j), \qquad j = 1, \ldots, \Omega \tag{25}$$

in turn. One must then combine the information generated in these Ω independent simulations. The traditional approach to our problem ("integration methods," Section IV.A) employs this strategy; its basic merit is that it is simple.

2. *Parallel Sampling*

Instead of exploring the path macrostates serially, we may choose to explore them in parallel. In this framework we simulate a set of Ω replicas of the physical system, with the jth member chosen to realize the conditions $\mathscr{C}_j$. The sampling distribution in the composite configuration space spanned by

$\{q\}^{(1)} \ldots \{q\}^{(\Omega)}$ is

$$P_{\mathscr{S}}(\{q\}^{(1)} \ldots \{q\}^{(\Omega)}) = \prod_{j=1}^{\Omega} P_0(\{q\}^{(j)} | \mathscr{C}_j) \tag{26}$$

This is not simply a way of exploiting the availability of parallel computing architectures to treat Ω tasks at the same time; more significantly it provides a way of breaking out of the straightjacket of local-update algorithms. The composite ensemble can be updated through interchanges of the coordinate sets associated with adjacent macrostates (j and $j+1$ say) to give updated coordinates

$$\{q'\}^{(j)} = \{q\}^{(j+1)} \quad \text{and} \quad \{q'\}^{(j+1)} = \{q\}^{(j)} \tag{27}$$

The chosen sampling distribution [Eq. (26)] is realized if such configuration exchanges are accepted with the appropriate probability [Eqs. (13) and (17)] reflecting the change incurred in the total energy of the composite system. This change is nominally "macroscopic" (scales with the size of the system), and so the acceptance probability will remain workably large only if the parameters of adjacent macrostates are chosen sufficiently close to one another. In practice this means utilizing a number Ω of macrostates that is proportional to $\sqrt{N}$ [17].

Interchanging the configurations of adjacent replicas is one instance (see Section IV.D for another) of a *global update* in which all the coordinates evolve simultaneously. The payoff, potentially, is a more rapid evolution around coordinate space—more formally a stronger mixing of the Markov chain. This is of course a general desideratum of any MC framework. Thus it is not surprising that algorithms of this kind have been independently devised in a wide variety of disciplines, applied in a correspondingly wide set of contexts and given a whole set of different names [17]. These include Metropolis Coupled Chain [18], Exchange Monte Carlo [19], and Parallel Tempering [20]. In Section IV.B we shall see how one variant has been applied to deal with the phase-coexistence problem in systems with a critical point.

3. *Extended Sampling*

We will use the term *extended sampling* (ES) to refer to an algorithm that allows exploration of a region of configuration space which is "extended" with respect to the range spanned by canonical Boltzmann sampling—specifically, one that assembles the statistical properties of our set of macrostates $\mathscr{C}_1 \ldots \mathscr{C}_\Omega$ within a *single* simulation. Again it is straightforward to write down the generic form of a sampling distribution that will achieve this end; we need only a "superposition" of the canonical sampling distributions for the set of macrostates,

$$P_{\mathscr{S}}(\{q\}) = W_0 \sum_{j=1}^{\Omega} P_0(\{q\} | \mathscr{C}_j) \tag{28}$$

where W_0 is a normalization constant. The superficial similarity between this form and those prescribed in Eqs. (25) and (26) camouflages a crucial difference. Each of the distributions $P_0(\{q\}|\mathscr{C}_j)$ involves a normalization constant identifiable as [Eqs. (1) and (5)]

$$w_j = Z(\mathscr{C}_j)^{-1} \tag{29}$$

We do not need to know the set of normalization constants $\{w\}$ to implement *serial* sampling [Eq. (25)] since each plays only the role of a normalization constant for a *sampling* distribution, which the MC framework does not require. Nor do we need these constants in implementing *parallel* sampling [Eq. (26)] since in this case they feature (through their product) only in the one overall normalization constant for the sampling distribution. But Eq. (28) is different. Writing it out more explicitly,

$$P_{\mathscr{S}}(\{q\}) = W_0 \sum_{j=1}^{\Omega} w_j e^{-\mathscr{E}(\{q\},\mathscr{C}_j)} \tag{30}$$

we see that the weights $\{w\}$ control the relative contributions that the macrostates make to the sampling distribution. While we are, in principle, at liberty to choose whatever mixture we please [we do not *have* to make the assignment prescribed by Eq. (29)], it should be clear intuitively (we shall develop this point in Section IV.C.1) that the choice should confer *roughly* equal probabilities on each macrostate, so that all are well-sampled. It is not hard to see that the weight-assignment made in Eq. (29) is in fact what we need to fulfill this requirement. Evidently, to make *extended* sampling work, we *do* need to "know" the weights $w_j = Z(\mathscr{C}_j)^{-1}$. There is an element of circularity here which needs to be recognized. Our prime objective is to determine the (relative) configurational weights of *two* macrostates (those associated with two different phases, under the same physical conditions [21]); to do so (somehow or other—we haven't yet said how) by extended sampling requires knowledge of the configurational weights of a *whole path's-worth* of macrostates. There *is* progress here nevertheless. While the two macrostates of interest are remote from one another, the path (by construction) comprises macrostates that are contiguous; it is relatively easy to determine the relative weights of pairs of contiguous macrostates, and therefore the relative weights of all in the set. In effect the extended sampling framework allows us to replace one hard problem with a large number of somewhat easier problems.

The machinery needed to turn this general strategy into a practical method ("building the extended sampling distribution" or "determining the macrostate weights") has evolved over the years from a process of trial and error to algorithms that are systematic and to some extent self-monitoring. The workings of the machinery is more interesting than it might sound; we will discuss some aspects of what is involved in Sections IV.C and IV.D. But we relegate more technical discussion (focused on recent advances) to Appendix A. Here we continue with a broader brush.

It would be hard to write a definitive account of the development of extended sampling methods; we will not attempt to do so. The seminal ideas are probably correctly attributed to Torrie and Valleau [22], who coined the terminology *umbrella sampling*. The huge literature of subsequent advances and rediscoveries may be rationalized a little by dividing it into two, according to how the macrostates to be weighted are defined.

If the macrostates are defined by a set of values [λ] of some generalized "field" λ, the sampling distribution is of the form

$$P_{\mathcal{S}}(\{q\}) = W_0 \sum_{j=1}^{\Omega} w_j e^{-\mathcal{E}(\{q\},\lambda_j)} \tag{31}$$

Extended sampling strategies utilizing this kind of representation appear in the literature with a variety of titles: *expanded ensemble* [23], *simulated tempering* [24], and *temperature scaling* [25].

On the other hand, if the macrostates are defined on some "macrovariable" M, the sampling distribution is of the form

$$P_{\mathcal{S}}(\{q\}) = W_0 \sum_{j=1}^{\Omega} w_j e^{-\mathcal{E}(\{q\})} \Delta_j[M] \tag{32}$$

where

$$\Delta_j[M] \equiv \begin{cases} 1, & M \in \text{range associated with } \mathcal{C}_j \\ 0, & \text{otherwise} \end{cases} \tag{33}$$

Realizations of this formalism go under the names *adaptive umbrella sampling* [26] and the *multicanonical ensemble* introduced by Berg and Neuhaus [27]. It seems right to attribute the recent revival in interest in extended sampling to the latter work.

In Sections IV.C and IV.D we shall see that extended sampling strategies provide a rich variety of ways of tackling the phase coexistence problem, including the distinctive problems arising when one (or both) of the phases is of solid form.

IV. PATH-BASED TECHNIQUES: A GUIDED TOUR

We now proceed to explore how the strategic options in regard to routing and sampling of paths (Sections III.B and III.C) can be fused together to form practical techniques for addressing the phase-coexistence problem. We do not set out to be exhaustive here: There are very many pairings of routes and sampling strategies (the choices are to some extent mutually independent, which is why we discussed them separately). We focus instead on a few key cases; we explain what information is gathered by the chosen sampling of the chosen path and how it is used to yield the desired free-energy comparison; and we assess the strengths and weaknesses of each technique as we go.

A. Keeping It Simple: Numerical Integration and Reference States

1. Strategy

The staple approach to the phase-coexistence problem involves serial sampling (Section III.C) along a path of the type depicted in Fig. 2a. Effectively the single-core problem (comparing the configurational weights—free energies—of the two physical macrostates) is split into two, each requiring comparison of a physical macrostate (c, α) with some suitable reference macrostate, $\mathscr{C}_{\alpha}^{\mathrm{REF}}$. A reference macrostate will be "suitable" if one can identify a path parameterized by some field λ linking it to the physical macrostate, with $\lambda = \lambda_1$ and $\lambda = \lambda_\Omega$ denoting respectively the physical and the reference macrostates. The sampling distribution at some arbitrary point, λ, on this path is then of the form [Eqs. (4), (5), and (25)]

$$P_{\mathscr{S}}(\{q\}) = P_0(\{q\}|\lambda) = \frac{1}{Z_\alpha(\lambda)} e^{-\mathscr{E}(\{q\},\lambda)} \tag{34}$$

with

$$Z_\alpha(\lambda) \equiv \int \prod_i dq_i \Delta_\alpha[M(\{q\})] e^{-\mathscr{E}(\{q\},\lambda)} \equiv e^{-\mathscr{F}_\alpha(\lambda)} \tag{35}$$

We have seen that free energies like $\mathscr{F}_\alpha(\lambda)$ are not themselves naturally expressible as canonical averages, but their *derivatives* with respect to field-parameters *are* expressible this way. Specifically,

$$\begin{aligned}
\frac{\partial \mathscr{F}_\alpha(\lambda)}{\partial \lambda} &= -\frac{1}{Z_\alpha(\lambda)} \frac{\partial Z_\alpha(\lambda)}{\partial \lambda} \\
&= \frac{1}{Z_\alpha(\lambda)} \int \prod_i dq_i \Delta_\alpha[M(\{q\})] \frac{\partial \mathscr{E}(\{q\}, \lambda)}{\partial \lambda} e^{-\mathscr{E}(\{q\},\lambda)} \\
&= \left\langle \frac{\partial \mathscr{E}(\{q\}, \lambda)}{\partial \lambda} \right\rangle_{\alpha,\lambda}
\end{aligned} \tag{36}$$

where the average is to be taken with respect to the canonical distribution for macrostate α, λ [Eq. (34)]. The free-energy difference between the physical and reference macrostates is thus formally given by

$$\mathcal{F}(\mathcal{C}_{\alpha}^{\mathrm{REF}}) - \mathcal{F}_{\alpha}(c) = \int_{\lambda_1}^{\lambda_\Omega} d\lambda \frac{\partial \mathcal{F}_{\alpha}(\lambda)}{\partial \lambda} = \int_{\lambda_1}^{\lambda_\Omega} d\lambda \left\langle \frac{\partial \mathcal{E}(\{q\}, \lambda)}{\partial \lambda} \right\rangle_{\alpha,\lambda} \tag{37}$$

A sequence of independent simulations, conducted at a set of points $[\lambda]$ spanning a path from λ_1 to λ_Ω, then allows one to estimate the free-energy difference by numerical quadrature:

$$\mathcal{F}(\mathcal{C}_{\alpha}^{\mathrm{REF}}) - \mathcal{F}_{\alpha}(c) \stackrel{eb}{=} \sum_{j=1}^{\Omega} \left\langle \frac{\partial \mathcal{E}(\{q\}, \lambda)}{\partial \lambda_j} \right\rangle_{\alpha,\lambda_j} \Delta\lambda \tag{38}$$

This takes us halfway. The entire procedure has to be repeated for the second phase $\tilde{\alpha}$, integrating along some path parameterized (in general) by some *other* field $\tilde{\lambda}$ running between the macrostate $\tilde{\alpha}, c$ and some *other* reference state $\mathcal{C}_{\tilde{\alpha}}^{\mathrm{REF}}$. Finally the results of the two procedures are combined to give the quantity of interest [Eq. (10)]

$$\begin{aligned}\Delta\mathcal{F}_{\alpha\tilde{\alpha}}(c) = \mathcal{F}(\mathcal{C}_{\alpha}^{\mathrm{REF}}) - \mathcal{F}(\mathcal{C}_{\tilde{\alpha}}^{\mathrm{REF}}) - \Delta\lambda \sum_{j=1}^{\Omega} \left\langle \frac{\partial \mathcal{E}(\{q\}, \lambda)}{\partial \lambda_j} \right\rangle_{\alpha,\lambda_j} \\ + \Delta\tilde{\lambda} \sum_{j=1}^{\tilde{\Omega}} \left\langle \frac{\partial \mathcal{E}(\{q\}, \tilde{\lambda})}{\partial \tilde{\lambda}_j} \right\rangle_{\tilde{\alpha},\tilde{\lambda}_j}\end{aligned} \tag{39}$$

We shall refer to this strategy as *numerical integration to reference macrostates* (NIRM). It is helpful to assess its strengths and weaknesses armed with an explicit example.

We shall consider what is arguably the archetypal example of the NIRM strategy: the Einstein Solid Method (ESM) [28]. The ESM provides a simple way of computing the free energies of crystalline phases and therefore addresses questions of the relative stability of competing crystalline structures. We describe its implementation for the simplest case where the interparticle interaction is of hard-sphere form; it is readily extended to deal with particles interacting through soft potentials [29].

The name of the method reflects the choice of the reference macrostate: a crystalline solid comprising particles which do not interact with one another, but which are bound by harmonic springs to the sites of a crystalline lattice, $\{\vec{R}\}_{\alpha}$, coinciding with that of the phase of interest.

The relevant path is constructed from sampling distributions of the general form prescribed in Eq. (34) with

$$\mathscr{E}(\{q\}, \lambda) = \mathscr{E}(\{\vec{r}\}, c) + \lambda \sum_{i=1}^{N} \left(\vec{r}_i - \vec{R}_i^{\alpha}\right)^2 \tag{40}$$

The first term on the right-hand side contains the hard-sphere interactions. The second term embodies the harmonic spring energy, which reflects the displacements of the particles from their lattice sites. In this case the point $\lambda_1 = 0$ locates the physical macrostate. With increasing λ, the energy cost associated with a given set of displacements increases, with a concomitant reduction in the size of the typical displacements. On further increasing λ, one ultimately reaches a point λ_Ω beyond which particles are so tightly bound to their lattice sites that they practically never collide with one another; the hard-sphere interaction term then plays no role, thus realizing the desired reference macrostate, whose free-energy may be computed exactly.

Some of the results of ESM studies of crystalline phases of hard spheres are shown in Table I. We shall discuss them below.

2. *Critique*

There is much to commend NIRM: It is conceptually simple; it can be implemented with only a modest extension of the simulation framework already needed for standard MC sampling; and it is versatile. It has been applied successfully in free-energy measurements of (*inter alia*) crystalline solids [28],

TABLE I
The Difference in the *Entropy* per Particle of the *fcc* and *hcp* Crystalline Phases of Hard Spheres[a]

ρ/ρ_{cp}	N	$\Delta s(10^{-5} \times k_B)$		Method	Reference
0.736	12,000	230	(100)	NIRM	36
0.736	12,096	87	(20)	NIRM	35
0.7778	216	132	(4)	ESPS	34
0.7778	1,728	112	(4)	ESPS	34
0.7778	1,728	113	(4)	NIRM	34
0.7778	216	133	(3)	ESPS	48
0.7778	1,728	113	(3)	ESPS	48
0.7778	5,832	110	(3)	ESPS	48
1.00	12,000	260	(100)	NIRM	36
1.0	216	131	(3)	ESPS	48
1.0	1,728	125	(3)	ESPS	48

[a]The associated uncertainties are in parentheses. For comparison we note that the excess entropy per particle at $\rho/\rho_{cp} = 0.7778, N = 1152$ is $-6.53\ldots$, with the phase-dependence showing in the fourth significant figure [28].

liquids [30], and liquid crystals [31]. It is probably regarded as *the* standard method for attacking the phase-coexistence problem, and it provides the benchmark against which other approaches must be assessed. Nevertheless (as one might guess from the persistence of attempts to develop "other approaches"), it is less than ideal in a number of respects. We discuss them in turn.

First, the NIRM method hinges on the *identification of a good path and reference macrostate.* A "good" path is short; but the reference macrostate (the choice of which is limited) may lie far from the physical macrostate of interest, entailing a large number of independent simulations to make the necessary link. In a sense the ESM provides a case in point: The reference macrostate is strictly located at $\lambda_\Omega = \infty$; corrections for the use of a finite λ_Ω need to be made [28]. This kind of problem is a nuisance, but no worse. A potentially more serious constraint on the path is that the derivative being measured should vary slowly, smoothly, and reversibly along it; if it does not, the numerical quadrature may be compromised. A phase transition en route (whether in the "real" space of the physical system or the extended-model space into which NIRM simulations frequently extend) is thus a particular hazard. The realization of NIRM known as the Single Occupancy Cell Method (SOCM) [32] provides an example where such concerns arise, and seem not to have been wholly dispelled [33].

The *choice of simulation parameters* also raises issues. Evidently one has to decide how many simulations are to be performed along the path and at which values of λ. In so doing, one must strike a suitable balance between minimizing computation time while still ensuring that no region of the path (particularly one in which the integrand varies strongly) is neglected. This may necessitate a degree of trial and error.

The *uncertainties* to be attached to NIRM estimates are problematic in a number of respects. Use of simple numerical quadrature [estimating the integral in Eq. (37) by the sum in Eq. (38)] will result in errors. One can reduce such errors by interpolation into the regions of λ between the chosen simulation values (for example, using histogram reweighting techniques discussed in Appendix B), but there will still be systematic errors associated both with the interpolation and with finite sampling times. No reliable and comprehensive prescription for estimating the magnitude of such errors has yet been developed.

These problems are exacerbated when one addresses the quantity of real interest—the *difference* between the free energies of two competing phases. The fact that NIRM treats this as *two* problems rather than *one* [Eq. (39)] is problematic in two respects.

First this aspect of the NIRM strategy *compounds* a problem inherent in all simulation studies of this problem (or, indeed, any other in many-body physics): finite-size effects. This issue deserves a section devoted to itself, and it gets one (Section VI.C). Here we note simply that such effects are harder to assess when one has to synthesize calculations on different phases, utilizing different reference states and different system sizes—and sometimes conducted by different authors.

Second, since the entire enterprise is constructed so as to locate points (of phase equilibrium) at which the free-energy difference *vanishes*, in NIRM one is inevitably faced with the task of determining some very small number by taking the difference between two relatively large numbers. This point is made more explicitly by the hard-sphere data in Table I. One sees that the *difference* between the values of the free energy [37] of the two crystalline phases is some *four orders of magnitude* smaller than the separate results for the two phases, determined by ESM. Of course one can see this as a testimony to the remarkable care with which the most recent recent ESM studies have been carried out [34]. Alternatively, one may see it as a strong indicator that another approach is called for.

B. Parallel Tempering; Around the Critical Point

1. Strategy

When the coexistence line of interest terminates in a critical point, the two phases can be linked by a single continuous path (Fig. 2b) that loops around the critical point, eliminating the need for reference macrostates while still avoiding the interphase region. In principle it is possible to establish the location of such a coexistence curve by integration along this route. But the techniques of parallel sampling (Section III.C) provide a substantially more elegant way of exploiting such a path, in a technique known as (hyper) parallel tempering (HPT) [38].

Studies of liquid–vapor coexistence are, generally, best addressed in the framework of an open ensemble; thus the state variables here comprise both the particle coordinates $\{\vec{r}\}$ and the particle number N. A path with the appropriate credentials can be constructed by identifying pairs of values of the chemical potential μ and the temperature T which trace out some rough *approximation* to the coexistence curve in the μ–T plane, but extend into the one-phase region beyond the critical point. Once again there is some circularity here to which we shall return. Making the relevant variables explicit, the sampling distribution [Eq. (26)] takes the form

$$P_{\mathcal{S}}(\{\vec{r}\}^{(1)}, N^{(1)} \ldots \{\vec{r}\}^{(\Omega)}, N^{(\Omega)}) = \prod_{j=1}^{\Omega} P_0(\{\vec{r}\}^{(j)}, N^{(j)} | \mu_j, T_j) \tag{41}$$

In the context of liquid–vapor coexistence the particle number N (or equivalently the number density $\rho = N/V$) plays the role of an order parameter. Estimates of the distribution $P_0(N|\mu, T)$ are available from the simulation for all the points

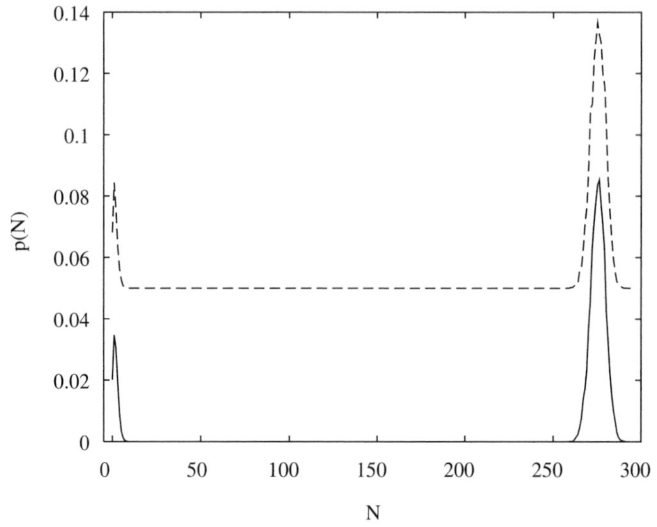

Figure 3. Probability distribution of the number N of particles in an LJ fluid. The simulations use the HPT method described in Section IV.B.1. The solid line shows the distribution for a replica whose μ–T parameters lie close to coexistence; the dashed line (offset) shows the distribution (for the same μ–T parameters) obtained by folding in (explicitly) the contributions of all replicas, using multihistogram reweighting. (Taken from Fig. 2 of Ref. 38.)

chosen to define the path. One may then identify the free-energy difference from the integrated areas of the branches of this distribution associated with each phase and proceed to search for coexistence using the criterion that these integrated areas should be equal [Eq. (10) et seq.]. Figures 3 and 4 show some explicit results [38] for a Lennard-Jones fluid.

2. *Critique*

The phase diagram shown in Fig. 4 is of course the ultimate objective of such studies; but it is the *distribution* shown in Fig. 3 that merits most immediate comment, lest its key feature be taken for-granted. Its "key feature" (which we shall come to recognize as the signature of success in this enterprise) is that it captures the contributions from *both* phases (the two distinct peaks) within the framework of a *single simulation*. The fact that both phases are well-visited shows that this strategy manages to break the ergodic block that dooms conventional MC sampling to explore only the particular phase in which the simulation happens to be launched (Section II.B).

The HPT method deals with the ergodic block by avoiding it. The configuration exchange between adjacent replicas [Eq. (27)] fuels a form of

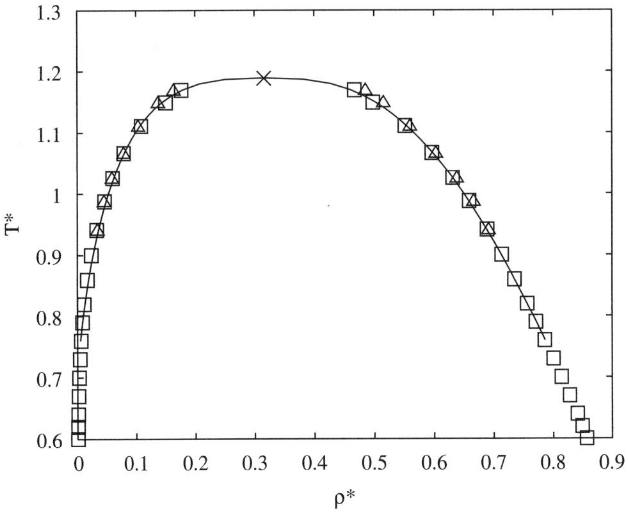

Figure 4. Phase diagram for the LJ fluid. The squares show results obtained by the HPT method described in Section IV.B.1 and illustrated in Fig. 3. The triangles show results obtained by the ESIT strategy described in Section IV.C.1 and illustrated in Fig. 5. (Taken from Fig. 3 of Ref. 38).

continuous tempering: a liquid phase configuration resident in a replica low down the coexistence curve may diffuse along the path and thus through the replicas, to a point (perhaps supercritical) at which ongoing local updating is effective in eroding memory of its liquid origins; that (ever-evolving) configuration may then diffuse downwards, to appear in the original replica as a vapor-phase configuration.

There is an elegant and powerful idea here, albeit one whose applicability (to the phase-coexistence problem) is limited to systems with critical points. But there is one respect in which it is less than satisfactory—the element of circularity already noted: The method will tell us *if* we have selected a point sufficiently close to coexistence for both phases to have observable probability; it does not in itself tell us what to do if our selection does not satisfy this criterion. How "close" we need to be depends sensitively on the size of the system simulated. The ratio of the two phase probabilities (at a chosen point in field space) varies exponentially fast with the system size [39]. Thus for a large system, unless we are very close to coexistence, the equilibrium (grand-canonical) PDF determined in HPT will show signs of only one phase *even when there is no ergodic block preventing access to the other phase*. If the initial choice is close enough to coexistence to provide at least *some* signature of the subdominant phase [40], then histogram reweighting techniques (Appendix B) can be used to give a better estimate of coexistence. But HPT provides no way of systematically improving bad initial estimates, because it provides no mechanism for dealing with the huge difference between the statistical weights of the two phases away from the immediate vicinity of coexistence. To address *that* issue, one must turn to extended sampling techniques.

C. Extended Sampling: Traversing the Barrier

1. Strategy

Viewed from the perspectives of configuration space provided by the caricature in Fig. 2, the most direct approach to the phase-coexistence problem calls for a full frontal assault on the ergodic barrier that separates the two phases. The extended sampling strategies discussed in Section III.C make that possible. The framework we need is a synthesis of Eqs. (10) and (32). We will refer to it generically as Extended Sampling Interface Traverse (ESIT).

Equation (10) shows that we can always accomplish our objective if we can measure the full canonical distribution of an appropriate order parameter. By "full" we mean that the contributions of both phases must be established *and calibrated on the same scale*. Of course it is the last bit that is the problem. (It is always straightforward to determine the two *separately* normalized distributions associated with the two phases, by conventional sampling in each phase in turn.) The reason that it is a problem is that the full canonical distribution of the (an) order parameter is typically vanishingly small at values intermediate between those characteristic of the two individual phases. The vanishingly small values provide a real, even quantitative, measure of the ergodic barrier between the phases. If the "full"-order parameter distribution is to be determined by a "direct" approach (as distinct from the circuitous approach of Section IV.B, or the "off the map" approach to be discussed in Section IV.D), these low-probability macrostates *must* be visited.

Equation (32) shows how. We need to build a sampling distribution that extends along the path of M-macrostates running between the two phases. To do its job, that sampling distribution must (Section III.C) assign roughly equal values to the probabilities of the different macrostates. More explicitly, the resulting measured distribution of M-values (following Ref. 27 we shall call it multicanonical)

$$P_{\mathcal{S}}(M_j) \equiv \int \prod_i dq_i P_{\mathcal{S}}(\{q\}) \Delta_j[M(\{q\})] \tag{42}$$

should be roughly flat. It needs to be "roughly flat" because the macrostate of lowest probability sets the size of the bottleneck through which interphase traverses must pass. It needs to be *no better* than "roughly" flat because of the way in which (ultimately) it is *used*. It is used to estimate the true canonical distribution $P_0(M)$. The two distributions are simply related by

$$P_0(M_j) \doteq w_j^{-1} P_{\mathscr{S}}(M_j) \tag{43}$$

where $\{w\}$ are the multicanonical weights that define the chosen sampling distribution [Eq. (32)] and $\doteq$ means equality to within an overall normalization constant. The procedure by which one uses this equation to estimate the canonical distribution from the measured distribution is variously referred to as "unfolding the weights" or "re-weighting"; it is simply one realization of the identity given in Eq. (24). The procedure eliminates any *explicit* dependence on the weights (hence the looseness of the criteria by which they are specified), but it leaves the desired legacy: The relative sizes of the two branches of the *canonical* distribution are determined with a statistical quality that reflects the number of interphase traverses in the *multicanonical* ensemble.

This strategy has been applied to the study of a range of coexistence problems, initially focused on lattice models in magnetism [41] and particle physics [42]. Figure 5 [43–45] shows the results of an application to liquid–vapor coexistence in a Lennard-Jones system with the particle number density chosen as an order parameter.

2. *Critique*

The ESIT strategy has clear advantages with respect to the NIRM and (to a lesser extent) HPT strategies discussed in preceding sections.

While HPT *requires* the existence of a critical point, ESIT does not (although its existence can be usefully *exploited* to assist in the task of weight generation [44]). Moreover, in contrast to HPT, ESIT does not *require* bootstrapping by a rather good guess of a point on the phase boundary. Extended sampling methods can take huge probability differentials in their stride (100 orders of magnitude is not uncommon) as one can see from the low probability of the macrostates in the interphase ravine in Fig. 5. Thus ESIT allows one to measure the free-energy difference between two phases relatively far from the phase boundary—in fact in any regime in which both phases are at least metastable [46].

ESIT has one further potential advantage with respect to HPT, which emerges once one appreciates what it entails at a *microscopic* level—in particular the nature of the intermediate macrostates that lurk in the ravine between the two peaks of the canonical distribution. To do so, one needs first to understand a *general* point about the weight generation procedures used to build the sampling distribution. These procedures do not *require* explicit specification of the configurations in the macrostates along the path; instead they *search* for

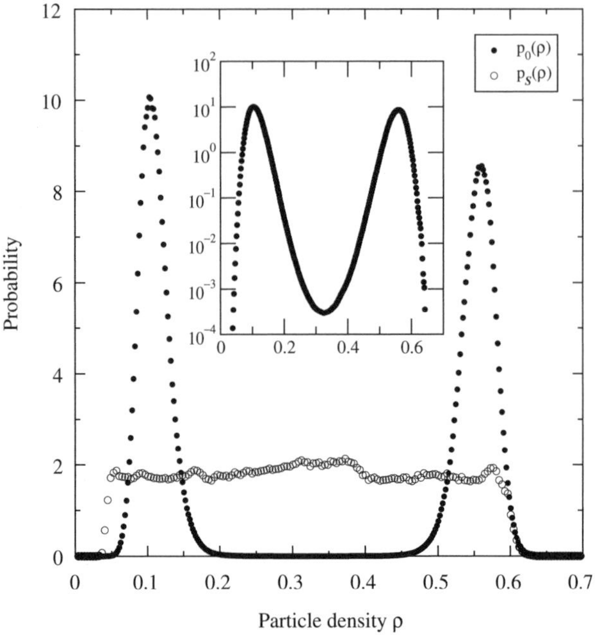

Figure 5. Results from a multicanonical simulation of the 3D Lennard-Jones fluid at a point on the coexistence curve. The figure shows both the multicanonical sampling distribution $P_{\mathcal{S}}(\rho)$ (symbol: ○) and the corresponding estimate of the equilibrium distribution $P_0(\rho)$ with $\rho = N/V$ the number density (symbol: •). The inset shows the value of the equilibrium distribution in the interfacial region [43].

the configurations that dominate those macrostates. The results of the "search" can sometimes be a little unexpected [47, 48]. But in the case of the liquid–vapor interphase path picked out by choosing the density as order parameter, the results of the search are clear *a priori*: The configurations dominating such a path will show two physically distinct regions, each housing one of the two phases, separated by an interface. The dominance of this kind of configuration is recognized in arguments that establish the convexity of the free energy in the thermodynamic limit and the way in which that limit is approached [49]; this picture also allows one to understand the depth of the probability ravine to be traversed [50]. It follows then that ESIT provides incidental access to these configurations, and thus to information about their dominant feature—the interface. Information of this kind has been exploited in a number of studies [51].

The broader and more far-reaching comparison to be made here is, however, between NIRM and strategies *like* ESIT which furnish canonical distributions spanning two phases, such as that shown in Fig. 5 [52]. Here it seems to us that ESIT wins in two respects. First, it is rather more transparent in regard to uncertainties (in free-energy differences). The error bounds emerging from ESIT (and family) represent purely statistical uncertainties associated with the measurement of the relative weights of two distribution-peaks. In contrast, NIRM error bounds have to aggregate the uncertainties (statistical and systematic) associated with different stages of the integration process. Second, it seems rather more satisfying to read-off the result for a free-energy difference directly from the likes of Fig. 5 than from a pair of numbers established by appeal to physically irrelevant reference states.

Nevertheless, the ESIT strategy is demanding in a number of respects. Generating the weights is a computationally intensive job, which is not yet fully self-managing. Subsequent sampling of the resulting multicanonical distribution is a slow process: The dynamics in M-space is a random walk in which visits to the macrostates of low equilibrium probability are secured only at the expense of repeated refusal of proffered moves to the dominant equilibrium macrostates. It helps (though it goes against the spirit of "one simulation") to break the space up into sections, whose length is chosen to reflect an interplay of the diffusion time *between* macrostates and the time associated with relaxation *within* macrostates [53].

These two reservations apply generically to ES methods; the following one is specific to ESIT itself. There are some phase coexistence problems in which an interphase path involving an *interface* is computationally fraught. In particular, if one of the phases is crystalline (as in the case of melting/freezing) or if both are crystalline (there is a potential *structural* phase transition), such a traverse will involve substantial, physically slow restructuring—vulnerable to further ergodic traps and compounding the intrinsic slowness of the multicanonical sampling process [54, 55]. In such circumstances it would be better if the interphase trip could be accomplished without encountering interfaces. This is possible.

D. From Paths to Wormholes: Phase Switch

We shall devote considerably more time to this fourth and final example of path-based strategies. We do so for two reasons. First it provides us with an opportunity to touch on a variety of other strands of thought about the free-energy estimation problem that should appear somewhere in this chapter, and can do so helpfully here. And second, we like it.

The strategy we shall discuss is a way of realizing the direct leap between phases represented in Fig. 2d. In the literature we have referred to it as "Lattice Switch Monte Carlo" [48, 56, 57] in the context of phase equilibrium between crystalline structures and "Phase Switch Monte Carlo" [58] in the context of solid–liquid coexistence. Here we shall develop the ideas in a general form, and we shall refer to the method as Extended Sampling Phase Switch (ESPS).

1. Strategy

Let us return to the core problem: the evaluation of the *ratio* [Eq. (10)] of configurational integrals of the form prescribed in Eq. (5). We can make the problem look different (and possibly make it easier) by choosing to express it in terms of coordinates that are in some sense "matched" to each phase. The simplest useful possibility is provided by an appropriate *linear* transformation

$$\{q\} = \{q\}_{\alpha}^{\text{REF}} + \tau_{\alpha}(\{u\}) \tag{44}$$

One can think of $\{q\}_{\alpha}^{\text{REF}}$ as some reference point in the configuration space of phase α and $\{u\}$ as a "displacement" from that point, modulo some rotation or dilation, prescribed by the operation τ_{α}. The single-phase partition function in Eq. (5) can then be written in the form

$$Z_{\alpha}(c) = \det \tau_{\alpha} \int \prod_{i} du_i e^{-\mathscr{E}_{\alpha}(\{u\},c)} \tag{45}$$

where $\det \tau_{\alpha}$ is the Jacobean [59] of the transformation (a configuration-independent constant, given the presumed linearity) and

$$e^{-\mathscr{E}_{\alpha}(\{u\},c)} = e^{-\mathscr{E}(\{q\},c)} \Delta_{\alpha}[\{q\}] \tag{46}$$

The form of the new energy function ("Hamiltonian") $\mathscr{E}_{\alpha}(\{u\}, c)$ reflects the *representation* chosen for the region of configuration space relevant to the phase; because this energy function carries a phase label, it can also be used to absorb the constraint [Eq. (6)] that restricts the integral to that region [60].

The difference between the free energies of the two phases [Eq. (10)] can now be written in the form

$$\Delta\mathscr{F}_{\alpha\tilde{\alpha}}(c) = \Delta\mathscr{E}^{0}_{\alpha\tilde{\alpha}} - \ln \det \mathbf{S}_{\alpha\tilde{\alpha}} - \ln \mathscr{R}_{\alpha\tilde{\alpha}} \tag{47}$$

where

$$\Delta\mathscr{E}^{0}_{\alpha\tilde{\alpha}} = \mathscr{E}(\{q\}_{\alpha}^{\text{REF}}, c) - \mathscr{E}(\{q\}_{\tilde{\alpha}}^{\text{REF}}, c) \tag{48}$$

is the difference between the Hamiltonians of the reference configurations while

$$\mathbf{S}_{\alpha\tilde{\alpha}} = \tau_{\alpha} \times \tau_{\tilde{\alpha}}^{-1} \tag{49}$$

These two contributions to Eq. (47) are computationally trivial; the computational challenge is now in the third term defined by

$$\mathcal{R}_{\alpha\tilde{\alpha}} = \frac{\int \prod_i du_i e^{-\mathcal{E}_\alpha(\{u\},c)}}{\int \prod_i du_i e^{-\mathcal{E}_{\tilde{\alpha}}(\{u\},c)}} \tag{50}$$

In the formulation of Eq. (10) we are confronted with a ratio of configurational integrals defined through the *same* energy function acting on *two explicitly different* regions of configuration space. In contrast, Eq. (50) features integrals defined through *different* energy functions acting on *one common* configuration space.

Configurational–integral ratios of the form of Eq. (50) appear widely in the free-energy literature, but the underlying physical motivation is not always the same. The spectrum of possible usages is covered by writing Eq. (50) in the more general form

$$\mathcal{R}_{AB} = \frac{\int \prod_i du_i e^{-\mathcal{E}_A(\{u\})}}{\int \prod_i du_i e^{-\mathcal{E}_B(\{u\})}} \equiv \frac{\mathcal{Z}_A}{\mathcal{Z}_B} \tag{51}$$

where A and B are two generalized macrostate labels, which identify two energy functions.

One meets this kind of ratio (perhaps most naturally) if one considers the difference between the free energies of two macrostates of *one* phase, corresponding to *different* choices of c—as in the influential work of Bennett [61]. And one meets it if one considers the difference between the free energies of a given model and some approximation to that model, as in the perturbation approach of Zwanzig [62]. Rahman and Jacucci [63] seem to have been amongst the first to consider this structure of problem with the kind of motivation we have given it here—that is, as a way of computing free energy differences between two *phases* directly.

To set the remainder of the discussion in as wide a context as we can, we shall develop it in the general notation of Eq. (51), reverting to the specific interpretation of the macrostate labels of primary interest here ($A \rightarrow \alpha, c; B \rightarrow \tilde{\alpha}, c$) as appropriate.

In common with most who have addressed the problem posed by this kind of configurational integral ratio, we shall exploit the statistics of an appropriate "order-parameter" defined on the common configuration space and of the general form

$$\mathcal{M}_{AB} \equiv M_A(\{u\}) - M_B(\{u\}) \tag{52}$$

There is considerable license in the choice of M_A and M_B. In the simplest cases it suffices to choose the two energy functions themselves. We make that choice explicitly here so as to expose connections with the work of others. Then the "order parameter" assumes the form

$$\mathcal{M}_{AB} = \mathcal{E}_A(\{u\}) - \mathcal{E}_B(\{u\}) \tag{53}$$

This quantity has the credentials of an "order parameter" in the sense that its behavior is qualitatively different in the two ensembles. To understand this, suppose that we sample from the canonical distribution of the B-ensemble, prescribed by the partition function $\mathcal{Z}_B$. A "typical point" $\{u\}$ will then characterize a typical configuration of B, in which—for example—no particle penetrates the core region of the potential of another; that same point $\{u\}$ will, however, describe a configuration of A which is *not* guaranteed to be typical of that ensemble, and *will* in general feature energy-costly regions of core penetration. The order parameter $\mathcal{M}_{AB}$ will thus generally be *positive* in ensemble B; by the same token it will be *negative* in ensemble A. (We shall see this explicitly in the examples that follow.)

One may measure and utilize the statistics of $\mathcal{M}$ in three strategically different ways.

First, Eqs. (50) and (53) lead immediately to the familiar Zwanzig formula [62],

$$\mathcal{R}_{AB} = \langle e^{-\mathcal{M}_{AB}} \rangle_B = \langle e^{-[\mathcal{E}_A(\{u\}) - \mathcal{E}_B(\{u\})]} \rangle_B \tag{54}$$

In principle then, *one single-ensemble-average* suffices to determine the desired ratio. However, this strategy (if unsupported by others) requires [63] that the two ensembles overlap in the sense that, loosely [64], the "dominant" configurations in the A ensemble are a subset of those in the B ensemble. This is a strong constraint; it will seldom, if ever, be satisfied [65].

The second generic strategy [61] utilizes *two single-ensemble-averages*—that is, averages with respect to the *separate* ensembles defined by $\mathcal{Z}_A$ and $\mathcal{Z}_B$. In particular, one may, in principle, measure the canonical probability distributions of the order parameter in each ensemble separately and exploit the relationship between them [63],

$$\mathcal{R}_{AB} P_0(\mathcal{M}_{AB} \mid A) = e^{-\mathcal{M}_{AB}} P_0(\mathcal{M}_{AB} \mid B) \tag{55}$$

again presupposing the choice prescribed in Eq. (53). One can reexpress this result in an alternative form:

$$\mathcal{R}_{AB} = \frac{\langle \mathcal{A}(\mathcal{M}_{AB}) \rangle_B}{\langle \mathcal{A}(\mathcal{M}_{BA}) \rangle_A} \tag{56}$$

where $\mathcal{A}$ is the Metropolis function defined in Eq. (19). Recalling the significance of that function, one can see each of the two terms on the right-hand side of Eq. (56) as the probability of acceptance of a Monte Carlo switch of the labels

A and B (and thus of the controlling "Hamiltonian") at a point in $\{u\}$-space, averaged over that space. In the numerator this switch is from $\mathscr{E}_B$ to $\mathscr{E}_A$ and the average is with respect to the canonical distribution in ensemble B; in the denominator the roles of A and B are reversed. This is Bennett's acceptance ratio formula [61].

Equation (55) shows that for this strategy to work, one needs the two ensembles to "overlap" in the sense [somewhat less restrictive than in the case of the Zwanzig formula, Eq. (54)] that the two single-ensemble PDFs are measurable at some *common* value of $\mathscr{M}$, the most obvious candidate being the $\mathscr{M} \simeq 0$ region intermediate between the values typical of the two ensembles. Equation (56) shows that this requirement is effectively equivalent to the condition that the probabilities of acceptance of a Hamiltonian switch can be measured, in *both* directions.

In the form described, this strategy will virtually always fail; to produce a generally workable strategy we need to introduce two further ingredients: The first essential, the second desirable. First we need to invoke ES techniques to extend the $\mathscr{M}$-ranges sampled in the single-ensemble simulations until they overlap; in principle this is enough to allow us to determine the desired ratio by matching up the two distributions in Eq. (55). But the information thus gathered is more fully and efficiently utilized by taking a further step. In the $\mathscr{M} \simeq 0$ regions, then accessed, "switches" (between the "Hamiltonians" of the two ensembles) can be implemented not just *virtually*, as envisaged in the acceptance ratio formula, but *actually* within a configuration space enlarged to include the macrostate label explicitly. One may then measure the full canonical PDF of $\mathscr{M} \equiv \mathscr{M}_{AB}$ and deduce the desired ratio from [Eq. (50)]

$$\mathscr{R}_{AB} = \frac{\int_{\mathscr{M}<0} d\mathscr{M} P_0(\mathscr{M}|c)}{\int_{\mathscr{M}>0} d\mathscr{M} P_0(\mathscr{M}|c)} \tag{57}$$

where we have used the fact that the sign of the order parameter acts as a signature of the ensemble.

This is the ESPS strategy. We have expressed it in a general way, as a switch between *any* nominated pair of "macrostates" or "ensembles." In summarizing it, we revert to the particular context of primary interest in which the two macrostates belong to *different* phases. The core idea is simple: to use ES methods to seek out regions of the configuration space of one phase which are such that a transition (switch, leap) to the other will be accepted with reasonable probability. The leap avoids mixed-phase configurations: The simulation explores both configuration spaces but is always to be found in one or the other.

The full machinery of ESPS is customizable in a number of respects, notably the choice of reference configurations, of order parameter and of transformation matrix. These issues are best explored in the context of specific examples.

2. *Examples*

We consider three examples of ESPS, ordered conceptually rather than chronologically.

Example A: hcp and fcc Phases of Lennard Jones Systems. In the case of crystalline phases it is natural to choose the reference configurations to represent the states of perfect crystalline order, described by the appropriate sets of lattice vectors

$$\{q\}_{\gamma}^{\mathrm{REF}} \longrightarrow \{\vec{R}\}_{\gamma}, \qquad \gamma = \alpha, \tilde{\alpha} \tag{58}$$

Note that the particle and lattice site indexing hidden in this notation mandates a one–one mapping between the lattice sites of the two systems; we are free to choose any of the $N!$ possibilities. In the case of *hcp* and *fcc* lattices [66] it is natural to exploit the fact that the one lattice can be made out of the other merely by translating close-packed planes, in the fashion depicted in Fig. 6.

The simplest choice for the τ operations in Eq. (44) is to take them as the identity operation; the operation **S** [Eq. (49)] is then also the identity.

The choices of $\{q\}^{\mathrm{REF}}$ and τ together define the geometry of the switch operation—the way in which a configuration of one structure is used to generate

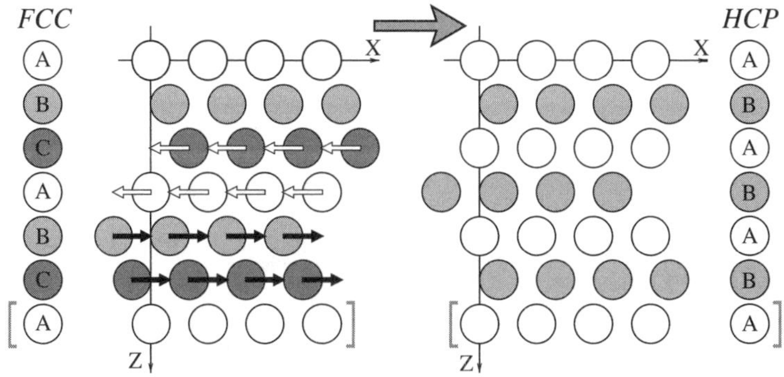

Figure 6. The simple transformation for switching between *fcc* and *hcp* lattices. The diagram shows six close-packed (x–y) layers. (The additional bracketed layer at the bottom is the periodic image of the layer at the top.) The circles show the boundaries of particles located at the sites of the two close-packed structures. In the lattice switch operation the top pair of planes are left unaltered, while the other pairs of planes are relocated by translations, specified by the black and white arrows. The switch operation is discrete: The relocations occur "through the wormholes." (Taken from Fig. 4 of Ref. 48.)

a configuration of the other: Here the switch exchanges one lattice for another, while conserving the physical displacements with respect to lattice sites.

The final choice to be made is the form of the order parameter; in this case the default (built out of the energy function) defined in Eq. (53) proves the right choice. Thus, making the phase labels explicit we take

$$\mathcal{M}_{\alpha\tilde{\alpha}} = \mathcal{E}_{\alpha}(\{u\}, c) - \mathcal{E}_{\tilde{\alpha}}(\{u\}, c) \tag{59}$$

Figures 7 and 8 show results for the Lennard Jones (LJ) crystalline phases established with ESIT, on the basis of these choices [57]. Commentary (on this and other results here) is deferred to the "critique" below.

Example B: hcp and fcc Phases of Hard Spheres. We have already noted that the order parameter in ESPS need not be constructed out of the true energy functions of the two phases. The defining characteristic of an ESPS order parameter is that it measures the difference between the values of some chosen function of the common coordinate set $\{u\}$ evaluated in the two phases, such that for some region (typically "sufficiently small values") of that quantity, an interphase switch can be successfully initiated. An ESPS "order parameter" merely provides a convenient thread that can be followed to the wormhole ends.

In the case of hard spheres the energy function does not provide a usefully graded measure of how far we are from a wormhole: The order parameter defined in Eq. (53) will generally be infinite because of hard-sphere overlap in the configurations created by the switch. But it is easy to find an alternative: All

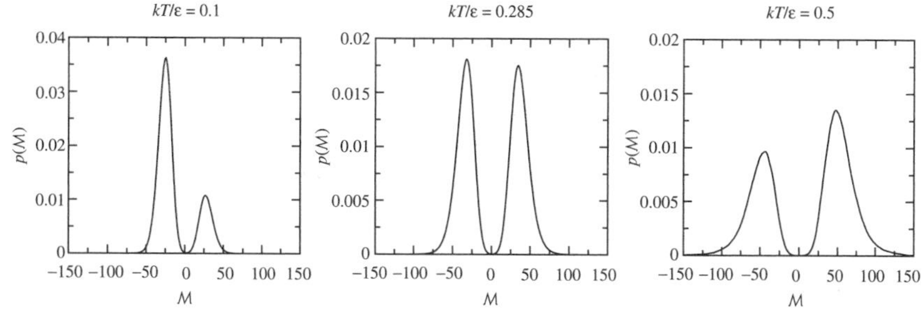

Figure 7. ESPS ("Lattice Switch") studies of the relative stability of *fcc* and *hcp* phases of the LJ solid at zero pressure, as discussed in Section IV.D. In this case the order parameter M [Eq. (59)] measures the difference between the energy of a configuration of one phase and the corresponding configuration of the other phase generated by the switch operation. The areas under the two peaks reflect the relative configurational weights of the two phases. The evolution with increasing temperature (from *hcp*-favored to *fcc*-favored behavior) picks out the *hcp–fcc* phase boundary shown in Fig. 8. (Taken from Fig. 7 of Ref. 57.)

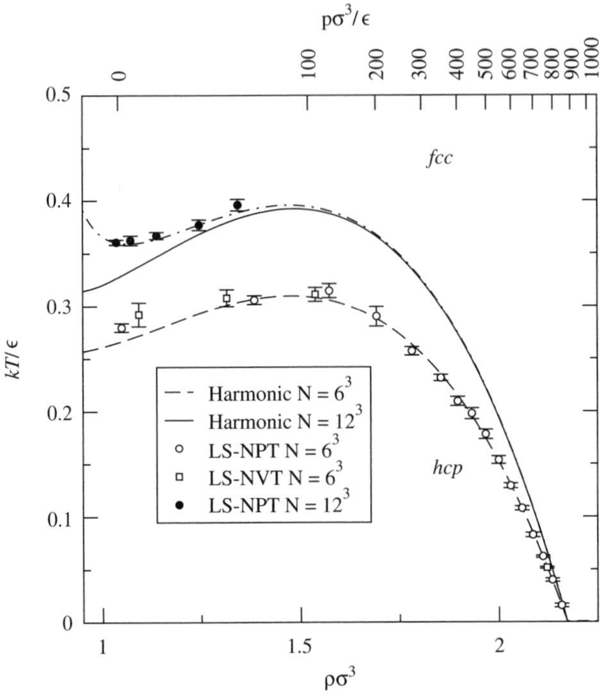

Figure 8. A variety of approximations to the classical Lennard-Jones phase diagram. The data points show the results of ESPS studies (discussed in Section IV.D), denoted here by "LS." The dashed and solid lines are the results of harmonic calculations (for the two system sizes). The dash–dotted line is a phenomenological parameterization of the anharmonic effects. The scale at the top of the figure shows the pressures at selected points on the (LS $N = 12^3$, *NPT*) coexistence curve. Tie-line structure is unresolvable on the scale of the figure. (Taken from Fig. 11 of Ref. 57.)

one has to do is build an order parameter out of a count of the number N^o of overlapping spheres. Instead of Eq. (59) we then have

$$\mathcal{M}_{\alpha\tilde{\alpha}} \equiv N^o_{\alpha}(\{u\}, c) - N^o_{\tilde{\alpha}}(\{u\}, c) \tag{60}$$

With the switch geometry chosen as for the LJ systems discussed above, the difference between the free energies of fcc and hcp hard-sphere systems can be determined precisely and transparently [48, 56]. Some of the results are included in Table I.

Example C: Liquid and fcc Phases of Hard Spheres. The ESPS strategy can also be applied when one of the phases is a liquid [58]. A configuration selected at random from those explored in canonical sampling of the liquid phase will serve as a reference state. Since the liquid and solid phases generally have significantly different densities, the simulation must be conducted at constant pressure [67]; the coordinate set $\{q\}$ then contains the system volume, and the switch must accommodate an appropriate dilation (and can do so easily through the specification of the volumes implicit in the reference configurations). While the overlap order parameter defined in Eq. (60) remains appropriate for simulations conducted in the solid phase, in the liquid phase it is necessary to engineer something a little more elaborate to account for the fact that the particles are not spatially localized. Such considerations also lead to some relatively subtle but significant finite-size effects. Figure 9 shows some results locating the freezing pressure of hard spheres this way [58].

3. *Critique*

The ESPS method draws on and synthesizes a number of ideas in the extensive free-energy literature, including the importance of *representations* and *space transformations between them* [63, 68, 69], the utility of *expanded ensembles* in turning virtual transitions into real ones [23], and the general power of *multicanonical methods* to seek out macrostates with any desired property [27].

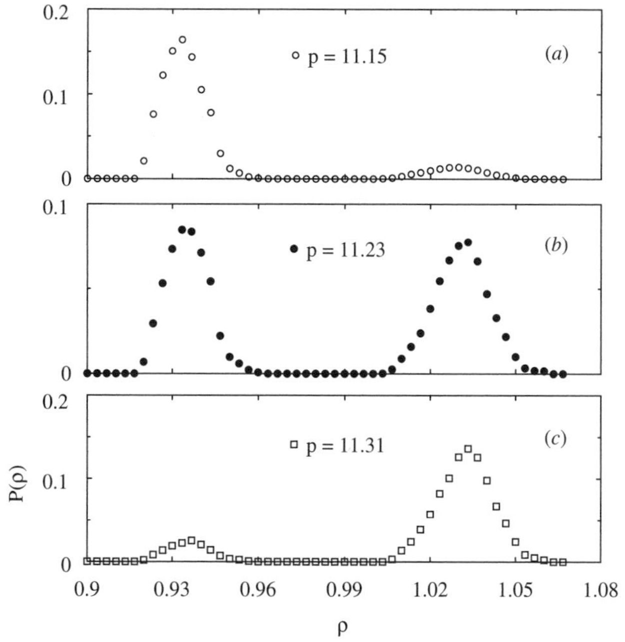

Figure 9. The distribution of the density of the system of $N = 256$ LJ particles in crystalline and liquid phases, as determined by ESPS methods. The three pressures are (a) just below, (b) at, and (c) just above coexistence for this N. (Taken from Fig. 1 of Ref. 58.)

Methodologically, ESPS has much in common with ESIT: like ESIT, it utilizes the paraphernalia of extended sampling to visit both phases in a single simulation; but in contrast to ESIT, it contrives to do this *without* having to traverse the interfacial configurations that make ESIT hard, probably impossible, to implement in problems involving solid phases.

ESPS thus shares a number of the advantages that ESIT has with respect to integration methods (Sections IV.A.2 and IV.C.2). It is pleasingly transparent: The evolution with temperature of the relative stability of *fcc* and *hcp* LJ crystals can be "read off" from Fig. 7; and the LJ freezing pressure can be "seen" in Fig. 9. Apart from finite-size effects, uncertainties are purely statistical. The fact that both phases are realized within the same simulation means that finite-size effects can be handled more systematically; this seems to be a particular advantage of the ESPS approach to the liquid–solid phase boundary.

ESPS remains a computationally intensive strategy, though not prohibitively so on the scale of its competitors: One explicit comparison (in the case hard-sphere crystals) indicates that ESPS and NIRM deliver similar precision for similar compute resource [34].

But the possibility of substantial improvements to ESPS remains. The idea of improved ("targeted") mappings between the configuration spaces has been discussed in general terms by Jarzynski [70]—albeit in the context of the Zwanzig formula [Eq. (54)], which will not generally work without ES props. In the ESPS framework, this mapping enters in the matrix **S**, reflecting the representations chosen for the two phases [Eqs. (44) and (49)]. One does not *have* to preserve the physical displacements in the course of the switch. By appropriate choice of the operations τ, it is possible [71] to implement a switch that, instead, conserves a set of Fourier coordinates, and therefore the *harmonic* contributions to the energy of the configurations of each phase; the determinant in Eq. (47) then captures the harmonic contribution to the free-energy difference, leaving the computational problem focused on the anharmonic contributions that (alone) are left in $\mathcal{R}$. This strategy greatly enhances the overlap between the two branches of the order parameter distribution; but the associated efficiency gains (resulting from the reduced length of "path" through $\mathcal{M}$-space) are offset by the greatly increased computational cost of the mapping itself [71].

V. DOING IT WITHOUT PATHS: ALTERNATIVE STRATEGIES

In this section we survey some of the strategic approaches to the phase-coexistence problem which do not fit comfortably into the path-based perspectives we have favored here.

A. The Gibbs Ensemble Monte Carlo Method

1. *Strategy*

Gibbs Ensemble Monte Carlo (GEMC) is an ingenious method introduced by Panagiotopoulos [72], which allows one to simulate the coexistence of liquid and vapor phases without having to deal with a physical interface between them.

GEMC utilizes *two* simulation subsystems ("boxes"); though physically separate, the two boxes are thermodynamically coupled through the MC algorithm, which allows them to exchange both volume and particles subject to the constraint that the total volume and number of particles remain fixed. Implementing these updates (in a way that respects detailed balance) ensures that the two systems will come to equilibrium at a common temperature, pressure, and chemical potential. The temperature is fixed explicitly in the MC procedure; but the procedure itself selects the chemical potential and pressure that will secure equilibrium.

If the overall number density and temperature are chosen to lie within the two-phase region, the system must phase separate; it does so through configurations in which each box houses one pure phase, since such arrangements avoid the free-energy cost of an interface. The coexistence densities can then be simply measured through a sample average over each box. By conducting simulations at a series of temperatures, the phase diagram in the temperature–density plane can be constructed. Details of the implementation procedure can be found in Ref. 29.

2. *Critique*

The GEMC method is elegant in concept and simple in practice; it seems fair to say that it revolutionized simulations of fluid-phase equilibria; it has been very widely used and comprehensively reviewed [73]. We note three respects in which it is less than ideal.

First, in common with any simulation of open systems, it runs into increasing difficulties as one moves down the coexistence curve to the region of high densities where particle insertion (entailed by particle exchange) has a low acceptance probability.

It also runs into difficulties of a different kind at the other end of the coexistence curve, as one approaches the critical point. GEMC effectively supposes that criticality may be identified by the coalescence of the two peaks in the separate branches of the density distribution captured by the two simulation boxes. The limiting critical behavior of the full density distribution in a system of finite size (Fig. 10, to be discussed below) shows that this is not so; the critical point cannot be reliably located this way. These difficulties are reflected in the strong finite-size dependence of the shape of the coexistence curve evident in GEMC studies [74]. To make sense of the GEMC behavior near the

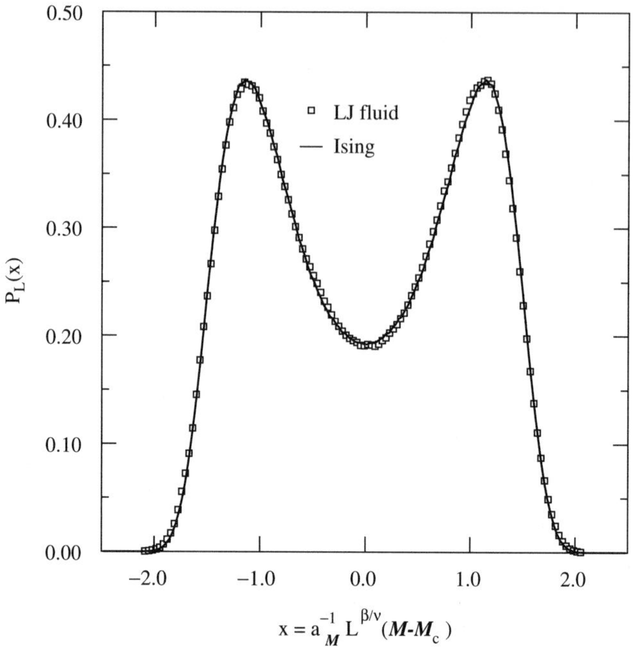

Figure 10. The distribution of the density of an LJ fluid at its critical point showing the collapse (given a suitable choice of scale) onto a form characteristic of the Ising universality class. (Taken from Fig. 3a of Ref. 44.)

critical point, therefore, one needs to invoke finite-size scaling strategies [75, 76]; but these are substantially less easy to implement than they are in the framework of the μVT ensemble, to be discussed in Section VI.C.

Finally we note that the efficiency of GEMC is reduced through the high computational cost of volume moves (each one of which requires a recalculation of *all* interparticle interactions). Comparisons show [73] that the strategy discussed in Section IV.C.1 (multicanonical methods and histogram reweighting, within a grand canonical ensemble) gives a better return in terms of precision per computational unit cost. But GEMC is undoubtedly easier to implement.

B. The NPT and Test Particle Method

1. *Strategy*

The NPT-TP method [77] locates phase coexistence at a prescribed temperature by finding that value of the pressure for which the chemical potentials of the two phases are equal. The chemical potential μ is identified with the difference between the Helmholtz free energies of systems containing N and $N-1$ particles. Then the Zwanzig formula ([62], Eq. (54)) shows that

$$\mu = \ln\langle e^{-[\mathcal{E}_N - \mathcal{E}_{N-1}]}\rangle_{N-1} = \mu^{ig} + \ln\langle e^{-\Delta U_{N,N-1}}\rangle_{N-1} \tag{61}$$

where $\Delta U_{N,N-1}$ is the additional configuration energy associated with the insertion of a "test" particle into a system of $N-1$ particles. Equation (61) is due to Widom [78] and forms the basis of the *test particle insertion method*. In contrast to other realizations of the Zwanzig formula, it works (or can do so) reasonably well, because the argument of the exponential is of O[1] rather than O[N], as long as the particle interactions are short-ranged.

The values of the chemical potentials in each phase, together with their pressure derivatives (available through the measured number densities), can be exploited to home in on the coexistence pressure. The method has been successfully applied to calculate the phase diagrams of a number of simple fluids and fluid mixtures [79, 80].

2. *Critique*

The NPT-TP method is obviously designed to deal with the coexistence of fluid phases: particle insertion into ordered structures is generally to be avoided. In this restricted context it is straightforward to implement, needing no more than the conventional apparatus of single-phase NPT simulation. However, in common with other strategies that involve particle insertion (such as Gibbs–Duhem integration, Section VI.B), it runs into difficulties at high densities; and it cannot readily handle the behavior in the critical region (Section VI.C.2).

C. Beyond Equilibrium Sampling: Fast Growth Methods

1. *Strategy*

The techniques discussed in this section might reasonably have been included in our collection of path-based strategies (Section IV). However, although the idea of a "path" features here too, it does so *without* the usual implications of equilibrium sampling.

At the heart of the techniques in question (we shall refer to them collectively as *Fast Growth*, FG) is a simple and beautiful result established by Jarzynski [81], which we write in the form [6]

$$\mathcal{R}_{AB} \equiv e^{-\Delta\mathcal{F}_{AB}} = \overline{e^{-W_{AB}^{\tau_S}}} \tag{62}$$

Here $W_{AB}^{\tau_s}$ is the work done in switching the effective energy function (through some time-dependent control parameter $\lambda(t)$) from $\mathcal{E}_B$ to $\mathcal{E}_A$, in time τ_s, while the system observes the dynamical or stochastic updating rules appropriate to the (evolving) energy function. The bar denotes an average over the ensemble of such procedures generated by choosing the initiating microstate randomly from macrostate B.

Equation (62) incorporates two more familiar claims as special cases. In the limit of *long* switching times the system has time to equilibrate at every stage of the switching procedure; then Eq. (62) reduces to the result of numerical integration along the path prescribed by the switching operation [cf. Eq. (37)]

$$\Delta\mathcal{F}_{AB} = \int_B^A d\lambda \left\langle \frac{\partial \mathcal{E}}{\partial \lambda} \right\rangle_\lambda \tag{63}$$

At the other extreme, if the switching time τ_s is short, the work done is just the energy cost of an instantaneous and complete Hamiltonian switch, and one recovers the Zwanzig formula [cf. Eq. (54)]

$$\Delta\mathcal{F}_{AB} = -\ln\langle e^{-[\mathcal{E}_A - \mathcal{E}_B]}\rangle_B \tag{64}$$

In fact, Eq. (62) holds *irrespective* of the switching time τ_s: The *equilibrium* free-energy difference is determined by the spectrum of a quantity, W_{AB}, associated with a *nonequilibrium* process. The "exponential average" of the work done in taking the system between the designated macrostates (at *any* chosen rate) thus provides an alternative estimator of the difference between the associated free energies.

The fact that Eq. (62) (in general) folds in nonequilibrium processes is made explicit through a third result that may be deduced from it. The convexity of the exponential [82] implies that

$$\overline{e^{-W_{AB}}} \geq e^{-\overline{W_{AB}}} \tag{65}$$

so that, from Eq. (62), we obtain

$$\Delta\mathcal{F}_{AB} \leq \overline{W_{AB}} \tag{66}$$

This is the Helmholtz inequality, a variant of the Second Law and thus an acknowledgment of the consequences of irreversible processes.

2. *Critique*

Our discussion of the Zwanzig formula shows that the FG representation is unlikely to be practically helpful if one chooses small τ_s—at least for the systems having large enough N to be of interest to us here. But given a choice of devoting a specified computational resource to *one long* switch [and appealing to the integration procedure, Eq. (63)] or *several short* switches [and appealing to the exponential averaging procedure, Eq. (62)], the latter seems to be the preferred strategy [81]: The two approaches are comparable in precision; but FG generates an estimate of its own uncertainties, and it is trivially implementable in parallel computing architectures. This seems a potentially fruitful avenue for further exploration.

One can avoid the issues associated with exponential-averaging if one uses the FG formula in the form of the inequality Eq. (66), in tandem with the corresponding inequality emerging from a reverse switch operation (from A to B). The two results together give upper and lower bounds on the difference between the two free energies; the bounds can be tightened by a variational procedure with respect to the parameters of the chosen switch [83]. Figure 11 shows the results of such a procedure [84] applied to the Bain switch operation [85] which maps an *fcc* lattice onto a *bcc* lattice by a continuous [86] deformation.

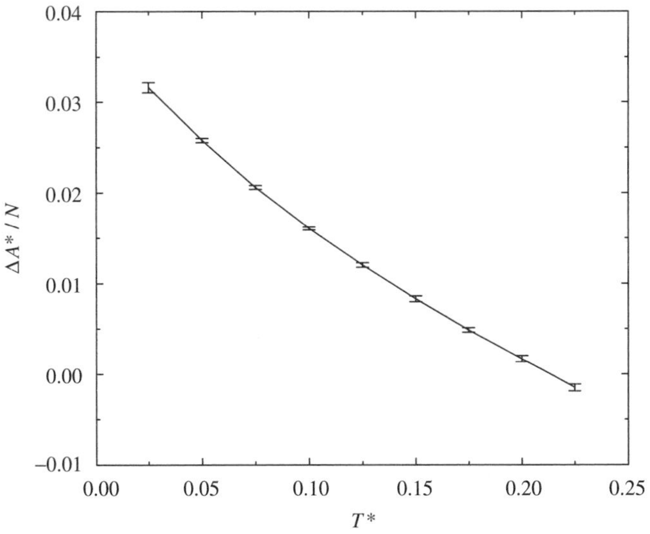

Figure 11. The difference between the free-energy densities of *fcc* and *bcc* phases of particles interacting through a Yukawa potential, as a function of temperature, determined through the FG methods discussed in Section V.C. The error bars reflect the difference between the upper and lower bounds provided by FG switches between the phases (along the Bain path [85]) in the two directions. The favored phase changes around $T^* = 0.21$. (Taken from Fig. 11 of Ref. 84.)

VI. DETERMINING THE PHASE BOUNDARY: EXTRAPOLATION, TRACKING, AND THE THERMODYNAMIC LIMIT

The path-based methods exemplified in the preceding section provide us with ways of estimating the difference between the free energies of the two phases, $\Delta\mathcal{F}_{\alpha\tilde{\alpha}}(\{\lambda\}, N)$ [Eq. (10)] at *some* point $c \equiv \{\lambda\}$ in the space of the controlling fields [88], for a system whose size N (which we shall occasionally make explicit in this section) is computationally manageable.

The practical task of interest here requires that we identify the set of points $c_x \equiv \{\lambda\}_x$ giving phase coexistence in the thermodynamic limit and defined by the solutions to the equation

$$\lim_{N\to\infty} \frac{1}{N} \Delta\mathcal{F}_{\alpha\tilde{\alpha}}(\{\lambda\}_x, N) = 0 \tag{67}$$

We divide this program into three parts. The first issue is how to use the data accumulated at our chosen state point to infer the location of *some* point c_x on the coexistence curve, for some finite N. The second is to map out the phase boundary emanating from that point. And the third is to deal with the corrections associated with the limited ("finite") size of the simulation system.

We shall restrict the discussion to a two-dimensional field space spanned by fields $\{\lambda\} \equiv \lambda_1, \lambda_2$ with conjugate macrovariables $\{M\} \equiv M_1, M_2$.

A. Extrapolation to the Phase Boundary

The simplest way of using the measurements at $\{\lambda\}$ to estimate the location of a point on the phase boundary is to perform a linear extrapolation in one of the fields using the result

$$\frac{\partial\Delta\mathcal{F}_{\alpha\tilde{\alpha}}(\{\lambda\})}{\partial\lambda} = \langle M\rangle_\alpha - \langle M\rangle_{\tilde{\alpha}} \tag{68}$$

Note that the quantities on the RHS of this equation represent separate single-phase expectation values, defined with respect to single-phase canonical distributions of the form given in Eq. (4); they are thus problem-free.

The implied estimate of the coexistence-value of the field λ_1 (say) is then

$$\lambda_{1x} = \lambda_1 - \frac{\Delta\mathcal{F}_{\alpha\tilde{\alpha}}(\{\lambda\})}{\langle M_1\rangle_\alpha - \langle M_1\rangle_{\tilde{\alpha}}} \tag{69}$$

Such extrapolations provide a simple, but possibly crude, way of correcting an initially poor estimate of coexistence.

One may be able to do better by appealing to histogram reweighting techniques (Appendix B). *If* the initial measurement of a free-energy difference is based on some form of extended sampling that establishes the full canonical distribution of some order parameter and *if* that order parameter is the conjugate of one of the fields spanning the phase diagram of interest, *then* HR allows one to scan through a range of values of that field to find the coexistence value [identified by the resulting equality of the areas of the two peaks in the canonical order parameter distribution, implied by Eq. (10)]. Such a process is easily automated.

B. Tracking the Phase Boundary

In principle, knowledge of a *single point* on the coexistence curve permits the *entire curve* to be traced without further calculation of free energies. The key result needed follows simply from Eq. (68), applied to each field λ_1 and λ_2 in turn. The *slope* of the coexistence curve $\Delta\mathcal{F}_{\alpha\tilde{\alpha}}(\{\lambda\}_x) = 0$ is as follows:

$$\left[\frac{d\lambda_1}{d\lambda_2}\right]_x = -\frac{\langle M_2\rangle_\alpha - \langle M_2\rangle_{\tilde{\alpha}}}{\langle M_1\rangle_\alpha - \langle M_1\rangle_{\tilde{\alpha}}} \tag{70}$$

This is the generalized Clausius–Clapeyron equation [89]. It expresses the slope entirely in terms of single-phase averages; the slope can be employed (in a predictor–corrector scheme) to estimate a nearby coexistence point. Fresh simulations performed at this new point yield the phase boundary gradient there, allowing further extrapolation to be made, and so on. In this manner one can in principle track the whole coexistence curve. This strategy is widely known as Gibbs–Duhem integration (GDI) [90].

GDI has been used effectively in a number of studies, most notably in the context of freezing of hard and soft spheres [91]. Its distinctive feature is simultaneously its strength and its weakness: Once bootstrapped by knowledge of one point on the coexistence curve, it subsequently requires only *single-phase* averages. This is clearly a virtue since the elaborate machinery needed for two-phase sampling is not to be unleashed lightly. But *without* any "reconnection" of the two configuration spaces at subsequent simulation state points, the GDI approach offers no feedback on integration errors. Since there will generally exist a *band* of practically stable states on each side of the phase boundary, it is possible for the integration to wander significantly from the true boundary with no indication that anything is wrong.

A more robust (though computationally more intensive) alternative to GDI is provided by a synthesis of extended (multicanonical) sampling and histogram reweighting techniques. The method is bootstrapped by an ES measurement of the full canonical distribution of a suitable order parameter, at some point on the coexistence curve [identified by the equal areas criterion specified in Eq. (10)].

HR techniques then allow one to map a region of the phase boundary close to this point. The range over which such extrapolations are reliable is limited (Appendix B), and it is not possible to extrapolate arbitrarily far along the phase boundary: Further multicanonical simulations will be needed at points that lie at the extremes of the range of reliable extrapolation. But there is no need to determine a new set of weights (a new extended sampling distribution) from *scratch* for these new simulations. HR allows one to generate a *rough estimate* of the canonical order parameter distribution (at these points) by extrapolation from the original measured distribution. The "rough estimate" is enough to furnish a usable set of weights for the new multicanonical simulations. Repeating the combined procedure (multicanonical simulation followed by histogram extrapolation), one can track along the coexistence curve. The data from the separate histograms can subsequently be combined self-consistently (through multihistogram extrapolation, as discussed in Appendix B) to yield the whole phase boundary. If one wishes to implement this procedure for a phase boundary that terminates in a critical point, it is advisable to start the tracking procedure nearby. At such a point the ergodic block presented to interphase traverses is relatively small (the canonical order parameter distribution is relatively weakly double-peaked), and so the multicanonical distribution (weights) required to initiate the whole process can be determined without extensive (perhaps without any) iterative procedures [92].

C. Finite-Size Effects

Computer simulation is invariably conducted on a model system whose size is small on the thermodynamic scale one typically has in mind when one refers to "phase diagrams." Any simulation-based study of phase behavior thus necessarily requires careful consideration of "finite-size effects." The nature of these effects is significantly different according to whether one is concerned with behavior close to or remote from a critical point. The distinction reflects the relative sizes of the linear dimension L of the system—the edge of the simulation cube, and the correlation length ξ—the distance over which the local configurational variables are correlated. By "noncritical" we mean a system for which $L \gg \xi$; by critical we mean one for which $L \ll \xi$. We shall discuss these two regions in turn and avoid the lacuna in between.

1. Noncritical Systems

In the case of noncritical systems the issue of finite-size effects is traditionally expressed in terms of the finite-size corrections to the free-energy densities of each of the two phases:

$$f_\alpha(c,\infty) - f_\alpha(c,N), \qquad \text{where} \quad f_\alpha(c,N) = \frac{1}{N}\mathcal{F}_\alpha(c,N) \tag{71}$$

In recent years, substantial efforts have been made to develop a theoretical framework for understanding the nature of such corrections [93]. In the case of *lattice models* (i.e., models of strictly localized particles) in the *NVT ensemble* with *periodic boundary conditions* (PBCs), it has been established *a priori* [94] and corroborated in explicit simulation [95] that the corrections are *exponentially small* in the system size [96].

However, these results do not immediately carry over to the problems of interest here where (while PBCs are the norm) the ensembles are frequently open or constant pressure, and the systems do not fit in to the lattice model framework. Even in the apparently simple case of crystalline solids in *NVT*, the free translation of the center of mass introduces N-dependent phase space factors in the configurational integral which manifest themselves as additional finite-size corrections to the free energy; these may not yet be fully understood [58, 97]. If one adopts the traditional stance, then, one is typically faced with having to make extrapolations of the free-energy densities in *each* of the two phases, *without* a secure understanding of the underlying form ($\frac{1}{N}$? $\frac{\ln N}{N}$? ...) of the corrections involved.

The problems are reduced if one shifts the focus of attention from the single-phase free energies to the quantities of real interest: the difference between the two free energies and the field values that identify where it vanishes. In both cases within an ES strategy that treats both phases together, there is only *one* extrapolation to do, which is clearly a step forward. If the two phases are of the same generic type (e.g., two crystalline solids), one can expect cancellations of corrections of the form $\frac{\ln N}{N}$—which should be identical in both phases—leaving presumably at most $\frac{1}{N}$ corrections. If the phases are not of the same generic type (a solid and a liquid), the logarithmic corrections will probably not cancel; reliable extrapolation will be possible only if they can be identified and allowed for explicitly, leaving only a pure power law.

Overall it seems that there is considerable room for progress here.

2. *Near-Critical Systems*

In the case of simulation studies of near-critical systems, the issues associated with "finite-size effects" have an altogether different flavor. First, they are no longer properly regarded as essentially small effects to be "corrected for"; the critical region is characterized by a strong and distinctive dependence of system properties on system size; the right strategy is to address that dependence head on and exploit it. Second, there is an extensive framework to appeal to here: The phenomenology of *finite-size scaling* [93] and the underpinning theoretical structure of the *renormalization group* [98] together show what to look for and what to expect. Third, the objectives are rather different, going well beyond the issues of phase-diagram mapping that we are preoccupied with here.

Our discussion will be substantially briefer than it might be; we will focus on the issue most relevant here (but not the most interesting): the location of the critical point in a fluid.

As in the coexistence curve problem, the key is the distribution of the order parameter, in this case the density. On the coexistence curve, remote from the critical point (i.e., in the region $\xi \ll L$) we have seen (Fig. 5) that this distribution comprises two peaks of equal area, each roughly centered on the corresponding single-phase average; the two peaks are narrow and near-Gaussian in form (the more so, the larger the system size) as one would expect from the Central Limit Theorem; the probability of interphase tunneling (an inverse measure of the ergodic barrier) is vanishingly small [50]. As one moves up the coexistence curve, simulations show more or less what one would guess: The peaks broaden, they become less convincingly Gaussian, and the tunneling probability increases; this is the natural evolution en route to the form that must be appropriate in the one-phase region beyond criticality—a single peak narrowing with increasing L, asymptotically Gaussian. Against this immediately intelligible backdrop, the behavior at criticality (Fig. 10) is comparatively subtle. Again (for large enough L, but still $L \ll \xi$) a limiting form is reached; however, that limiting form comprises neither one Gaussian nor two, but something in between. The distribution narrows with increasing L (while preserving the shape of the "limiting form"); however, here the width varies not as $1/L^{d/2} = 1/\sqrt{N}$ (familiar from Central Limit behavior) but as $1/L^{\beta/\nu}$, where β and ν are the critical exponents of the order parameter and the correlation length respectively. There is good reason to believe that the shape of the distribution shares the distinctive quality of the critical exponents—all are universal signatures of behavior common to a wide range of physically disparate systems. The idea (of long-standing [99]) that fluids and Ising magnets belong to the same universality class is corroborated (beyond the exponent level) by the correspondence between their critical-point order parameter distributions [100, 100a]. Once that correspondence has been established to one's satisfaction, one is at liberty to *exploit* it to refine the assignment of fluid critical-point parameters. The form of the density distribution depends sensitively on the choice for the controlling fields (chemical potential μ and temperature T); *demanding* correspondence with the form appropriate for the Ising universality class (studied extensively and known with considerable precision) sets narrow bounds on the location of the fluid critical point [44].

Finally we should not overlook one—perhaps unexpected—feature of the critical distribution: It shows that clear signatures of the two (incipient) phases persist along the coexistence curve and *right through to the critical point* in a finite system. Failure to appreciate this will lead (as it has in the past) to substantial overestimates of critical temperatures [100b].

VII. DEALING WITH IMPERFECTION

Real substances often deviate from the idealized models employed in simulation studies. For instance, many complex fluids, whether natural or synthetic in origin, comprise mixtures of *similar* rather than *identical* constituents. Similarly, crystalline phases usually exhibit a finite concentration of defects that disturb the otherwise perfect crystalline order. The presence of imperfections can significantly alter phase behavior with respect to the idealized case. If one is to realize the goal of obtaining quantitatively accurate simulation data for real substances, the effects of imperfections must be incorporated. In this section we consider the state-of-the-art in dealing with two kinds of imperfection, polydispersity and point defects in crystals.

A. Polydispersity

Statistical mechanics was originally formulated to describe the properties of systems of *identical* particles such as atoms or small molecules. However, many materials of industrial and commercial importance do not fit neatly into this framework. For example, the particles in a colloidal suspension are never strictly identical to one another, but have a range of radii (and possibly surface charges, shapes, etc.). This dependence of the particle properties on one or more continuous parameters is known as polydispersity. One can regard a polydisperse fluid as a mixture of an infinite number of distinct particle species. If we label each species according to the value of its polydisperse attribute, σ, the state of a polydisperse system entails specification of a density *distribution* $\rho(\sigma)$, rather than a finite number of density variables. It is usual to identify two distinct types of polydispersity: *variable* and *fixed*. Variable polydispersity pertains to systems such as ionic micelles or oil–water emulsions, where the degree of polydispersity (as measured by the form of $\rho(\sigma)$) can change under the influence of external factors. A more common situation is fixed polydispersity, appropriate for the description of systems such as colloidal dispersions, liquid crystals, and polymers. Here the form of $\rho(\sigma)$ is determined by the synthesis of the fluid.

Computationally, polydispersity is best handled within a grand canonical (GCE) or semi-grand canonical ensemble in which the density distribution $\rho(\sigma)$ is controlled by a conjugate chemical potential distribution $\mu(\sigma)$. Use of such an ensemble is attractive because it allows $\rho(\sigma)$ to fluctuate as a whole, thereby sampling many different realizations of the disorder and hence reducing finite-size effects. Within such a framework, the case of variable polydispersity is considerably easier to tackle than fixed polydispersity: The phase behavior is simply obtained as a function of the width of the prescribed $\mu(\sigma)$ distribution. Perhaps for this reason, most simulation studies of phase behavior in polydisperse systems have focused on the variable case [90, 101–103].

Handling fixed polydispersity is computationally much more challenging: One wishes to retain the efficiency of the GCE, but to do so, a way must be found to adapt the imposed form of $\mu(\sigma)$ such as to realize the prescribed form of $\rho(\sigma)$. This task is complicated by the fact that $\rho(\sigma)$ is a *functional* of $\mu(\sigma)$. Recently, however, a new approach has been developed which handles this difficulty. The key idea is that the required form of $\mu(\sigma)$ is obtainable iteratively by functionally minimizing a cost function quantifying the deviation of the measured form of $\rho(\sigma)$ from the prescribed "target" form. For efficiency reasons, this minimization is embedded within a histogram reweighting scheme (Appendix B), obviating the need for a new simulation at each iteration. The new method is efficient, as evidenced by tests on polydisperse hard spheres [104] where it permitted the first direct simulation measurements of the equation of state of a polydisperse fluid.

B. Crystalline Defects

Defects in crystals are known to have a potentially major influence on phase behavior. For instance, dislocation unbinding is believed to be central to the 2D melting transition, while in 3D there is evidence to suggest that defects can act as nucleation centers for the liquid phase [105]. In superconductors, defects can pin vortices and influence "vortex melting" [106].

Almost all computational studies of defect free energies (and their influence on phase transitions) have been concerned with point defects. Polson et al. [97] used the results of early calculations [107] of the vacancy free energy of a hard-sphere crystal, to estimate the equilibrium vacancy concentration at melting. Comparison with the measured free energies of the perfect hard-sphere crystal (obtained from the Einstein Crystal NIRM Method discussed in Section IV.A) allowed them to estimate the effect of vacancies on the melting pressure, predicting a significant shift. A separate calculation for interstitials found their equilibrium concentration at melting to be three orders of magnitude smaller than that of vacancies. In follow-up work, Pronk and Frenkel [108] used a technique similar to the Widom particle insertion method (Section V.B.1) to calculate the vacancy free energy of a hard-sphere crystal. For interstitial defects they employed an extended sampling technique in which a tagged "ghost particle" is grown reversibly in an interstitial site.

To our knowledge there have been no reported measurements of equilibrium defect concentrations in soft-sphere models. Similarly, relatively few measurements have been reported of defect free energies in models for real systems. Those that exist rely on integration methods to connect the defective solid to the perfect solid. In *ab initio* studies the computational cost of this procedure can be high, although results have recently started to appear, most notably for vacancies and interstitial defects in silicon. For a review see Ref. 109.

VIII. OUTLOOK

Those who read this chapter may share with its authors the feelings expressed in Ref. 110: The dynamics in this particular problem space seems to have been rather more diffusive than ballistic. It is therefore wise to have some idea of where the ultimate destination is and to be familiar with the strategies that are most likely to take us there.

The long-term goal is a computational framework that will be grounded in electronic structure as distinct from phenomenological particle potentials; that will predict global phase behavior *a priori*, rather than simply decide between two nominated candidate phases; and that will handle quantum behavior, in contrast to the essentially classical framework on which we have focused here. That goal is distant but not altogether out of sight. Integrating *ab initio* electronic structure calculations with the statistical mechanics of phase behavior has already received some attention [109, 111]. The WL algorithm ([112], Appendix A) offers a glimpse of the kind of self-monitoring configuration-space search algorithm that one needs to make automated *a priori* predictions of phase behavior possible. And folding in quantum mechanics requires only a dimensionality upgrade [113].

There are of course many other challenges, a little less grand: the two space and time scales arising in asymmetrical binary mixtures [114]; the fast attrition (exponential in the chain length) for the insertion of polymers in NPT-TP or grand-canonical methods [115]; the long-range interactions in coulombic fluids [116]; the extended equilibration times for dense liquids near the structural glass transition [117]; and the extreme long-time dynamics of the escape from metastable states in nanoscale ferromagnets [118].

As regards the strategies that seem most likely to take us forward, we make three general observations:

Making the most of the information available in simulation studies requires an understanding of finite-size effects; it also requires awareness of the utility of quantities that one would not naturally consider were one restricted to pen and paper.

We will surely need new algorithms; they come from physical insight into the configurational core of the problem at hand.

One needs to match formulations to the available technology: parallel computing architectures give some algorithms a head start.

APPENDIX A: BUILDING EXTENDED SAMPLING DISTRIBUTIONS

In contrast to canonical sampling distributions whose form can be *written down* [Eq. (1)] the Extended Sampling (ES) distributions discussed in Section III.C have to be *built*. There is a large literature devoted to the building techniques, extending back at least as far as Ref. 22. We restrict our attention to relatively recent developments (those that seem to be reflected in current practices), and we shall focus on those aspects that are most relevant to ES distributions facilitating two-phase sampling.

In the broad-brush classification scheme offered in Section III.C, the domain of an ES distribution may be prescribed by a range of values of one or more *fields*, or one or more *macrovariables*. We shall focus on the latter representation, it seems simpler to manage. The generic task then is to construct a ("multicanonical") sampling distribution that will visit all (equal-sized) intervals within a chosen range of the nominated macrovariable(s) with roughly equal probability: The multicanonical distribution of the macrovariable(s) is essentially flat over the chosen range.

In formulating a strategy for building such a distribution, most authors have chosen to consider the particular case in which the macrovariable space is one-dimensional and is spanned by the configurational *energy*, E [119]. The choice is motivated by the fact that a distribution that is multicanonical in E samples configurations typical of a range of *temperatures*, providing access to the simple (reference-state) behavior that often sets in at high or low temperatures. For the purposes of two-phase sampling, we typically need to track a path defined on some macrovariable other than the energy (ideally, in *addition* to it; we will come back to this). The hallmarks of a "good" choice are that in some region of the chosen variable (inevitably one with intrinsically low equilibrium probability) the system may pass (has a workably large chance of passing) from one phase to the other. In discussing the key issues, then, we shall have in mind this kind of quantity; we shall continue to refer to it as an order parameter and denote it by M.

One can easily identify the generic structure of the sampling distribution we require. It must be of the form

$$P_{\mathcal{S}}(\{q\}) \doteq \frac{P_0(\{q\} \mid c)}{\hat{P}_0(M(\{q\}))} \tag{A1}$$

Here $\hat{P}_0(M)$ is an *estimate* of the true canonical M-distribution. Appealing to the sampling identity 24, the resulting M-distribution is

$$P_{\mathcal{S}}(M) = \langle \delta[M - M(\{q\})] \rangle_{\mathcal{S}} \doteq \left\langle \frac{P_{\mathcal{S}}}{P_0} \delta[M - M(\{q\})] \right\rangle_0 \doteq \frac{P_0(M \mid c)}{\hat{P}_0(M)} \tag{A2}$$

and is multicanonical (flat) to the extent that our estimate of the canonical M-distribution is a good one.

We can also immediately write down the prescription for generating the ensemble of configurations defined by the chosen sampling distribution. We need a simple MC procedure with acceptance probability [Eq. (13)], with the presumption of Eq. (16)]

$$P_{\mathscr{A}}(\{q\} \to \{q'\}) = \min\left\{1, \frac{P_{\mathscr{S}}(\{q'\})}{P_{\mathscr{S}}(\{q\})}\right\} \tag{A3}$$

In turning this skeleton framework into a working technique, one must make choices in regard to three key issues:

1. How to *parameterize* the estimator $\hat{P}_0$.
2. What *statistics* of the ensemble to use to guide the update of $\hat{P}_0$.
3. What *algorithm* to use in updating $\hat{P}_0$.

The second and third issues are the ones of real substance; the issue of parameterization is important only because the proliferation of different choices that have been made here may give the impression that there are more techniques available than is actually the case. That proliferation is due, in some measure, to the preoccupation with building ES distributions for the *energy*, E. There are many "natural parameterizations" here because there are many ways in which E appears in canonical sampling. Thus Berg and Neuhaus [27] employ an E-dependent *effective temperature*; Lee [120] utilizes a *microcanonical entropy* function; Wang and Landau [112] focus on a *density of states* function. Given our concern with macrovariables other than E, the most appropriate parameterization of the sampling distribution here is through a *multicanonical weight function* $\hat{\eta}(M)$, in practice represented by a discrete set of multicanonical weights $\{\hat{\eta}\}$. Thus we write [121]

$$\hat{P}_0(M \mid c) \doteq e^{-\hat{\eta}(M)} \tag{A4}$$

implying [through Eq. (A1)] a sampling distribution

$$P_{\mathscr{S}}(\{q\}) \doteq \sum_{j=1}^{\Omega} e^{-\mathscr{E}(\{q\})+\hat{\eta}_j} \Delta_j[M] \doteq e^{-\beta E(\{q\})+\hat{\eta}[M(\{q\})]} \tag{A5}$$

which is of the general form of Eq. (32), with $w_j \equiv e^{\hat{\eta}_j}$.

There are broadly two strategic responses to the second of the issues raised above: To drive the estimator in the right direction, one may appeal to the *statistics of visits to macrostates* or to the *statistics of transitions between macrostates*. We divide our discussion accordingly.

Statistics of Visits to Macrostates

The extent to which any chosen sampling distribution (weight function) meets our requirements is reflected most directly in the M-distribution it implies. One can estimate that distribution from a histogram $H(M)$ of the macrostates visited in the course of a set of MC observations. One can then use this information to refine the sampling distribution to be used in the next set of MC observations. The simplest update algorithm is of the form [122]

$$\hat{\eta}(M) \rightarrow \hat{\eta}(M) - \ln[H(M) + 1] + k \tag{A6}$$

The form of the logarithmic term serves to ensure that macrostates registering no counts (of which there will usually be many) have their weights incremented by the same finite amount (the positive constant k [123]); macrostates that *have* been visited are (comparatively) down-weighted.

Each successive iteration comprises a fresh simulation, performed using the weight function yielded by its predecessor; since the weights attached to unvisited macrostates is enhanced (by k) at every iteration that fails to reach them, the algorithm plumbs a depth of probabilility that grows exponentially with the iteration number. The iterations proceed until the sampling distribution is roughly flat over the entire range of interest.

There are many tricks of the trade here. One must recognize the interplay between signal and noise in the histograms: The algorithm will converge only as long as the signal is clear. To promote faster convergence, one can perform a linear extrapolation from the sampled into the unsampled region. One may bootstrap the process by choosing an initial setting for the weight function on the basis of results established on a smaller (computationally-less-demanding) system. To avoid spending excessive time sampling regions in which the weight function has already been reliably determined, one can adopt a multistage approach. Here one determines the weight function separately within slightly overlapping windows of the macrovariable. The individual parts of the weight function are then synthesized using multihistogram reweighting (Appendix B) to obtain the full weight function. For further details the reader is referred to Ref. 124.

The strategy we have discussed is generally attributed to Berg and Neuhaus (BN) [27]. Wang and Landau (WL) have offered an alternative formulation [112]. To expose what is different and what is not, it is helpful to consider first the case in which the macrovariable is the energy, E. Appealing to what one knows *a priori* about the canonical energy distribution, the obvious parameterization is

$$\hat{P}_0(E) \doteq \hat{G}(E)e^{-\beta E} \tag{A7}$$

where $\hat{G}(E)$ is an estimator of the *density of states* function $G(E)$. Matching this parameterization to the multicanonical weight function $\hat{\eta}(E)$ implied by choosing $M = E$ in Eq. (A4), one obtains the correspondence

$$\hat{\eta}(E) = \beta E - \ln \hat{G}(E) \tag{A8}$$

There is thus no major difference here. The differences between the two strategies reside instead in the procedure by which the parameters of the sampling distribution are *updated*, and the point at which that procedure is *terminated.*

Like BN, WL monitors visits to macrostates. But, while BN updates the weights of *all* macrostates after *many* MC steps, WL updates its "density of states" for *the current* macrostate after *every* step. The update prescription is

$$\hat{G}(E) \rightarrow f\hat{G}(E) \tag{A9}$$

where f is a constant, greater than unity. As in BN, a visit to a given macrostate tends to reduce the probability of further visits. But in WL this change takes place *immediately* so the sampling distribution evolves on the basic time scale of the simulation. As the simulation proceeds, the evolution in the sampling distribution irons out large differences in the sampling probability across E-space, which is monitored through a histogram $H(E)$. When that histogram satisfies a nominated "flatness criterion," the entire process is repeated (starting from the current $\hat{G}(E)$, but zeroing $H(E)$) with a smaller value of the weight-modification factor, f.

Like BN, then, the WL strategy entails a two-time-scale iterative process. But in BN the aim is only to generate a set of weights that can be utilized in a further, final multicanonical sampling process; the iterative procedure is terminated when the weights are sufficiently good to allow this. In contrast, in WL the iterative procedure is pursued further—to a point [125] where $\hat{G}(E)$ may be regarded as a definitive approximation to $G(E)$, which can be used to compute any (single phase) thermal property at any temperature through the partition function

$$Z(\beta) = \int dE\, G(E) e^{-\beta E} \tag{A10}$$

In the context of energy sampling, then, BN and WL achieve essentially the same ends, by algorithmically different routes. Both entail choices (in regard to their update schedule) that have to rest on experience rather than on any deep understanding. WL seems closer to the self-monitoring ideal, and it may scale more favorably with system size.

The WL procedure *can* be applied to any chosen macrovariable, M. But while a good estimate $\hat{G}(E)$ is sufficient to allow multicanonical sampling in E [and a definitive one is enough to determine $Z(\beta)$, Eq. (A10)], the M-density of states does not itself deliver the desired analogues; we need, instead, the *joint* density of states $G(E, M)$ which determines the restricted, single-phase partition functions through

$$Z_\alpha(\beta) = \int dE \int dM\, G(E, M) e^{-\beta E} \Delta_\alpha[M] \tag{A11}$$

The WL strategy *does* readily generalize to a 2D macrovariable space. The substantially greater investment of computational resources is offset by the fact that the relative weights of the two phases can be determined at *any* temperature. Reference 126 provides one of the few illustrations of this strategy, which seems simple, powerful, and general.

Statistics of Transitions Between Macrostates

The principal general feature of the algorithms based on visited macrostates is that the domain of the macrovariable they explore expands relatively slowly into the regions of interest. The algorithms we now discuss offer significant improvement in this respect. Although (inevitably, it seems) they exist in a variety of guises, they have a common core that is easily established. We take the general detailed balance condition [Eq. (11)] and sum over configurations $\{q\}$ and $\{q'\}$ that contribute (respectively) to the macrostates M_i and M_j of some chosen macrovariable M. We obtain immediately

$$P_{\mathcal{S}}(M_i)\overline{P_{\mathcal{S}}(M_i \to M_j)} = P_{\mathcal{S}}(M_j)\overline{P_{\mathcal{S}}(M_j \to M_i)} \tag{A12}$$

The terms with overbars are macrostate transition probabilities (TP). Specifically, $\overline{P_{\mathcal{S}}(M_i \to M_j)}$ is the probability (per unit time, say) of a transition from some nominated configuration in M_i to *any* configuration in M_j, ensemble-averaged over the configuration in M_i. Adopting a more concise (and suggestive) matrix notation, we have

$$p^M_{\mathcal{S}}[i]\rho^M_{\mathcal{S}}[ij] = p^M_{\mathcal{S}}[j]\rho^M_{\mathcal{S}}[ji] \tag{A13}$$

This is a not-so-detailed balance condition; it holds for any sampling distribution and any macrovariable [122]. The components of the eigenvector of the TP matrix (of eigenvalue unity) thus identify the macrostate probabilities. This is more useful than it might seem. One can build up an approximation of the transition matrix by monitoring the transitions [126a, 126b] that follow when a simulation is launched from an *arbitrary* point in configuration space. The "arbitrary" point can be judiciously sited in the heart of the interesting region; the subsequent simulations then carry the macrovariable right though the chosen region, allowing one to accumulate information about it from the outset. With a sampling distribution parameterized as in Eq. (A5), the update scheme is simply

$$\hat{\eta}(M) \rightarrow \hat{\eta}(M) - \ln[\hat{P}_{\mathcal{S}}(M)] + k \tag{A14}$$

where $\hat{P}_{\mathcal{S}}(M)$ is the estimate of the multicanonical M-distribution that is *deduced* from the measured TP matrix [127]. Reference 55 describes the application of this technique to a structural phase transition.

One particular case of Eq. (A12) has attracted considerable attention. If one sets $M = E$ and considers the infinite temperature limit, the probabilities of the macrostates E_i and E_j can be replaced by the associated values of the density-of-states function $G(E_i)$ and $G(E_j)$. The resulting equation has been christened *the broad-histogram relation* [128]; it forms the core of extensive studies of transition probability methods referred to variously as "flat histogram" [129] and "transition matrix" [130]. Applications of these formulations seem to have been restricted to the situation where the energy is the macrovariable, and the energy spectrum is discrete.

Methods utilizing macrostate transitions do have one notable advantage with respect to those that rely on histograms of macrostate visits. In transition methods the results of separate simulation runs (possibly initiated from different points in macrovariable space) can be straightforwardly combined: One simply aggregates the contributions to the transition-count matrix [122]. Synthesizing the information in separate histograms (see Appendix B) is less straightforward. The easy synthesis of data sets makes the TP method ideally suited for implementation in parallel architectures. Whether these advantages are sufficient to offset the fact that TP methods are undoubtedly more complicated to implement is, perhaps, a matter of individual taste.

APPENDIX B: HISTOGRAM REWEIGHTING

Without loss of generality the effective configurational energy may always be written in the form

$$\mathcal{E}(\{q\}, \{\lambda\}) = -\sum_{\mu} \lambda^{(\mu)} M^{(\mu)}(\{q\}) \tag{B1}$$

where $\{\lambda\}$ and $\{M\}$ are sets [88] comprising one or more mutually conjugate fields and macrovariables [131]. We consider two ensembles that differ only in the values of one or more of the fields $\{\lambda\}$. The canonical sampling distributions of the two ensembles are then related by

$$P_0(\{q\}|\{\lambda'\})e^{-\sum_\mu \lambda'^{(\mu)} M^{(\mu)}(\{q\})} \doteq P_0(\{q\}|\{\lambda\})e^{-\sum_\mu \lambda^{(\mu)} M^{(\mu)}(\{q\})} \tag{B2}$$

where, again, $\doteq$ signifies equality to within a configuration-independent constant. Performing the configurational sum for *fixed* values of the macrovariables $\{M\}$ then yields the relationship

$$P_0(\{M\}|\{\lambda'\})e^{-\sum_\mu \lambda'^{(\mu)} M^{(\mu)}} \doteq P_0(\{M\}|\{\lambda\})e^{-\sum_\mu \lambda^{(\mu)} M^{(\mu)}} \tag{B3}$$

so that

$$P_0(\{M\}|\{\lambda'\}) \doteq P_0(\{M\}|\{\lambda\})e^{-\sum_\mu [\lambda^{(\mu)} - \lambda'^{(\mu)}] M^{(\mu)}} \tag{B4}$$

In principle then, the $\{M\}$-distribution for *any* values of the fields can be inferred from the $\{M\}$-distribution for *one* particular set of values [132]. This is the basis of the Histogram Reweighting (HR) Technique [133], also known as Histogram Extrapolation [134]. It can be seen as a special case of the sampling identity given in Eq. (24) [135]. Like that identity, its formal promise is not matched by what it can deliver in practice. The measurements in the $\{\lambda\}$-ensemble (from which the extrapolation is to be made) determine only an *estimate* of the $\{M\}$-distribution for that ensemble. The estimate will be relatively good for the most probable $\{M\}$ values (around the "peak" of that distribution) which are "well-sampled" and will be relatively poor for the less probable values (in the "wings"), which are sampled less well. This trade-off (desirable if one wants only properties of the $\{\lambda\}$ ensemble) limits the range of field-space over which extrapolation will be reliable. The distribution associated with a set $\{\lambda'\}$ remote from $\{\lambda\}$ may peak in a region of $\{M\}$-space far from the peak of the measured distribution, lying instead in its poorly sampled wings. In such circumstances the estimate provided by the extrapolation prescribed by Eq. (B4) will be unreliable (and will typically reveal itself as such, through its ragged appearance).

In the scheme we have discussed, extrapolations are made on the basis of a histogram determined at a *single* point in field-space. The multihistogram method [133] extends this framework. It entails a sequence of separate simulations spanning a range of the field (or fields) whose conjugate macrovariable(s) are of interest. The intervals are chosen so that the tails of the histograms of the macrovariable accumulated at neighboring state points overlap. It is clear from the discussion above that in principle *every* histogram will provide some information about *every* region and that the most reliable information about any *given* region will come from the histogram which samples that region most effectively. These ideas can be expressed in an explicit prescription for synthesizing all the histograms to give an estimate of the canonical distribution of the macrovariable across the whole range of fields [133, 136].

References

1. R. Evans, in *Fundamentals of Inhomogeneous Fluids*, D. Henderson, ed., Marcel Dekker, New York, 1992.
2. C. Caccamo, *Phys. Rep.* **274**, 1 (1996).
3. R. A. Cowley, *Adv. Phys.* **12**, 421 (1963).
4. H. N. V. Temperley, J. S. Rowlinson, and G. S. Rushbrooke, *Physics of Simple Liquids*, North-Holland, Amsterdam, 1968.
5. N. B. Wilding (unpublished).
6. In the arguments of exponentials we shall generally absorb factors of $kT \equiv 1/\beta$ into effective energy functions.
7. K. Binder and D. W. Heermann, *Monte Carlo Simulation in Statistical Physics*, Springer, New York, 1998.
8. D. P. Landau and K. Binder, *A Guide to Monte Carlo Simulations in Statistical Physics*, Cambridge University Press, New York, 2000.
9. O. Narayan and A. P. Young, *Phys. Rev. E* **64**, 021104 (2001).
10. N. Metropolis, A. W. Rosenbluth, M. N. Rosenbluth, A. H. Teller, and E. Teller, *J. Chem. Phys.* **21**, 1087 (1953).
11. It would be more circumspect to say that the partition function cannot in general be *naturally or usefully* expressed as a canonical average since that average inevitably falls into the category for which canonical sampling is inadequate [13].
12. L. D. Fosdick, *Methods Comput. Phys.* **1**, 245 (1963).
13. In general the expectation value of an observable which is an *exponential* function of some *extensive* ("macroscopic") property cannot be reliably estimated from its sample average if one chooses $P_{\mathcal{S}} = P_0$.
14. Throughout this section we use the notation introduced in Eq. (7) in which the phase label is absorbed into an extended macrostate label $\mathcal{C}$. The path defined by the set of macrostates $\{\mathcal{C}\}$ may (and sometimes will) extend from one phase to another.
15. We shall refer to a field λ and a macrovariable M as conjugate if the λ-dependence of the configurational energy is of the form $\mathcal{E}(\{q\}, c, \lambda) = \mathcal{E}(\{q\}, c) - \lambda M(\{q\})$.
16. It would be more conventional to refer to "densities," which are *intensive*. But we prefer to deal with their *extensive* counterparts, which we shall refer to as *macrovariables*. The preference reflects our focus on simulation studies which are necessarily conducted on systems of some given (finite) size.
17. Y. Iba, *Int. J. Mod. Phys. C* **12**, 623 (2001).
18. C. J. Geyer and E. A. Thompson, *J. R. Stat. Soc. B* **54**, 657 (1992).
19. K. Hukushima, H. Takayama, and K. Nemoto, *Int. J. Mod. Phys. C.* **3**, 337 (1996).
20. E. Marinari, Optimized Monte Carlo Methods, in *Advances in Computer Simulation*, J. Kertesz and I. Kondor, eds., Springer-Verlag, New York, 1997).
21. More explicitly, $\mathcal{C}_1 \equiv \alpha, c$ and $\mathcal{C}_2 \equiv \tilde{\alpha}, c$.

22. G. M. Torrie and J. P. Valleau, *Chem. Phys. Lett.* **28**, 578 (1974); G. M. Torrie and J. P. Valleau, *J. Comp. Phys.* **23**, 187 (1977).
23. A. P. Lyubartsev, A. A. Martsinovski, S. V. Shevkunov, and P. N. Vorontsov-Velyaminov, *J. Chem. Phys.* **96**, 1776 (1992).
24. E. Marinari and G. Parisi, *Europhys. Lett.* **19**, 451 (1992).
25. J. P. Valleau, *Adv. Chem. Phys.*, **105** 369 (1999).
26. M. Mezei, *J. Comp. Phys.*, **68**, 237 (1987).
27. B. A. Berg and T. Neuhaus, *Phys. Lett. B* **267**, 249 (1991); B. A. Berg and T. Neuhaus, *Phys. Rev. Lett.* **68**, 9 (1992).
28. D. Frenkel and A. J. C. Ladd, *J. Chem. Phys.* **81**, 3188 (1984).
29. D. Frenkel and B. Smit, *Understanding Molecular Simulation*, Academic Press, New York, 1996.
30. J. P. Hansen and L. Verlet, *Phys. Rev.* **184**, 151 (1969).
31. D. Frenkel, H. N. W. Lekkerkerker, and A. Stoobants, *Nature* **332**, 822 (1988).
32. W. G. Hoover and F. H. Ree *J. Chem. Phys.* **49**, 3609 (1968).
33. H. Ogura, H. Matsuda, T. Ogawa, N. Ogita, and A. Ueda, *Prog. Theor. Phys.* **58**, 419 (1977).
34. S. Pronk and D. Frenkel, *J. Chem. Phys.* **110**, 4589 (1999).
35. P. G. Bolhuis, D. Frenkel, S.-C. Mau, and D. A. Huse, *Nature* **388**, 235 (1997).
36. L. V. Woodcock, *Nature* **384**, 141 (1997); **388**, 236 (1997).
37. For hard spheres the Helmholtz free energy is purely entropic. Thus Table I shows the difference between the entropy per particle of the two phases, $[S_{fcc} - S_{hcp}]/N$.
38. Q. Yan and J. J. de Pablo, *J. Chem. Phys.* **111**, 9509 (1999).
39. Equation (10) shows that the *logarithm* of the probability ratio measures the free-energy difference, which is *extensive*.
40. In fact, HPT seems to require that *all* the points chosen for replicas lie suffiently close to coexistence that both phases have detectable signatures.
41. B. A. Berg, U. H. E. Hansmann, and T. Neuhaus, *Z. Phys. B* **90**, 229 (1993).
42. B. Grossmann, M. L. Laursen, T. Trappenberg, and U. J. Wiese, *Phys. Lett. B* **293**, 175 (1992).
43. Reference 44 and N. B. Wilding (unpublished).
44. N. B. Wilding, *Phys. Rev. E* **52**, 602 (1995).
45. Some of the results from Ref. 44 are included in the LJ phase diagram shown in Fig. 4.
46. Operationally we may regard a phase as "metastable" and assign it a "free energy" (in relation to some other, also metastable) if the conditions for single-phase sampling [see Eq. (21) and accompanying discussion] are met.
47. The configurations unearthed by the ES algorithm used in Lattice Switch studies of hard spheres (Section IV.D.2) have some distinctive features [48].
48. A. D. Bruce, A. N. Jackson, and G. J. Ackland, *Phys. Rev. E* **61**, 906 (2000).
49. See, for example, M. S. S. Challa and J. H. Hetherington, *Phys. Rev. A* **38**, 6324 (1988).
50. When the interphase traverse proceeds by way of such two-phase configurations the free-energy barrier to be surmounted must scale with the area of the interface supporting two phases in equal measure. The probability of the ravine macrostates (and thus the probability of tunneling between the two phases *at coexistence*) must vanish as e^{-F_I} where the free-energy cost of the interface is of order $F_I \sim \sigma_s L^{d-1}$, with σ_s the interface tension and $L \sim N^{1/d}$ the edge of the d-dimensional simulation cube.

51. See, for example, Ref. 41.
52. The method to be discussed in the next section belongs in this family; so does HPT, but less compellingly because of the limitations we have already touched on in Section IV.B.2.
53. See Ref. 29, p. 176.
54. The traverse might work acceptably well if the dynamics of the interface between the two phases is favorable: systems with martensitic phase transitions may fall into this category: Z. Nishiyama, *Martensitic Transformations*, Academic Press, New York, 1978. Note also that the special case in which the structural phase transition involves no change of symmetry can be handled within the standard multicanonical framework [55].
55. G. R. Smith and A. D. Bruce, *Phys. Rev. E* **53**, 6530 (1996).
56. A. D. Bruce, N. B. Wilding, and G. J. Ackland, *Phys. Rev. Lett.* **79**, 3002 (1997).
57. A. D. Bruce, A. N. Jackson, G. J. Ackland, and N. B. Wilding, *Phys. Rev. E* **65**, 36710 (2002).
58. N. B. Wilding and A. D. Bruce, *Phys. Rev. Lett.* **85**, 5138 (2000).
59. In general the partition function Eq. (2) should already incorporate a Jacobean reflecting the transformation from cartesian coordinates to the generalized coordinates $\{q\}$. It may always be absorbed into the configurational energy and we have adopted that convention.
60. The constraint is rarely made explict in MC studies: it is tacitly assumed that the conditions for "single-phase sampling" [Eq. (21) and accompanying discussion] are met.
61. C. H. Bennett, *J. Comp. Phys.* **22**, 245 (1976).
62. R. W. Zwanzig, *J. Chem. Phys* **22**, 1420 (1954).
63. A. Rahman and G. Jacucci, *Il Nuovo Cimento D* **4**, 357 (1984).
64. One sees this more clearly when one views this relationship as the result of integrating Eq. (55).
65. But a judicious choice of the *representations* chosen for the two phases, reflected in the transformation τ [Eq. (44)], can bring one closer to this attractive limit. See Refs. 70 and 71 and further remarks below.
66. We use the term "lattice" rather loosely. The set of vectors $\{\vec{R}\}_\alpha$ identifies the orthodox lattice convoluted with the orthodox basis.
67. This will frequently be true for solid–solid phase behavior too; in the systems studied with Lattice Switch methods to date, the effects have proved rather small.
68. M. C. Moody, J. R. Ray and A. Rahman, *J. Chem. Phys.* **84**, 1795 (1985).
69. A. F. Voter, *J. Chem. Phys.* **82**, 1890 (1985).
70. C. Jarzynski, *Phys. Rev. E* **65**, 046122 (2002).
71. A. R. Acharya, A. D. Bruce, and G. J. Ackland, in preparation.
72. A. Z. Panagiotopoulos, *Mol. Phys.* **61**, 813 (1987).
73. A. Z. Panagiotopoulos, *J. Phys. Cond. Matt.* **12**, R25 (2000).
74. J. P Valleau, *J. Chem. Phys.* **108**, 2962 (1998).
75. K. K. Mon and K. Binder, *J. Chem. Phys.* **96**, 6989 (1992).
76. A. D. Bruce, *Phys. Rev. E* **55**, 2315 (1997).
77. D. Möller and J. Fischer, *Mol. Phys.* **69**, 463 (1990); *ibid* **75**, 1461 (1992).
78. B. Widom, *J. Chem. Phys.* **39**, 2808 (1963).
79. A. Lotfi, J. Vrabec, and J. Fischer, *Mol. Phys.* **76**, 1319 (1992).
80. J. Vrabec and J. Fischer, *Mol. Phys.* **85**, 781 (1995).
81. See, for example, H. Falk, *Am. J. Phys.* **38**, 858 (1970).
82. C. Jarzynski, *Phys. Rev. Lett.* **78**, 2690 (1997).

83. W. P. Reinhardt and J. E. Hunter III, *J. Chem. Phys.* **97**, 1599 (1992).

84. M. A. Miller and W. P. Reinhardt, *J. Chem. Phys.* **113**, 7035 (2000).

85. A. G. Khatchaturyan, *Theory of Structural Phase Transformations in Solids*, John Wiley & Sons, New York, 1983).

86. A *continuous* space transformation like this can also be implemented in the lattice switch framework described in Section IV.D. By the same token, one can generally devise a continuous version of any LS transformation. Thus, for example, the obvious continuous counterpart of the *fcc-hcp* LS transformation depicted in Fig. 6 entails a *gradual shear* of the close-packed planes. Since the LS implementation remains always in the space of one or other of the crystalline structures, one might expect it to prove the safer choice. Where an explicit comparison of the two strategies has been made [87], this was indeed the conclusion.

87. S.-C. Mau and D. A. Huse, *Phys. Rev. E* **59**, 4396 (1999).

88. We use $\{\lambda\}$ to denote a set of physically different fields $\lambda_1, \lambda_2 \ldots$. In contrast, $[\lambda]$ (appearing in our discussion of paths) is shorthand for a set of different values $\lambda^{(1)}, \lambda^{(2)} \ldots$ of some (particular) field λ. We use $\{M\}$ and $[M]$ in a similar way.

89. It reduces to the familiar form with the macrovariables M_1 and M_2 chosen as respectively the enthalpy and the volume whose conjugate fields are $1/\mathrm{T}$ and P/T.

90. D. A. Kofke, *J. Chem. Phys.* **98**, 4149(1993); R. Agrawal and D. A. Kofke, *Phys. Rev. Lett.* **74**, 122 (1995).

91. P. G. Bolhuis and D. A. Kofke, *Phys. Rev. E* **54**, 634 (1996); D. A. Kofke and P. G. Bolhuis, *Phys. Rev. E* **59**, 618 (1999); M. Lisal and V. Vacek, *Molecular Simulation* **18**, 75 (1996); F. A. Escobedo and J. J. de Pablo, *J. Chem. Phys.* **106**, 2911 (1997).

92. Recall that knowing the weights that define a multicanonical distribution of some quantity M over some range is equivalent to knowing the true canonical M-distribution *throughout the chosen range*.

93. For a review of FSS see *Finite Size Scaling and Numerical Simulation of Statistical Systems*, V. Privman, ed., World Scientific Publishing, Singapore, 1990.

94. C. Borgs and R. Kotecký, *J. Stat. Phys.* **61**, 79 (1990); C. Borgs, R. Koteck, and S. Miracle-Sol, *J. Stat. Phys.* **62**, 529 (1991).

95. C. Borgs and W. Janke, *Phys. Rev. Lett.* **68**, 1738 (1992).

96. As an immediate corollary, the field values that satisfy the equal-areas criterion will identify the coexistence curve of two such ensembles to within exponentially small corrections [95].

97. J. M. Polson, E. Trizac, S. Pronk, and D. Frenkel, *J. Chem. Phys.* **112**, 5339 (2000).

98. J. J. Binney, N. J. Dowrick, A. J. Fisher, and M. E. J. Newman, *The Theory of Critical Phenomena: An Introduction to the Renormalization Group*, Oxford University Press, New York, 1992.

99. See, for example, A. L. Sengers, R. Hocken, and J. V. Sengers, *Phys. Today* **30**, 42 (1977).

100. A. D. Bruce and N. B. Wilding, *Phys. Rev. Lett.* **68**, 193 (1992); N. B. Wilding and A. D. Bruce, *J. Phys: Condens. Matter* **4**, 3087 (1992).

100a. The treatment of scaling-field mixing in Ref. 100 has recently been corrected: M. E. Fisher and G Orkoulas, *Phys. Rev. Lett.* **85**, 696 (2000); G. Orkoulas, M. E. Fisher and C. Üstün, *J. Chem. Phys* **113**, 7530 (2000).

100b. In a number of recent papers the ideas outlined here have been developed to yield a framework in which the liquid-vapor coexistence curve can be tracked with remarkable precision deep into the critical region: E. Luijten, M. E. Fisher and A. Z. Panagiotopoulos, *Phys. Rev. Lett.* **88**, 185701 (2002); Y. C. Kim, M. E. Fisher and E Luitjen, cond-mat/0304032.

101. M. R. Stapleton, D. J. Tildesley, and N. Quirke, *J. Chem. Phys.* **92**, 4456 (1990).
102. T. Kristof and J. Liszi, *Mol. Phys.* **99**, 167 (2001).
103. P. V. Pant and D. N. Theodorou, *Macromolecules* **28**, 7224 (1995).
104. N. B. Wilding and P. Sollich, *J. Chem. Phys.* **116**, 7116 (2002).
105. L. Gomez, A. Dobry, and H. T. Diep, *Phys. Rev. B* **63**, 224103 (2001).
106. I. A. Rudnev, V. A. Kashurnikov, M. E. Gracheva, and O. A. Nikitenko, *Physica C* **332**, 383 (2000).
107. C. H. Bennett and B. J. Alder, *J. Chem. Phys.* **54**, 4796 (1971).
108. S. Pronk and D. Frenkel, *J. Phys. Chem. B* **105**, 6722 (2001).
109. D. Alfe, G. A. De Wijs, G. Kresse, and M. J. Gillan *Int. J. Quant. Chem.* **77**, 871 (2000).
110. Ecclesiastes 1:9. [*The Holy Bible*, King James Version, American Bible Society, New York, 1999).]
111. G. J. Ackland, *J. Phys. Condens. Matter* **14**, 2975 (2002).
112. F. Wang and D. P. Landau, *Phys. Rev. Lett.* **86**, 2050 (2001); *Phys. Rev. E*, **64** 056101 (2001).
113. F. H. Zong, D. M. Ceperley, *Phys. Rev. E* **58**, 5123 (1998); C. Rickwardt, P. Nielaba, M. H. Muser, and K. Binder, *Phys. Rev. B* **63**, 045204 (2001).
114. M. Dijkstra, R. van Roij, and R. Evans, *Phys. Rev. Lett.* **82**, 117 (1999).
115. D. Frenkel, G. C. A. M. Mooij, B. Smit, *J. Phys. Condens. Matter* **4**, 3053 (1992); S. Consta, N. B. Wilding, D. Frenkel, and Z. Alexandrowicz, *J. Chem. Phys.* **110**, 3220 (1999).
116. M. E. Fisher and A. Z. Panagiotopoulos, *Phys. Rev. Lett.* **88**, 185701 (2002).
117. W. Kob, C. Brangian, T. Stuhn, and R. Yamamoto, in *Computer Simulation Studies in Condensed Matter Physics XIII*, D. P. Landau, S. P. Lewis, and H. B. Schttler, eds., Springer, Berlin, 2000.
118. M. Kolesik, M. A. Novotny, and P. A. Rikvold, cond-mat 0207405.
119. In this section we refer to E rather than its dimensionless counterpart $\mathscr{E} \equiv \beta E$, since it seems advisable to make the measure of temperature $\beta = 1/kT$ explicit.
120. J. Lee, *Phys. Rev. Lett.* **71**, 211 (1993); Erratum: *Phys. Rev. Lett.* **71**, 2353 (1993).
121. Different authors have made different choices of notation and adopted different sign conventions at this point. We adopt those of Ref. 122.
122. G. R. Smith and A. D. Bruce, *J. Phys. A* **28**, 6623 (1995).
123. The constant k merely serves to assign a convenient lower (or upper) bound to the weights; the value chosen is absorbed in the normalization of the sampling distribution.
124. B. A. Berg, *Int. J. Mod. Phys. C* **4**, 249 (1993); *Fields Inst. Commun.* **26**, 1 (2000).
125. The WL scheme is terminated when the histogram is "sufficiently" flat *and* the weight-modification factor f is "sufficiently" close to unity.
126. Q. Yan, R. Faller, and J. J. de Pablo, *J. Chem. Phys.* **116**, 8745 (2002).

126a. One may count the macrostate-transitions that are actually implemented, as in Ref. 55 or (retaining somewhat more information) store the acceptance probabilities of all attempted transitions, as in Ref. 126b.

126b. M. Fitzgerald, R. R. Picard and R. N. Silver, *Europhys. Lett* **46**, 282 (1999); *J. Stat. Phys.* **98**, 321 (2000).

127. There is a caveat here: The traverse through the interesting region needs to be slow enough to allow a rough local equilibrium in each macrostate to be established; this condition is satisfied relatively poorly at the outset (where the MC trajectory heads rapidly for the equilibrium states); but it gets better the closer the sampling distribution comes to the multicanonical limit; one requires no more than this.

128. P. M. C. de Oliveira, T. J. P. Penna and H. J. Herrmann, *Braz. J. Phys.* **26**, 677 (1996); P. M. C. de Oliveira, T. J. P. Penna, and H. J. Herrmann, *ibid Eur. Phys. J. B* **1**, 205 (1998); P. M. C. de Oliveira, *Eur. Phys. J. B* **6**, 111 (1998).

129. J. S. Wang and L. W. Lee, *Comput. Phys. Commun.* **127**, 131 (2000).

130. J. S. Wang, T. K. Tay, and R. H. Swendsen, *Phys. Rev. Lett.* **82**, 476 (1999); J. S. Wang and R. H. Swendsen, cond-mat/0104418.

131. We subsume into the definitions of the macrovariables any parameters that do not have the credentials ("linearity") of the fields $\{\lambda\}$.

132. The unspecified constant in Eq. (B3) is prescribed by normalization.

133. A. M. Ferrenberg and R. H. Swendsen, *Phys. Rev. Lett.* **61**, 2635 (1989); A. M. Ferrenberg and R. H. Swendsen, *Phys. Rev. Lett.* **63**, 1195 (1989); R. H. Swendsen, *Physica A* **194**, 53 (1993).

134. The name seems to reflect no more than the fact that the resulting estimates of PDFs are as always based on (normalized) histograms accumulated in the sampling process.

135. The observable Q is then a delta function that picks out prescribed values for the macrovariables in the set $\{M\}$.

136. To do so, one must determine a set of weights that moderate the relative contributions of each histogram. There is more than a passing similarity here to the expanded ensemble framework discussed in Section III.C.

MOLECULAR MODELS FOR CALCULATION OF DIELECTRIC/FAR-INFRARED SPECTRA OF LIQUID WATER

VLADIMIR I. GAIDUK and BORIS M. TSEITLIN

Institute of Electronics and Radio Engineering, Russian Academy of Sciences, Fryazino, 141190, Moscow Region, Russia

CONTENTS

List of Conventional Symbols
- I. Introduction
  - A. Short Survey of This Chapter
  - B. Prehistory of This Review
- II. Review of the ACF Method
  - A. Spectral Function and Complex Susceptibility (General Expressions)
    1. Connection of Complex Refraction Index with Current Density
    2. Representation of Current Density $J_E(t)$ by Using the t_0 Theorem
    3. Application of Statistical Distributions
    4. Average Perturbation (AP) Theorem and Its Consequences
    5. Spectral Function as a Dipole Autocorrelator
    6. Expression of Complex Susceptibility Through Spectral Function
  - B. Spectral Function of a Dipole Reorienting in a Local Axisymmetric Potential
    1. General Properties
    2. Ensemble Average
    3. Representation of Spectral Function in Terms of Fourier Amplitudes
  - C. About Correlation Between Dielectric Spectra and Parameters of a Molecular Model
    1. Localization of a Dipolar Ensemble
    2. Estimation of the Absorption-Peak Frequency
    3. About Absorption Bandwidth
    4. Characteristics of the Relaxation Spectrum
- III. Protomodel: Dipole in Rectangular Potential Well
  - A. A Dipole in a Conical Cavity
    1. Geometry and Law of Motion
    2. Configuration of Phase Space

Advances in Chemical Physics, Volume 127, Edited by I. Prigogine and Stuart A. Rice
ISBN 0-471-23583-0

B. Spectral Function (Rigorous Expression)
1. Representation by Series
2. Analytical Representation
C. Planar Librations–Regular Precession (PL–RP) Approximation
1. Calculation of the Librational Spectral Function $L_{PL}(z)$
2. Calculation of the Precessional Spectral Function $L_{RP}(z)$
3. Resulting Spectral Function. The Small β Approximation
4. Absorption Frequency Dependence
D. Statistical Parameters
E. Summation of Series in Expression for the Spectral Function
IV. Hat-Flat Model
A. Classification of Trajectories
B. Spectral Function of Librators
1. Representation in Form of a Series
2. Analytical Representation
3. Planar Libration–Regular Precession (PL–RP) Approximation
C. Spectral Function of Rotators
1. Small β Approximation
2. Rotation in Space Angle Close to Spherical Angle
3. Interpolation for Arbitrary Cone Angle β
D. Statistical Parameters
1. Proportions of Subensembles
2. Mean Localization
3. Mean Number of Reflections. Estimation of Absorption-Peak Frequency
E. Hybrid Model
F. Qualitative Spectral Dependences
G. Application to Polar Fluids
1. General Expressions for Spectral Characteristics
2. Application to Water H_2O
3. Application to Strongly Absorbing Nonassociated Liquid
V. Hat-Curved Model and Its Application for Polar Fluids
A. Problem of Unspecific Interactions
1. Evolution of Rectangular-Like Models
2. Evolution of "Field" Models
3. Hat-Curved Model as Symbiosis of Rectangular-Well and Parabolic-Well Models
B. Basic Assumptions of a Linear Response Theory
C. Molecular Subensembles and Spectral Function
1. Form of Potential and Classification of Particles' Subensembles
2. "Partial" Spectral Functions
3. Qualitative Description of Absorption Spectrum
D. Calculated Spectra of Polar Fluids
1. Dielectric and FIR Spectra of Liquid H_2O
2. Dielectric/FIR Spectra of Nonassociated Liquid (CH_3F)
3. Concluding Remarks
E. Spectral Function (Derivation)
1. Norm C of Boltzmann Distribution $W = C\exp[-h(\Gamma)]$
2. Phase Regions of Subensembles
3. General Representation of Spectral Functions
4. Spectral Function of Librators

5. Spectral Function of Precessors
6. Spectral Function of Rotators
7. Statistical Averages

Appendix 1. Calculation of Fourier Amplitudes b_{2n-1} for Librators
Appendix 2. Transformation of Integral for Spectral Function of Precessors
Appendix 3. Optical Constants of Liquid Water

VI. Specific Interactions in Water
A. Problem of Specific Interactions
B. Modified Spectral Function $R(z)$
C. Calculated Spectra
D. Discussion
1. Role of Specific Interactions
2. Role of Unspecific Interactions
3. Next Step: Composite Model Characterized by Two Relaxation Times
E. Modified Spectral Function (Derivation)
1. Purely Harmonically Changing $\tilde{\mu}(T)$
2. Account of Decaying Term $\exp(-t/\tau_{\text{vib}})$

VII. Composite Models: Application to Water
A. Hat-Curved–Harmonic Oscillator Composite Model
1. About Two Mechanisms of Dielectric Relaxation
2. Two Components of Complex Permittivity
3. Reorientation Process
4. Vibration Process
5. Results of Calculations
6. HC–HO Model: Discussion
B. Hat-Curved–Cosine-Squared Potential Composite Model
1. Theory and Calculated Spectra of H_2O and D_2O
2. HC–CS Model: Discussion
3. Conclusions
C. Nonrigid Oscillator: Linear-Response Theory for the Parabolic Potential
1. Equation of Motion of Harmonic Oscillator
2. Spectral Function for Back-and-Forth Motion of Two Charged Particles
3. Frequency Dependences Pertinent to Different Collision Models
4. About History of the Question

VIII. Dielectric Response of Aqueous Electrolyte Solutions
A. Schemes of Ion's Motion in Water Surroundings
1. Motion Along a Line
2. Motion Inside a Sphere
B. The Frequency-Dependent Ionic Conductivity
C. Effect of Ions on Wideband Spectra
D. Discussion

Appendix. Relation of Ionic Susceptibility to Conductivity

IX. Structural–Dynamical (SD) Model
A. Problem of Elastic Interactions
B. Dynamics of Elastic Interactions (Account of Stretching Force)
1. Equation of Motion and Its Solution
2. Dynamics of Transverse Vibrations
3. Mean Frequencies/Amplitudes and Corresponding Density Distributions. Form of Rotational Absorption Band

4. Numerical Estimations
C. Dynamics of Elastic Interactions (Generalized Consideration)
1. Calculation Scheme
2. Results of Calculations
D. Application of the ACF Method
1. Spectral Function of Restricted Rotators
2. Dielectric Response of Restricted Rotators
3. Composite HC–CD Model: Frequency Dependence of Wideband FIR Absorption
X. Conclusions and Perspectives
A. Effect of Temperature on Wideband Dielectric Spectra
B. Far-Infrared Dielectric Response of Ice
C. About Evolution of Molecular Models
D. Hat-Curved–Elastic Bond (HC–EB) Model as a Simplest Composite Model
Acknowledgments
References

LIST OF CONVENTIONAL SYMBOLS

a.c.	Alternating current (time-varying field)
ACF	Autocorrelation function
CS	Cosine squared (model)
FIR	Far infrared
GT	Ref. 1
HC	Hat-curved (model)
PL–RP	Planar libration–regular precession approximation
R–band	Band around $\nu = 200\ \text{cm}^{-1}$
RR	Restricted rotators
SD	Structural-dynamical (model)
SF	Spectral function
VIG	Ref. 2

Symbol	Value	Example	Symbol	Value	Example
i	$\sqrt{-1}$		$\breve{f}$	Symbol referring to librators	
$*$	Complex conjugation	$\chi^* = \chi' + i\chi''$	$\mathring{f}$	Symbol referring to rotators	
$\hat{E}$	Complex amplitude		$\in$	"is element of" sign	$x \in [a, b]$
$\langle f \rangle$	Ensemble average of quantity $f(\Gamma)$	$\langle f \rangle \equiv \int_{\Gamma} f\, W\, d\Gamma$	$\propto$	Proportionality sign	$G \propto N$

Symbols Common to All Sections

$c = 2.9979 \times 10^{10}$ cm/s	Speed of light
$C(t)$, $\tilde{C}(\omega)$	Single-particle dipolar ACF and its spectrum
$d\Gamma$	Element of phase-space volume
$e = 4.803242$ units of CGSE	Charge of electron
$E = \mathrm{Re}[\hat{E}e^{i\omega t}]$	Alternating current (a.c.) electric field
$\hat{E}$, E_m	Complex, real amplitude of radiation field
F	Induced distribution
$\{F_\mathrm{B}, F_\mathrm{G}, F_\mathrm{T}\}$	Boltzmann, Gross, isothermal-induced distributions
f	Form factor of the hat-curved model
$G = \mu^2 N / (3k_\mathrm{B}T)$	Normalized concentration of molecules
g	Kirkwood correlation factor
H	Steady-state energy (Hamiltonian) of a dipole
$h = H/(k_\mathrm{B}T)$	Dimensionless energy of a dipole
I	Moment of inertia of a molecule
$K_\parallel(z), K_\perp(z)$	Longitudinal and transverse components of the spectral function
$k = (\omega/c)\sqrt{\varepsilon} = k' - \mathrm{i}\, k''$	Complex propagation constant
k	Elasticity constant (in Section IX)
$k_\mathrm{B} = 1.3807 \times 10^{-16}$ erg/K	Boltzmann constant
$k\mu$	Factor correcting μ value in a liquid
$L(z)$	Spectral function pertinent to an isotropic medium
$\breve{L}$, $\mathring{L}$	Spectral function pertinent to librating and rotating particles
l	Projection of the normalized angular momentum on the symmetry axis
l	Maximal shift of an ion
M	Molecular mass (gram · mole)
m	Mass of a molecule
$m_\mathrm{H} = 1.6726 \times 10^{-24}$ g	Mass of proton
m_ion	Mass of ion
N	Concentration
$N_\mathrm{A} = 6.0220 \times 10^{23}$ (g · mole)$^{-1}$	Avogadro constant
$n^* = \mathrm{Re}\,[\sqrt{\varepsilon^*}] = n + i\kappa$	Complex refraction index
n_∞	Optical refraction index
P	Polarization
$q = \mu_\mathrm{E}/\mu$	Normalized dipole moment projection
r	Radius vector of a charged particle
$\breve{r}$, $\mathring{r}$	Proportions of librators and rotators

$S(Z) = \sigma/\omega_p$	Normalized complex conductivity
T	Temperature
t, t_v	Time, lifetime (random quantity)
U_0	Well depth
$u = U_0/(k_B T)$	Dimensionless well depth
v	Velocity of a particle
$W(\mathrm{H}) = C \exp[-H/(k_B T)]$	Steady-state distribution function
$X = \omega/\omega_p$	Dimensionless frequency (for electrolyte solutions)
$x = \omega\eta = 2\pi c \nu \eta$	Normalized frequency of a.c. field
$x_D = \eta/\tau_D = 2\pi c \nu_D \eta$	Normalized frequency of the loss maximum $\varepsilon''(\omega)$
$Y = (\omega_p \tau)^{-1}$	Dimensionless collision frequency (for electrolyte solutions
$y = \eta/\tau = \sqrt{I/(2k_B T)}/\tau$	Normalized collision frequency
$Z = X + iY$	Normalized complex frequency (for electrolyte solutions)
$z = x + iy$	Normalized complex frequency
$\alpha = \omega\epsilon''/cn = 4\pi\nu \mathrm{Im}(\sqrt{\varepsilon^*})$	Absorption coefficient (in cm^{-1})
β	Libration amplitude, cone angle
Γ	Phase-space volume
γ	Phase of a.c. field
δq	Charge "attached" to a water molecule
$\varepsilon = \varepsilon' + i\varepsilon''$, ε_s	Complex, static dielectric permittivity (dimensionless quantity)
$\eta = \sqrt{I/(2 k_B T)}$	Normalizing time parameter
Θ	Angle between a.c. field vector and the symmetry axis of the potential (for dipoles)
Θ	Angle between a.c. field vector and the normal to the plane of motion (for ions)
λ	Wavelength of electromagnetic radiation in vacuum
λ	Precessional shift of a dipole moment during a half period
$\lambda = r/L$	Normalized covalent OH length
$\mu = \mu_0 k_\mu (n_\infty^2 + 2)/3$	Dipole moment of a molecule in liquid
μ_0	Dipole moment of a molecule in vacuum
$\nu = \omega \cdot (2\pi c)^{-1} = 1/\lambda$	Wave number (in cm^{-1}), termed "frequency"
$\Pi = \int_{-\infty}^{\infty} \omega\chi''(\omega) d\omega$	Integrated absorption
ρ	Density
σ	Conductivity (in aqueous solutions of electrolytes)

τ, τ_{or}	Mean lifetime (for reorientation)
τ_{ion}	Mean lifetime (for ions)
τ_{vib}	Mean lifetime (for vibration)
τ_D	Debye relaxation time
Φ	Dimensionless period of the $q(t) = \mu_E(t)/\mu$ function
$\chi^* = \chi' + i\chi''$	Complex susceptibility
ψ	Current precessional shift of a dipole moment
$\omega = 2\pi c\nu$	Angular frequency of radiation
$\omega_p = 2e\sqrt{\pi N/m}$	Plasma frequency
$\Im$	Period of $q(t)$ function (dimensional quantity)

Symbols for Sections II–IV

w	Effective potential
θ	Angle between the symmetry axis and the normal to the plane of dipole's rotation
ϑ	Current deflection of a dipole moment from the symmetry axis
ξ	Mean dimensionless inverse period

Symbols for Sections V–VII

S	Normalized steepness in the hat-curved model
V	Effective potential
θ	Current deflection of a dipole moment from the symmetry axis
$\bar{\theta}$	The position of θ, at which the effective potential V attains its minimum value
Σ and $\breve{\Sigma}$, $\tilde{\Sigma}$, $\mathring{\Sigma}$	Statistical integral and its part, corresponding to contributions of librators, precessors, rotators
$\varphi = t/\eta$	Dimensionless (normalized, reduced) time
ϕ	Azimuthal coordinate (turn of a dipole-moment vector around the symmetry axis of a potential)

I. INTRODUCTION

One may consider this chapter to be a continuation of the previous work [1] published in ACP by the same authors and denoted further GT. In this work the theory was applied to strongly polar nonassociated liquids. Now we consider more important aqueous media, mostly liquid water. We shall present an analytical description of spectra of the complex dielectric permittivity and absorption based on a modern linear-response theory. The book [2] by one of the authors (denoted VIG) and the article GT will be used as a basis of our review, but we avoid repetition of the materials from Refs. 1 and 2. Only recent publications [3–12c] are included into the review.

In our approach [1, 2] termed the *dynamic method* the complex susceptibility $\chi = \chi' - i\chi''$ is determined by a law of *undamped* motion of a dipole in a given potential well and by *dissipation mechanism* often described as *stosszahlansatz* in the underlying kinetic or Boltzmann equation. In this review we shall refer to this (dynamic) method as the *ACF method*, since it is actually based on calculation of the *spectrum* of the dipolar autocorrelation function (ACF). Actually we use a one-particle approximation, in which the form of an employed potential well (being in many cases *rectangular or close* to it) is taken *a priori*. Correlation of the particles coordinates is characterized *implicitly* by the Kirkwood correlation factor g, its value being taken from the experimental data. The ACF method is *simple* and *effective*, because we do not employ the stochastic equations of motions. This feature distinguishes our method from other well-known approaches—for example, from those described in books [13, 14].

In the main group of molecular models studied here the dielectric response of a linear molecule, characterized by a moment of inertia I, is examined. A molecule librates/rotates in a conservative *intermolecular* potential. Besides, in the harmonic oscillator model such a response is determined by the elastic constant k and by the masses of two vibrating particles.

In the review also more complex systems of particles/charges are considered in an *additivity approximation.* Namely, a few specific types of motion are treated in terms of *quasi-independent* molecular ensembles, each being characterized by its own potential well.

Our theoretical group succeeded (sometimes in collaboration with other scientists) to elaborate models capable of describing spectra of water and aqueous media. Substantial progress, achieved during a period of several years, is reflected in the present review, where we consider the following new topics:

(i) Rigorous consideration of spatial motion of dipoles under influence of a nonhomogeneous conservative potential
(ii) Effect of vibrations of the H-bonded molecules on a wideband spectrum of liquid water (in other words, a polarization model of water is also considered)
(iii) Role of dispersion of the ionic conductivity pertinent to aqueous electrolyte solutions

(iv) Perspectives of modeling based on a simple pattern of the water structure

We do not know theoretical descriptions other than ours of the dielectric/FIR spectra applicable for water in the range from 0 to 1000 cm^{-1}, which were made on a molecular basis *in terms of complex permittivity* $\varepsilon(\omega)$.

It should, however, be noted that there exist rather complex and nontransparent descriptions made [15] in terms of the *absorption* vibration spectroscopy of water. This approach takes into account a multitude of the vibration lines calculated for a few water molecules. However, within the frames of this method for the wavenumber[1] $\nu < 1000\ cm^{-1}$, it is difficult to get information about the time/spatial scales of molecular motions and to calculate the spectra of *complex*-permittivity or of the complex refraction index—in particular, the *low-frequency* dielectric spectra of liquid water.

Our rather voluminous chapter could *conditionally* be divided into two parts (a possibility exists to read them independently one from another). In the first part (Sections II–IV) written mostly by B.M.T., a brief review of the ACF method is given and two basic rectangular-well models are described. The other part (Sections I,V–X), written mostly by V.I.G., concerns substantial complication of these models and their application for description, sometimes *quantitative*, of wideband dielectric and far-infrared (FIR) spectra of strongly polar fluids. A schematic diagram (Fig. 1) illustrates the main topics of both above-mentioned parts, which are here marked **A** and **B**.

Besides, the review could conditionally be divided in accord with another criterion. **(a)** In Sections III–V and VII we discuss so-called *unspecific interactions*, which take place in a local-order structure of *various* polar liquids. **(b)** In Sections VI–IX we also consider *specific interactions* [16]. These are *directly* determined by the *hydrogen bonds* in water, are reflected in the band centered at 200 cm^{-1}, which is termed here the R-band, and is characterized by some spectral features in the submillimeter wavelength range (from 10 to 100 cm^{-1}). Note that sometimes in the literature the R-band is termed the "translational" band, since the peak frequency of this band does not depend on the moment of inertia I of a water molecule.

Accounting for such composition of the chapter, before the second part we include the list of formulas useful for calculations (Section IV.G.1) and in Section V.B—the list of main assumptions employed in the ACF method. With more details this method is considered in Section II.

The liquid-state theory—in particular, that capable of describing the wideband spectra $\varepsilon = \varepsilon' - i\varepsilon''$ of the complex permittivity and of absorption

[1]The wavenumber ν is termed "frequency" throughout this chapter.

A	Sections	B	Sections
ACF METHOD		**IMPROVED MODELS**	
• Complex susceptibility as a consequence of Maxwell equations	II.A–II.C	**Hat–curved model with a rigid dipole**	
• Kubo-like dipolar autocorrelator (spectral function)	II.A.4–II.A.5	• The problem	V.A
• Complex permittivity relevant to a wideband spectrum	II.A.6	• Basic assumptions of the ACF method	V.B
• Features of the dipolar autocorrelator	II.B	• The model	V.C
• Correlation between the spectra and model parameters	II.C	• Application to H_2O, CH_3F	V.D
BASIC MOLECULAR MODELS		• Calculation of the spectral function	V.E
Protomodel		**Hat–curved model with a non-rigid dipole**	
• Rigorous spectral function	III.A–III.B	• Problems of specific interactions	VI.A–VI.B
• Planar libration–regular precession approximation	III.C	• Application to H_2O, D_2O	VI.C–VI.D
• Statistical parameters	III.D	• Modified spectral function (derivation)	VI.E
Hat-flat model		**Hat–curved–harmonic oscillator model**	
• Rigorous spectral function	IV.A–V.C	• Response mechanisms and calculation schemes	VII.A.1–VII.A.4
• Statistical parameters	IV.D	• Application to H_2O, D_2O	VII.A.5–VII.A.6
• The spectral function of the hybrid model	IV.E	• Hat-curved–cosine-squared potential model	VII.B
• The form of wideband spectra	IV.F	• Harmonic oscillator model (derivations)	VII.C
• A list of main formulas	IV.G.1	**Ionic models**	
• Preliminary application to water H_2O	IV.G.2	• Models of one-dimensional and spatial motion of ions	VIII.A
• Application to liquid CH_3F	IV.G.3	• Autocorrelator and dispersion of the complex conductivity:	VIII.B
		• Wideband spectra of NaCl– and KCl–water solutions	VIII.C,D
		Potential well based on water structure	
		• Rotational and transverse motions of a dipole	IX.A–IX.C
		• Mean frequencies, estimation of the form of the band	IX.A–IX.C
		• Application of the ACF method	IX.D

Figure 1. Main topics described in the present chapter.

coefficient α of liquid water H_2O and D_2O—is far from being completed. New ideas constantly appear concerning improvement of existing models and generation of new models/calculation schemes. In this chapter an attempt has been undertaken to emphasize *evolution* of the proposed molecular models, moving from a rather primitive infinitely deep rectangular potential well to a rather exotic "*self-consistent*" well described in the Section IX. Evolution of *forms* of intermolecular potentials employed in the review is illustrated in Fig. 2.

A. Short Survey of this Chapter

Section II. The purpose of this section is to write down the expressions that generally allow us to calculate the complex susceptibility $\chi^*(\omega)$ due to motions of dipoles in an arbitrary conservative potential well, which is assumed to be known. (The asterisk denotes the complex-conjugation symbol.) Starting from nonhomogeneous Maxwell equations comprising the current-density term, we come to the Kubo-like dipolar autocorrelator $L(\omega)$ termed the *spectral function* (SF). The derivation is based on two theorems. The first one allows us to express $\chi^*(\omega)$ through the integral determined by the trajectories and the statistical distributions *disturbed* by a.c. electric field. The second theorem transforms this integral to another one, in which the integrand depends only on a steady-state trajectory of a dipoles and includes the Maxwell–Boltzmann distribution function. The spectral function $L(\omega)$ is linearly related to the spectrum $\tilde{C}(\omega)$ of the autocorrelation function (ACF) determined by *undamped motion* of a dipole in a potential U during the lifetime t_v. The exponential distribution $\exp(-t_v/\tau)$, characterized by the mean lifetime $\tau \equiv \langle t_v \rangle$, is used for calculation of the spectral function, where τ^{-1} plays the role of a friction coefficient. Involving the so-called "collision model," one may express the susceptibility $\chi^*(\omega)$ as a rational function of $L(\omega)$, of the Kirkwood correlation factor g, and of the mean lifetime τ. (In our chapter the Gross collision model is used.) The factor g is assumed to be known from the experiment, while τ and the quantities, which determine the form of a potential well, present the fitting parameters of the model.

In this chapter, dielectric response of only *isotropic medium* is considered. However, *in a local-order scale*, such a medium is actually *anisotropic*. The anisotropy is characterized by a local axially symmetric potential. *Spatial* motion of a dipole in such a potential can be represented as a superposition of oscillations (librations) in a symmetry-axis plane and of a dipole's precession about this axis. In our theory this anisotropy is revealed as follows. The spectral function presents a linear combination of the transverse ($K_\perp$) and the longitudinal ($K_\parallel$) spectral functions, which are found, respectively, for the parallel and the transverse orientations of the potential symmetry axis with

Model	Potential well Type of interaction	Properties of the model
(a) Protomodel (Section III)		Qualitative description of the FIR spectra . *Poor description of the Debye relaxation region.*
(b) Hat-flat (Section IV)	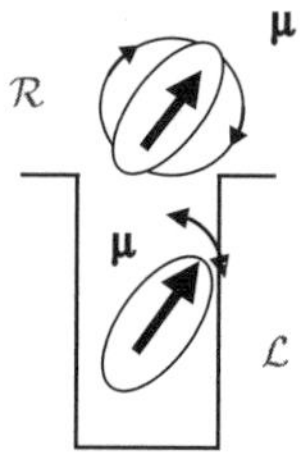	Reasonable description of wideband spectra of H_2O and of non-associated liquids. *The model cannot be applied to liquid D_2O.*
(c) Hat-curved (HC) (Sections V–VII) (c1) HC with a rigid dipole (c2) HC with a non-rigid dipole (c3) Composite HC – harmonic oscillator model	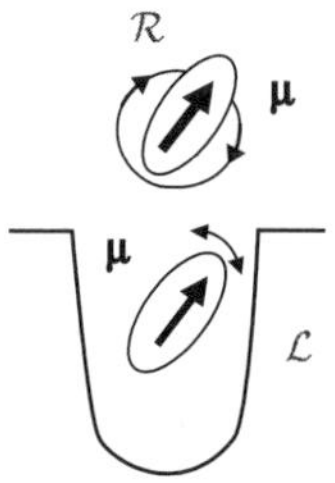 • (c1) pure HC • (c2) HC + $\tilde{\mu}(t)$ • (c3) HC + 	Satisfactory description of the FIR and of the Debye spectra (c1) *Poor description of liquid D_2O and of submillimeter spectra.* (c2) *Poor description of submillimeter spectra.* (c3) Good agreement of the theoretical and experimental spectra
(d) Hybrid + ionic (Section VIII)	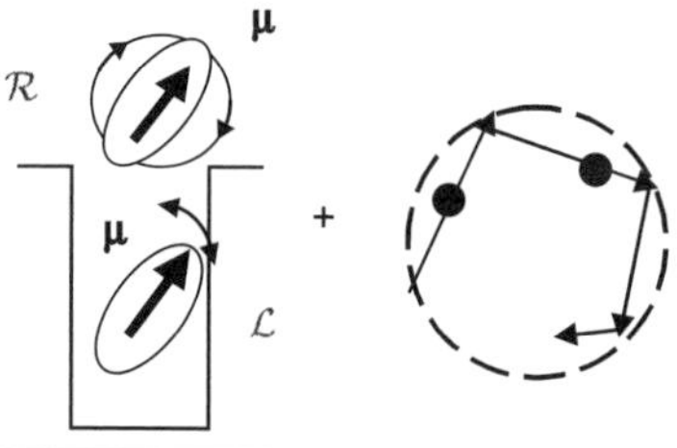	Frequency dependence of the conductivity and effect of ions on wideband spectra of a solution.
(e) Structural–dynamical (Section IX)	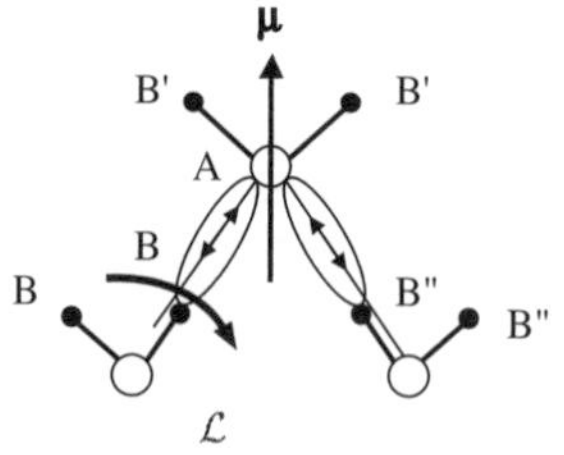	Estimated absorption curve resembles the R-band in water.

respect to a.c. electric-field vector **E**. When we employ rather narrow potential wells, the spectral function actually is determined only by the $K_{\perp}$ term. The latter is expressed in terms of the Fourier amplitudes of the periodic component of a dipoles' motion. Our method allows us to consider the potential wells of a rather complex form, such as depicted in Fig. 2 a–d. Eventually, application of the Fourier expansions allows us to get rather simple analytical formulas for the spectral function and thus provides the success of our work.

At the end of this section some useful interrelations are established between the free parameters of the model and the features of the calculated spectra, such as the absorption-peak frequency, bandwidth, and position of the low-frequency loss maximum.

Section III. We consider *infinitely deep* potential well shown in Fig. 2a. Due to spatial motion of dipoles, such potential geometrically can be presented as a *conical cavity with perfectly reflecting walls.* The *rectangular* "walls" of this well (just as other walls considered in this chapter) present a *rough phenomenological* model of molecular surroundings. During time intervals between elastic reflections from such walls, the particles execute free rotation in the planes inclined at different angles to the symmetry axis of the potential. Analytical expression in the form of a double integral over the phase variables of a reorienting dipole is derived for the spectral function $L(\omega)$. Unlike GT and VIG, the infinite series, describing $L(\omega)$, are summed up. We term this model "protomodel," since dielectric behavior found for this rather simple well presents a basis for further consideration. In the next section the protomodel will be compared with the model elaborated for a finite-well potential.

Using the so-called "planar libration–regular precession" (PL–RP) approximation, it is possible to reduce the double integral for the spectral function to a simple integral. The interval of integration is divided in the latter by two intervals, and in each one the integrands are substantially simplified. This simplification is shown to hold, if a *qualitative* absorption frequency dependence should be obtained. Useful simple formulas are derived for a few "statistical parameters" of the model expressed in terms of the cone angle β and of the lifetime τ. A small β approximation is also considered, which presents a basis for the "hybrid" model. The latter is employed in Sections IV and VIII, as well as in other publications (VIG).

Figure 2. Forms of the potential wells described in the chapter and the features pertinent to the corresponding molecular models. $\mathscr{L}$ and $\mathscr{R}$ denote, respectively, ensembles of librating and rotating dipoles; $\boldsymbol{\mu}$ denotes a dipole moment; VIB refers to the charges $\pm\delta q$ of a nonrigid dipole vibrating along the H-bond; $\tilde{\mu}(t)$ denotes a given harmonically changing component of a dipole moment.

Section IV. The "hat-flat" potential employed here is shown in Fig. 2b. This potential differs from the rectangular well introduced in Section III by a *finite well depth* U_0. The dipoles with small and high energy (compared with some barrier energy commensurable with U_0) constitute two groups, conventionally termed "librators" and "rotators." The law of motion of the librators is the same as that considered in Section III, namely, for U_0 tending to ∞. Rotators perform free or hindered *complete rotation* during the time interval between strong collisions.

We derive an analytical expression for the spectral function in terms of a double integral, which differs from the formulas given in Section III by account of finiteness of the well depth. Two important approximations are also given, in which the spectral function is represented by simple integrals. These are:

(i) The planar libration–regular precession (PL–RP) approximation modified in comparison with that described in Section III
(ii) Approximation of a small cone angle β termed the *hybrid model*

The formulas for the statistical characteristics are given as a generalization of the expressions obtained in Section III. These characteristics depend now on three model parameters: angular well width β, mean lifetime τ, and the reduced well depth $u = U_0/(k_B T)$, where k_B is the Boltzmann constant and T is the temperature.

For a reasonable set of the parameters the calculated far-infrared absorption frequency dependence presents a *two-humped* curve. The absorption peaks due to the librators and the rotators are situated at higher and lower frequencies with respect to each other. The absorption dependences obtained rigorously and in the above-mentioned approximations agree reasonably. An important result concerns the low-frequency (Debye) relaxation spectrum. The hat-flat model gives, *unlike the protomodel*, a reasonable estimation of the Debye relaxation time τ_D. The negative result for τ_D obtained in the protomodel is explained as follows. The subensemble of the rotators vanishes, if $u \to \infty$.

A list of useful formulas is given, which facilitates further calculations. In this section we briefly compare the dielectric/FIR spectra of ordinary water H_2O and of a strongly absorbing nonassociated liquid (CH_3F). We discuss the principal differences of these spectra and of the parameters of the model, which describe them.

The rectangular form of the well, being crude, nevertheless allows us to obtain a rather adequate description of local-order intermolecular forces arising in a liquid, which generally presents a state intermediate between solid body and gas. We emphasize that popular [13, 14] parabolic, cosine, or cosine-squared potential wells generally give *poor description* of the wideband dielectric spectra of nonassociated liquids and are *not* at all applicable for associated liquids.

The drawbacks of the hat model are discussed, so that a necessity for improvement of this model becomes evident.

Section V. Starting from this section and until Section IX we try to reach also a *quantitative* agreement of the theoretical and experimental spectra. Real progress in theoretical studies aimed at calculation of wideband spectra in polar fluids depends on a possibility to consider nontrivial forms of the potential wells, which determine trajectories of the particles and specific spectral features of various liquids. A model termed the hat-curved (HC) model is considered, where a hat-like potential, shown in Fig. 2c, differs from the rectangular one described in the previous section. The hat-curved model, being more realistic than the hat-flat model, allows us to control the width of the far-infrared absorption band by changing the form factor f of the well. To simplify final analytical expressions, it is assumed that the well is rather deep and narrow. We describe the wideband frequency dependences $\varepsilon^*(\nu)$ and $\alpha(\nu)$ of the complex permittivity and of the absorption coefficient in the Debye relaxation and FIR ranges—that is, from 0 to 1000 cm^{-1}. These dependences are calculated for water H_2O at two temperatures and for a typical strongly absorbing nonassociated liquid (CH_3F) at low temperature. Application to water of the hat-curved model is shown to *radically* improve description of the main quasi-resonance (librational) band with the absorption peak placed near 700 cm^{-1}. The spatial and time scales of molecular events, pertinent to water and to a nonassociated liquid, are compared.

A brief list of basic assumptions used in the ACF method precedes the detailed analysis of the results of calculations. The derivation of the formula for the spectral function is given at the end of the section. The calculations demonstrate a substantial progress as compared with the hat-flat model but also reveal two drawbacks related to disagreement with experiment of (i) the form of the FIR absorption spectrum and (ii) the complex-permittivity spectrum in the submillimeter wavelength region. We try to overcome these drawbacks in the next two sections, to which Fig. 2c refers.

Section VI. It is possible to unblock the first drawback (i), if to assume a *nonrigidity* of a dipole—that is, to propose a polarization model of water. This generalization roughly takes into account *specific interactions* in water, which govern hydrogen-bond vibrations. The latter determine the absorption R-band in the vicinity of 200 cm^{-1}. A simple modification of the hat-curved model is described, in which a dipole moment of a water molecule is represented as a sum of the constant ($\bar{\mu}$) and of a small quasi-harmonic time-varying part $\tilde{\mu}(t)$.

The latter is determined by the oscillation frequency, decaying coefficient, and vibration lifetime. This nonrigid dipole moment stipulates a Lorentz-like addition to the correlation function. As a result, the form of the calculated R-band substantially changes, if to compare it with this band described in terms of the "pure" hat-curved model. Application to ordinary and heavy water of the so-corrected hat-curved model is shown to improve description (given in terms of a simple analytical theory) of the far-infra red spectrum comprising superposition of the R- and librational bands.

At the end of the section, derivation of a modified spectral function is given with account of the $\tilde{\mu}(t)$-part of a dipole moment.

However, the so-corrected hat-curved model still does *not* give a perfect agreement with the experiment, since it does not allow us to eliminate the second drawback (ii), namely, disagreement with experiment of the calculated complex permittivity in the *submillimeter* wavelength region.

Section VII. To make away with this drawback (ii), as well as with the first one (i), an alternative polarization model of water is suggested.

First, two charged H-bonding molecules vibrating harmonically along the H-bond direction are considered. Thus the employed harmonic oscillator (HO) model yields a dielectric response pertinent to this type of motion. The existence of such nonrigid dipole with charges $\pm\,\delta q_{\mathrm{vib}}$ is *postulated*. The charge δq_{vib} is shown to be commensurable with charge e of electron. This HO model is applied for calculations additionally with the hat-curved model. The complex permittivity ε^* and the absorption coefficient α are calculated in the range from 0 to 1000 cm^{-1} for water H_2O at temperatures 22.2°C and 27°C and for water D_2O at 22.2°C. The composite HC–HO model is shown to avoid the drawbacks relevant to application (given in Sections V and VI) of the only hat-curved model. The composite model allows us to explain on a molecular basis the difference between the far-infrared spectra of liquid H_2O and D_2O. An interpretation is given of the well-known results by Liebe et al. [17] based on empirical double Debye/double Lorentz representation of the complex permittivity.

Second, an alternative hat-curved–cosine-squared potential (HC–CS) model is also considered, which, as it seems, is more adeuate than the HC–HO model. The CS potential is assumed to govern angular deflections of H-bonded *rigid* dipole from equilibrium H-bond direction. The HC–CS model agrees very well with the experimental spectra of water.

Third, the expression for the spectral function pertinent to the HO model is derived in detail using the ACF method. Some general results given in GT and VIG (and also in Section II) are confirmed by calculations, in which an undamped *harmonic* law of motion of the bounded charged particles is used *explicitly*. The complex susceptibility, depending on a type of a collision model, is found in form of the Lorentz, Gross, and "isothermal" lines by using compact and elegant derivations as compared with those given previously [18].

Section VIII. We consider dielectric response of an ion moving inside a hydration sheath. Assuming perfect reflections of the ion from the sheath, we model the latter by means of infinitely deep potential well (a like approach is used in Section III for the protomodel). So, no charges are proposed to penetrate through the sheath. Two variants of the ionic model are described. In the first one a charge (ion) moves *along a line* between two reflection points, and in the second model an ion moves in a space *inside a spherical cavity* with elastically reflecting walls (see Fig. 2d).

Using the theory presented in Sections II and VII, we find in analytic form the frequency dependence of the ionic complex conductivity. The features of this dependence are as follows.

(a) In the *low-frequency* region the real part S' of the normalized conductivity S is almost independent of frequency, while the imaginary part S'' decreases with the rise of the normalized frequency X from zero value, so that it is negative. The noticeable change of S' with ω occurs for ω about 1/10 of the plasma frequency ω_p.
(b) In the high-frequency region, S' and S'' have the form of fading oscillations. This "ionic" model presents a basis for description of the wideband spectra in aqueous electrolyte solutions.

Employing the *additivity approximation*, we find dielectric response of a *reorienting single* dipole (of a water molecule) in an intermolecular potential well. The corresponding complex permittivity $\varepsilon^*_{\text{dip}}$ is found in terms of the hybrid model described in Section IV. The *ionic* complex permittivity $\Delta\varepsilon^*_{\text{ion}}$ is calculated for the above-mentioned types of one-dimensional and spatial motions of the charged particles. The effect of ions is found for low concentrated NaCl and KCl aqueous solutions in terms of the resulting complex permittivity $\varepsilon^*_{\text{dip}} + \Delta\varepsilon^*_{\text{ion}}$. The calculations are made for long ($\tau_{\text{ion}} \gg \tau$) and rather short ($\tau_{\text{ion}} = \tau$) ionic lifetimes.

Section IX. We discuss a *principally* new way, initiated only recently [12, 12a], of modeling of intermolecular interactions. Namely, we start from a very simplified *molecular structure* of liquid water depicted in Fig. 2e. An approximate solution of the equations of motion of a single polar molecule is found. The key aspect of this approach is that the H-bond resembles a *string* characterized by stretching and bending *elasticity constants* k and k_α [15, 16, 19]. We consider "pure" *rotational and pure transverse deflections* of a water molecule with respect to an equilibrium position, which are accompanied by *expanding and bending* of the H-bond. Since the elastic force tends to restore equilibrium, oscillations should arise in such structures. The solution of a dynamical problem, given in the section, shows that:

(i) It is possible to find the "*self-consistent*" form of the potential well determined by water structure and force constants.
(ii) The estimated distribution of *rotational frequences* falls into the FIR region and resembles the form of the absorption R-band in water.
(iii) About an order of magnitude less is the mean frequency of *transverse* vibrations of a water molecule, occurring perpendicular to the H-bond equilibrium direction. This frequency falls into the submillimeter wavelength range. Therefore, in a first approximation both above-mentioned types of motion could be studied independently one from another.
(iv) The application of this structural–dynamical (SD) model in terms of the ACF method yields a good description of the R-band. Moreover, the composite HC–SD model is applicable for the range from 0 to $1000\,\text{cm}^{-1}$.

Section X. In this final section the conclusions are made and some perspectives of our work are discussed.

First, we emphasize importance of studies of the *effect of temperature* T on the wideband spectra of water. A new step could be made in terms of the composite model described in Section VII. We may try to employ the following property discovered in this section. The parameters of the model, pertinent to intermolecular-potential *geometry* and to *cooperative* motions of the H-bonded molecules, exhibit only a *small* dependence on T. Hopefully, this result may facilitate describing or, better, *predicting* by means of analytical formulas the dependence on T of the wideband spectra of water and of aqueous solutions.

Second, the composite hat-curved–harmonic oscillator model provides a good perspective for a spectroscopic investigation of ice I (more precisely, of ice Ih), which is formed at rather low pressure near the freezing point (0°C). The molecular structure of ice I evidently resembles the water structure. Correspondingly, well-known experimental data show a similarity of the *FIR spectra* (unlike the low-frequency spectra) recorded in liquid water and in ice Ih. This similarity suggests an idea that *rotational mobility* does not differ much in these two phase states of H_2O. On the other hand, (i) the FIR absorption spectra are more narrow in ice and (ii) the peaks of the R- and librational bands are situated at lower frequencies than in the case of water. It seems therefore that if we shall apply the hat-curved model also for ice Ih, then the form factor f should be less than for water.

Third, the success of the composite HC–SD model described in Section IX implies the idea that liquid water presents as if a solution of two components. The main one comprises ~95% of molecules (librators), which reorient rather freely in a deep potential well and are characterized by a *broken H-bond*. The second component comprises ~5% of molecules, which are H-bonded and perform fast vibration. Molecules of the first group live much longer than those of the second group. Thus a physical sense of the HC model is clarified in Section X as that describing dielectric response of dipoles with broken H-bonds.

It would be important to find analogous mechanism also for description of the main (librational) absorption band in water. After that it would be interesting to calculate for such molecular structures the *spectral function/complex dielectric permittivity* in terms of the ACF method. *If* this attempt will be successful, a new level of a "nonheuristic" molecular modeling of water and, generally, of aqueous media could be accomplished. We hope to convincingly demonstrate in the future that even a drastically simplified local-order structure of water could constitute a basis for a satisfactory description of the wideband spectra of water in terms of an analytical theory.

B. Prehistory of This Review

Our linear-response dielectric-relaxation theory (the ACF method) was born on a domain bordered by physical electronics and by molecular and chemical physics. The method had started from Refs. 20 and 21. In our study of polar fluids we may distinguish the following three periods.

1. *In the first period*, which ended with a review [18], the complex susceptibility $\chi^*(\omega)$ was expressed through the law of motion of the particles *perturbed* by a.c. external field $E(t)$. The results of these calculations *rigorously* coincide with those obtained, for example, in Refs. 22 and 23, respectively, for the planar and spatial extended diffusion model (compare with our Ref. 18, pp. 65 and 68). The most important results of this period are (i) the planar confined rotator model [17, p. 70; 20], which has found a number of applications in our and other [24–31] works; (ii) the composite so-called confined rotator–extended diffusion model. However, this approach had no perspectives because of troublesome calculations of the susceptibility $\chi^*(\omega)$.

2. *In the second period*, which was ended by review GT after the average perturbation theorem was proved, it became possible to get the Kubo-like expression for the spectral function $L(z)$ (GT, p. 150). This expression is applicable to any axially symmetric potential well. Several collision models were also considered, and the susceptibility was expressed through the same spectral function $L(z)$ (GT, p. 188). The law of motion of the particles should now be determined only by the steady state. So, calculations became much simpler than in the period (1). The best achievements of the period (2) concern the cone-confined rotator model (GT, p. 231), in which the dipoles were assumed to librate in space in an infinitely deep rectangular well, and applications of the theory to nonassociated liquids (GT, p. 329).

3. In the *third period*, which ended in 1999 after the book VIG was published, various fluids had been studied: strongly polar nonassociated liquids, liquid water, aqueous solutions of electrolytes, and a solution of a nonelectrolyte (dimethyl sulfoxide). Dielectric behavior of water bound by proteins was also studied. The latter studies concern hemoglobin in aqueous solution and humidified collagen, which could also serve as a model of human skin. In these investigations a simplified but effective approach was used, in which the susceptibility $\chi^*(\omega)$ of a complex system was represented as a superposition of the contributions due to several quasi-independent subensembles of molecules moving in different potential wells (VIG, p. 210). (The same approximation is used also in this chapter.) On the basis of a small-amplitude libration approximation used in terms of the cone-confined rotator model (GT, p. 238), the "hybrid model" was suggested in Refs. 32–34 and in VIG, p. 305. This model was successfully employed in most of our interpretations of the experimental results. Many citations of our works appeared in the literature.

4. *Fourth period.* The authors of the present chapter propose to initiate their next (fourth) period devoted to investigation of dielectric relaxation of polar fluids and ensembles of charged particles. We have now enough materials for a detailed description of an important but intricate problem concerning dielectric response of water molecules with account of *stretching and bending vibrations of H-bonded molecules*. Moreover, Sections VII–X could also be useful for investigation of biological systems. Here we mean, first of all, hops of the protons [35] and "fast resonant intermolecular transfer of OH-stretched excitations over many water molecules;... liquid water may play an important role in transporting vibrational energy between OH groups located on either different biomolecules or along extended biological structures" [36]. These processes could be closely related to the second relaxation region in water discussed in Section VII.

II. REVIEW OF THE ACF METHOD

A. Spectral Function and Complex Susceptibility (General Expressions)

1. Connection of Complex Refraction Index with Current Density

We start from the first pair of Maxwell equations written with account of electric charges moving in vacuum. Let vector **J** be the electric current density produced by these charges. Combining the above-mentioned equations, we get the second-order differential equation for electric field vector **E**:

$$(\text{graddiv} - \nabla^2)\mathbf{E} = -\frac{1}{c^2}\frac{\partial}{\partial t}\left(\frac{\partial \mathbf{E}}{\partial t} + 4\pi\mathbf{J}\right) \tag{1}$$

where c is the speed of light in vacuum. We represent **E** and **J** in a form of a transverse homogeneous plane wave:

$$\{\mathbf{E}(t), \mathbf{J}(t)\} = \frac{1}{2}\{\hat{\mathbf{E}},\hat{\mathbf{J}}\}\exp[i(\omega t - \mathbf{kr})] + \frac{1}{2}\{\hat{\mathbf{E}}^*,\hat{\mathbf{J}}^*\}\exp[-i(\omega t - \mathbf{k}^*\mathbf{r})] \tag{2}$$

Let the wave vector $\boldsymbol{\kappa}$ be normal to electric field ($\mathbf{k}\perp\mathbf{E}$), $\hat{\mathbf{E}}$ and $\hat{\mathbf{J}}$ are the corresponding complex amplitudes, * is a complex-conjugation symbol, $i = \sqrt{-1}$, and ω is angular frequency of radiation. Since in the case of transverse wave div $\mathbf{E} = 0$, Eq. (1) in representation (2) reduces to the following equation for the complex amplitudes:

$$\left(k^{*2} - \frac{\omega^2}{c^2}\right)\hat{\mathbf{E}}^* = 4\pi i\frac{\omega}{c^2}\hat{\mathbf{J}}^* \tag{3}$$

Next, we neglect[2] spatial dispersion by setting $\mathbf{kr} = 0$ *on the right-hand side* of Eq. (2). Combining it with the identity

$$\hat{\mathbf{J}}^* = \frac{\omega}{\pi}\int_0^{2\pi/\omega} \mathbf{J}(t, \mathbf{r})\exp(i\omega t)]\, dt$$

[2] This means that the change of phase/amplitude of the wave [Eq. (2)] can be neglected on the space scale about a mean path passed by particles between strong collisions. This approximation holds for rather low frequencies, at which the wavelength of radiation in a medium is much larger than the above-mentioned microscopic distance.

which follows from Eq. (2), we get from Eq. (3) the scalar equation

$$\hat{n}^{*2} - 1 = \frac{8i}{\hat{E}^{*}\omega}\frac{\omega}{2\pi}\int_0^{2\pi/\omega} J_E(t)\exp(i\omega t)]\, dt \tag{4}$$

for the complex refraction index

$$\hat{n}^{*} = (c/\omega)k^{*}$$

In Eq. (4), J_E is projection of vector $\mathbf{J}$ onto direction of field $\mathbf{E}$.

2. *Representation of Current Density $J_E(t)$ by Using the t_0 Theorem*

We transform the integral in Eq. (4) by using the method, which is described in detail in GT, Section II or in VIG, Chapter IV. This method comprises the following steps.

Step 1. We represent the current density vector at the point $\mathbf{r}$ and at the instant t as a sum of the contributions produced by multiple moving *point* charges q_m, $m- = 1, 2, \ldots, M$

$$\mathbf{J}(r,t) = \sum_{m=1}^{M} q_m \mathbf{v}_m(\mathbf{r},t)\delta(\mathbf{r} - \mathbf{r}_m(\mathbf{r},t))$$

where $\mathbf{r}_m(\mathbf{r},t)$ is the *law of motion* of the mth charge, $\mathbf{v}_m(\mathbf{r},t)$ is its velocity, and $\delta(\cdot)$ is delta function. We substitute this sum into Eq. (4) and average the result over a unit volume containing N particles, so that N is a number density. Thus we express the right-hand side of Eq. (4) through some complex effective current density $\bar{J}$:

$$\hat{n}^{*2} - 1 = \frac{8i\bar{J}}{\hat{E}^{*}\omega} \tag{5a}$$

$$\bar{J} \equiv \frac{\omega}{2\pi}\int_0^{2\pi/\omega} \sum_{m=1}^{N} j_m(t)\, dt \tag{5b}$$

$$j_m(t) \equiv q_m v_{Em}(t)\exp(i\omega t) \tag{5c}$$

Each term in the sum (5b) is, in general, a *nonperiodic* function of time, while the sum itself varies in time *periodically*, since it describes a response of the medium to the *harmonic* radiation field $\boldsymbol{E}(t)$.

Step 2. It is hardly possible to calculate the r.h.s. of Eq. (5b) *directly*. The t_0 theorem formulated and proved in GT and VIG allows us to replace the sum $\bar{J}$ of integrals by a double integral of one function

$$j(t) \equiv q v_E(t)\exp(i\omega t) \tag{5d}$$

in such a way that

$$\bar{J} = \frac{\omega}{2\pi} \int \delta N \int_{t_0}^{t_0+t_v} j(t)\, dt \tag{6}$$

The time dependence $j(t)$ is determined (a) by the equations of motions, (b) by the initial conditions (which are pertinent to a point Γ in phase space and to an instant t_0, when a particle is produced by a strong collision), and (c) by the lifetime[3] t_v between two successive strong collisions). Thus,

$$j(t) \equiv j(t, t_0, \Gamma, t_v)$$

The inner integration in Eq. (6) gives the contribution to $\bar{J}$ of one charged particle due to its interactions with a.c. field during lifetime t_v. This contribution is specified by an "initial" instant t_0. The second integration gives the effect of *all* particles.

Explanation of the Theorem. The idea of the theorem is following. Let us choose some instant $t = t_0$ and consider a chain of the particles termed the t_0 *chain.* These particles are characterized by the same phase point Γ and lifetime t_v but have phases ωt_0 differing by $2\pi n$, where n is integer. The first particle of this chain starts interaction with field $E(t)$ after a strong collision occurring just at the instant t_0; the second particle started its interaction at a period $2\pi/\omega$ of the field $E(t)$ earlier; the third particles started at two such periods earlier, and so on, until the last particle of this chain (before it had exerted a strong collision). Because of a periodicity of a.c. field, *each* particle moves along the same trajectory.

At another instant $t = t_0 + 2\pi/\omega$ the first particle, mentioned above, will take the place of the second particle, and so on, so the previous chain seems as if it shifts at one link but actually conserves its form. (Note that the last particle of the previous chain have now quitted the present chain.) Taking the integral $\frac{\omega}{2\pi} \int_0^{t_v} j(t)\, dt$, we find the contribution to $\bar{J}$ due to one t_0 chain, and the second integration in Eq. (6) sums up the effect of all chains. It is clear that the instants t_0 should be varied in the interval $[t_0,\ t_0 + 2\pi/\omega]$; or likewise, the phases $\gamma \equiv \omega t_0 \in [0,\ 2\pi]$.

The normalizing condition for the relation (6) can be found; if we replace j by 1, then

$$\int t_v \delta N = \frac{2\pi}{\omega} N \tag{7}$$

[3]Subscript of the quantity t_v refers to the word "vive," namely, to a lifetime.

If all the particles are *identical*—that is, if they differ only by the phase ωt_0—then we may set $\delta N = C_N\, dt_0$. After integration over t_0 in the interval $[0, 2\pi/\omega]$ we find from Eq. (7): $C_N = \frac{N}{\tau}$, so that $\bar{J} = \frac{N}{\tau}\int_0^{t_v} j(t)\, dt$. Substituting this into Eq. (5a), we have

$$\hat{n}^{*2} - 1 = \frac{8\pi i}{\omega \hat{E}^*}\frac{N}{\tau}\int_{t_0}^{t_0+t_v} j(t)\, dt, \qquad \text{if all the particles are identical} \tag{8}$$

3. *Application of Statistical Distributions*

To generalize Eq. (8), we return to Eqs. (6) and (7) and take into account that the particles differ by initial conditions and lifetimes. Introducing the steady-state distribution $W(H)$ over particles' energies and the distribution $F(\gamma, \Gamma)$ induced by a.c. field, we represent δN as

$$\delta N(\gamma) = \frac{2\pi}{\omega\tau} N W(\Gamma)\, d\Gamma\; F(\gamma,\Gamma)\frac{d\gamma}{2\pi}\Psi_v(t_v)\, dt_v \tag{9}$$

where $\Psi_v(t_v)$ is the distribution over lifetimes t_v between instants of strong collisions.[4] This expression satisfies the condition (7), if we normalize $W(\Gamma)$ and $\Psi_v(t_v)$ as follows:

$$\int_\Gamma W(H)\, d\Gamma = 1 \tag{10a}$$

$$\int_0^\infty \Psi(t_v)\, dt_v = 1 \tag{10b}$$

$W(\Gamma)$ is the Maxwell–Boltzmann distribution and Ψ is supposed to be the exponential one:

$$W \propto \exp[-H/(k_B T)] \tag{11a}$$

$$\Psi_v(t_v) = \tau^{-1}\exp(-t_v/\tau) \tag{11b}$$

For simplicity we further write $F(\gamma,\Gamma) \equiv F(\gamma)$, $\gamma \equiv \omega t_0$. This distribution accounts for the nonconservative addition to the energy H. The quantity $\tau = \int_0^\infty t_v \Psi_v(t_v)\, dt_v$ is the mean lifetime; furthermore, it is termed simply *lifetime*. Thus Eq. (8) should be replaced by

$$\hat{n}^{*2} - 1 = \frac{8iN}{\omega\,\tau\,\hat{E}^*}\int_\Gamma W(H)\, d\Gamma \int_0^{2\pi} F(\gamma)e^{i\gamma}\frac{d\gamma}{2\pi}\int_0^\infty e^{-t_v/\tau}\frac{dt_v}{\tau}\int_{t_0}^{t_0+t_v} j(t)\, dt \tag{12}$$

[4] $\Psi(t_v)$ is taken by analogy with the distribution used in the studies of gas state.

Equation (12) is *directly* related to ensemble of *free charges*. In the case of a polar medium, positive q and negative $-q$ charges of each particle are bound one to another; therefore we may represent the function (5d) as

$$j(t) = q\frac{d}{dt}\left[r_E^+(t) - r_E^-(t)\right]\exp(i\omega t) = \frac{d\mu_E}{dt}\exp(i\omega t) \tag{13}$$

$\mathbf{r}^+, \mathbf{r}^-$ are coordinates of positively and negatively charged particles, and μ_E is a projection of a dipole moment $\boldsymbol{\mu} = q(\mathbf{r}^+ - \mathbf{r}^-)$ of a pair of coupled charges onto the a.c. field direction.

Furthermore, we consider dielectric response of a *rigid* dipole in such frequency interval whereby one may assume that a polar particle reorients *as a whole*. Nonrigidity of a dipole becomes important in an optical region.[5] We replace 1 on the left-hand side of Eq. (12) by the optical permittivity n_∞^2, which approximately accounts for this nonrigidity. In the frequency range under consideration we set n_∞^2 to be independent on frequency.[6] Substituting Eq. (13) into the right-hand side of this formula, we get

$$\hat{n}^{*2} = n_\infty^2 + 4\pi\chi^*(\omega) \tag{14a}$$

$$\chi^*(\omega) = \frac{2iN}{\omega\tau\hat{E}^*}\int_\Gamma W(\Gamma)\,d\Gamma\int_0^{2\pi} F(\gamma)\frac{d\gamma}{2\pi}\int_0^\infty e^{-t_v/\tau}\,\frac{dt_v}{\tau}\int_{t_0}^{t_0+t_v} e^{i\omega t}\frac{d\mu_E(t)}{dt}\,dt \tag{14b}$$

We have introduced the *effective complex susceptibility* $\chi^*(\omega) \equiv \chi'(\omega) + i\chi''(\omega)$, stipulated by reorienting dipoles. This *scalar* quantity plays a fundamental role in subsequent description, since it connects the properties and parameters of our molecular models with the frequency dependences of the complex permittivity $\varepsilon^*(\nu)$ and the absorption coefficient $\alpha^*(\nu)$ calculated for these models.

Changing the time variable ($t + t_0 \to t$), we simplify Eq. (14b) by partial integration over t. Finally, instead of Eq. (14b) we write

$$\chi^* = \frac{2iN}{\omega\tau\hat{E}^*}\int_\Gamma W(h)\,d\Gamma\int_0^{2\pi} F(\gamma)e^{i\gamma}\frac{d\gamma}{2\pi}\int_0^\infty \frac{d\mu_E(t)}{dt}e^{i\hat{\omega}t}\,dt \tag{14c}$$

We have introduced here a dimensionless energy $h = H/(k_B T)$ and the complex frequency

$$\hat{\omega} \equiv \omega + i/\tau \tag{15}$$

[5]In the case of water, this property should be taken into account also in the far-infrared region (see Sections VI and VII).

[6]For instance, in the case of water we set $n_\infty^2 = 1.7$ near the FIR edge of the librational band, namely, at $\nu \approx 1000\,\mathrm{cm}^{-1}$.

4. *Average Perturbation (AP) Theorem and Its Consequences*

The drawback of Eq. (14c) is that the integrands comprise a dynamic quantity $\mu_E(t)$ and an induced distribution $F(\gamma)$. Both are *perturbed by radiation field and therefore depend on its complex amplitude* $\hat{E}^*$. Because of that further calculations become cumbersome [18]. It is possible to overcome this drawback on the basis of a linear-response approximation. The field-induced difference $\delta\mu_E$ of the law of motion $\mu_E(t)$ from the steady-state law $\mu_E^0(t)$ is proportional to the field amplitude $\hat{E}^*$. The same is supposed with respect to the difference $F(\gamma) - 1$ of induced and homogeneous distributions (for the latter $F \equiv 1$). A steady-state dipole's trajectory does not depend on phase γ and therefore does not contribute (at $F = 1$) to the integral (14c). Then in a linear approximation we may represent the average[7] of $\mu_E(t)F(\gamma)$ over γ as a sum

$$\langle \mu_E(t)\, F(\gamma)\rangle_\gamma = \mu_E^0(t)\, \langle [F(\gamma) - 1]\rangle_\gamma + \langle \delta\mu_E(t)\rangle_\gamma \tag{16}$$

Correspondingly, the right-hand side of Eq. (14c) can be represented as the sum

$$\chi^*(\omega) = \chi^*_{\text{st}}(\omega) + \chi^*_{\text{dyn}}(\omega) \tag{17}$$

Equation (14c) is reduced to the first term of Eq. (17), if one replaces $\mu_E(t)$ by the function $\mu_E^0(t)$ determined for the steady-state law of motion but for a *rigorous* induced distribution F (differing from 1). On the contrary, $\chi^*(\omega)$ is reduced to the second term, if one set in Eq. (14c) a *perturbed* $\mu_E(t)$ dependence but assumes the distribution F to be homogeneous ($F = 1$). Thus, the first and second terms of Eqs. (16) and (17) are stipulated, respectively, by effect of radiation on a statistic distributions and on dynamics of an individual particle. That is why we use the subscripts "st" and "dyn" in Eq. (17).

To find the component $\chi^*_{\text{dyn}}(\omega)$, we apply the average perturbation (AP) theorem, formulated and proved in GT, pp. 373–376 (see also VIG, pp. 82–87). This theorem allows us to express the ensemble average value for $\delta\mu_E$, induced by a.c. electric field, through an integral, including *unperturbed* time dependence $\mu_E^0(t)$. The formulation of the theorem is[8]

$$\int_\Gamma \delta q(t) W(h)\, d\Gamma = (k_\text{B}T)^{-1} \int_\Gamma q(t, p_i, s_i) A(t, p_i, s_i) W(h)\, d\Gamma \tag{18}$$

Here p_i and s_i are the canonically conjugated integrals of motion corresponding to the ith degree of freedom; $A(t, p_i, s_i)$ is the work depending on the same

[7] The average of $\mu_E(t)F(\gamma)$ over γ is denoted by subscript γ.

[8] The Reader should not mix a dynamical variable δq with notation of charge q, involved in Eq. (5d) and in other equations given in Section II.A.2.

parameters and produced in the interval [0, t] by some time-varying FORCE(t), which is proportional to the a.c. electric field $E(t)$. In our case

$$q = \mu_E^0/\mu \quad \text{and} \quad A(t) = \int_0^t E(t')\frac{d}{dt'}\mu_E^0(t')\,dt' \tag{19}$$

Generally, the AP theorem is applicable to *any steady-state dynamic quantity* $q(t)$ [of the sort discussed above, *viz.* like $\mu_E^0(t)$]. The theorem permits to calculate an increment δq induced *by any external force*, if the following conditions hold:

(a) The value $|A(t)/(k_B T)|$ is small at any time t, so that the increment $\delta q(t)$ is proportional to the FORCE(t), which performs the work $A(t)$.
(b) The canonically conjugated quantities s_i, p_i, which generally may differ from the integrals of motion, are used as the phase variables of a relevant ensemble of particles.
(c) The one-particle Boltzmann distribution function determined by these variables is used, so that $W(h) \propto \exp[-h(s_i, p_i)]$.

Thus we set the variable $q = \mu_E^0/\mu$ in Eq. (18) and express through this quantity the work $A(t)$ given by Eq. (19), where we represent $E(t')$ in view of Eq. (2). We take into account that: (i) $\mathbf{kr} = 0$; (ii) the variable t in Eq. (14b) is replaced by $t - t_0$. Then Eq. (2) transforms to

$$\mathbf{E}(t) = \frac{1}{2}\{\hat{E}\exp[i(\omega t + \gamma)] + \hat{E}^* \exp[-i(\omega t + \gamma)]\} \tag{19a}$$

The first term in $\{\cdot\}$ does not contribute into the average over γ in Eq. (14c). Then, substituting Eq. (19) in Eq. (18) we have

$$\int_\Gamma \delta q(t) W(h)\,d\Gamma = \frac{1}{2k_B T}\int_\Gamma e^{i\hat{\omega}t}\mu_E^0(t)W(h)\,d\Gamma \int_0^t e^{-(i\omega t' + \gamma)}\frac{d}{dt'}\mu_E^0(t')\,dt' \tag{20}$$

To extract the susceptibility component $\chi_{\text{dyn}}^*(\omega)$ from Eq. (14c), we make with account of Eq. (20) the following replacement:

$$F\int_\Gamma \dots \frac{d\mu_E}{dt} W\,d\Gamma \to \int_\Gamma \dots \frac{d}{dt}\delta\mu_E W\,d\Gamma$$
$$= \frac{1}{2k_B T}\int_\Gamma \dots e^{i\hat{\omega}t}\frac{d\mu_E^0}{dt} W\,d\Gamma \int_0^t e^{-(i\omega t' + \gamma)}\frac{d}{dt'}\mu_E^0(t')\,dt'$$

It is convenient to represent the result by introducing the *spectral function* (SF) $L(z)$ as follows:

$$\chi_{\text{dyn}}(\omega) = \frac{\hat{\omega}}{\omega} GL(z) \tag{21}$$

where

$$L(z) \equiv \frac{3i}{\hat{\omega}\tau} \int_{\Gamma} D(z) W(h)\, d\Gamma \tag{22a}$$

$$D(z) \equiv \int_0^{\infty} \frac{dq(t)}{dt} \exp\left(i\frac{zt}{\eta}\right) \int_0^t \frac{dq(t')}{dt'} \exp\left(-i\frac{xt'}{\eta}\right) dt' \tag{22b}$$

Here we introduce reduced quantities: the complex frequency z, related to $\hat{\omega}$ by a microscopic time scale η, and the normalized concentration G:

$$z = x + iy \equiv \eta\hat{\omega}, \qquad \eta = \sqrt{\frac{I}{2k_B T}}, \qquad G \equiv \frac{\mu^2 N}{3k_B T} \tag{23}$$

I being is the moment of inertia of a linear polar molecule.[9]

It is important that our ACF method suggests *undamped reorientation* of a molecule between the instants of strong collisions. Therefore a dimensionless *strong-collision frequency* y introduced in Eq. (23) plays a role of a *friction coefficient*: the more y, the larger the latter.

The spectral function (22a) is a main object of further calculations given in Sections III–VI and VII–IX. In the next subsection we shall derive a more convenient expression for $L(z)$.

5. *Spectral Function as a Dipole Autocorrelator*

We may simplify the formula (22b) for $D(z)$, if we employ an equality[10]

$$\int_{-\Im/2}^{\Im/2} \frac{d}{dt}[q(t)q(t')]\, dt_0 = 0 \tag{24}$$

where t_0 is the integral of motion canonically conjugated with an energy H. This quantity is involved in the time dependence $q(t)$ additively with t. In this and further subsections we consider a *conservative* system, unlike Sections II.A.2, and II.A.3, where the constant $t_0 = \gamma/\omega$ is related to a.c. field. The meaning of

[9] Below in Sections III–V we consider *linear rigid polar molecules* (rotators), while in Sections VI and VII a dipole is assumed to be nonrigid.

[10] Equation (24) is equivalent to Eq (3.15) in GT.

Eq. (24) is follows. In our microscopic models a particle's trajectory is characterized by some periodicity:

$$q(t, t_0) = q(t, t_0 + \Im) \tag{25}$$

The meaning of the quantity $\Im$ will be clarified in Section II.B. Since t_0 could be chosen as *one of the phase variables*, the ensemble average of the integrand in Eq. (24) also vanishes:

$$\left\langle \frac{d}{dt}[q(t)q(t')] \right\rangle = 0 \tag{26}$$

It was proved in GT, p. 150, with account of Eq. (26) that integration over t' in Eq. (21) is equivalent to multiplication by τ of the integrand taken at $t = 0$. Thus Eqs. (20) and (21) are transformed to

$$L(z) = \frac{3i}{\hat{\omega}} \int_\Gamma \left(\frac{dq}{dt}\right)_{t=0} W(h)\, d\Gamma \int_0^\infty \frac{dq(t)}{dt} \exp\left(\frac{izt}{\eta}\right) dt \tag{27a}$$

Next, (a) we use again Eq. (26); (b) take into account that the distribution W is an odd function of the derivative $\left(\frac{dq}{dt}\right)_{t=0}$; (c) carry out integration in Eq. (27a) by parts. Then we have

$$\begin{aligned} L(z) &= \frac{3i}{\hat{\omega}} \int_\Gamma \left(\frac{dq}{dt}\right)_{t=0} W(h)\, d\Gamma \left[q(0) + iz/\eta \int_0^\infty q(t) \exp\left(iz\frac{t}{\eta}\right) dt\right] \\ &= 3 \int_\Gamma \left(\frac{dq}{dt}\right)_{t=0} W(h)\, d\Gamma \int_0^\infty q(t) \exp\left(\frac{izt}{\eta}\right) dt \\ &= -3 \int_\Gamma q(0) W(h)\, d\Gamma \int_0^\infty \frac{dq(t)}{dt} \exp\left(\frac{izt}{\eta}\right) dt \end{aligned}$$

Finally, we derive the formula for the spectral function convenient for calculations:

$$L(z) = 3iz/\eta \int_\Gamma q(0) W(h)\, d\Gamma \int_0^\infty [q(t) - q(0)] \exp(izt/\eta)\, dt \tag{27b}$$

This expression is obtained by using *only* the Maxwell equations and the equations of classical mechanics and the condition (26) for $q(t)$. Because of this, Eq. (27b) is valid *irrespectively* of the type of the projection $\mu_E^0 = |\boldsymbol{\mu}|q$. This projection may correspond to reorientation of a rigid dipole or to vibration of bound charges of a nonrigid dipole.

6. *Expression of Complex Susceptibility Through Spectral Function*

Now we turn to calculation of the susceptibility component $\chi^*_{\rm st}(\omega)$ in Eq. (17). To extract it from Eq. (14c), one should replace there μ_E for μ_E^0 and account for an inhomogeneity of the induced distribution $F(\gamma)$. The latter is determined by a chosen collision model. Such models are described in detail in GT, Section IV.B, and in VIG, Section VI, where they are separated into the self-consistent and non-self-consistent models. For one simple example they are considered also in Section VII.C

We shall start with the Boltzmann collision model, which belongs to the last group. The Boltzmann-induced distribution $F_{\rm B}$, established at the instant $t = 0$ of the strong collision, in the case of an isotropic medium, is given by

$$F_{\rm B} = \exp\left[-\frac{\mu_E(0)E(0)}{k_{\rm B}T}\right] \simeq 1 - \frac{\mu(0)E(0)}{k_{\rm B}T} = 1 - \frac{\mu(0)}{k_{\rm B}T}\left(\hat{E}\,e^{i\gamma} + \hat{E}^* e^{-i\gamma}\right) \tag{28a}$$

We shall show further that at $F = F_{\rm B}$ the complex susceptibility is given by

$$\chi^* = GL(z) \tag{28b}$$

Now (and further) we shall employ an important self-consistent *Gross* collision model. We shall represent the Gross induced distribution $F = F_{\rm G}$ as

$$F_{\rm G}(\gamma) - 1 \simeq \frac{\mu_E(0)}{2k_{\rm B}T}[\hat{A}^*(z)\hat{E}e^{i\gamma} + \hat{A}(z)\hat{E}^* e^{-i\gamma}] \tag{28c}$$

An unknown complex factor $\hat{A}(z)$is found from the standard relation

$$\hat{P} = \chi\hat{E} \tag{29}$$

between the complex polarization $\hat{P}$ of an isotropic polar medium and the complex field amplitude, the factor of proportionality being equal to the complex susceptibility. On the other hand, by definition, the polarization at an instant $t = 0$ of a strong collision is determined by the total dipole moment of a unit volume comprising N polar molecules:

$$P(0) = \mathrm{Re}\left(\hat{P}e^{i\gamma}\right) \equiv Ng\int_{\Gamma} \mu_E(0) WF(\gamma)\, d\Gamma \tag{30}$$

Up to now we neglected *interaction between the particles of a medium.* To take it roughly into account, we have introduced in Eq. (30) the Kirkwood correlation factor g. The latter, in general, differs from 1, since the total dipole moment of a unit volume could be larger (if $g > 1$) or less (if $g < 1$) than the dipole moment found as sum of electric moments of noninteracting particles. The coefficient g determined by experimental data[11] approximately accounts for the fact that the statistical distribution should somehow differ from a *one-particle distribution function WF*, which is employed in this work.

Substituting (28c) into (14b) with account of (27b), we have the following relation between two components of the complex susceptibility:

$$\chi_{\text{st}}^* = -\frac{iy}{z}\hat{A}(z)\chi_{\text{dyn}}^* = -\frac{iy}{x}G\hat{A}(z)L(z)$$

whence the resulting complex susceptibility is connected with the complex factor $\hat{A}$ by

$$\chi^* = \frac{G}{x}[z - iy\hat{A}(z)L(z)]. \tag{31}$$

Now we shall obtain the formulas for the Boltzmann and Gross distributions F_{B} and F_{G}. The first one, given by Eq. (28a), is obtained from Eq. (31) if we set $\hat{A} = 1$. Then we get Eq. (28b). To find $\hat{A}$ at $F = F_{\text{G}}$, we take into account (i) the definitions (19) of q and (20) of G and (ii) the equality $\langle q^2 \rangle = 1/3$ pertinent to an isotropic medium. Thus we have from Eqs. (28c) and (30)

$$P(0) = \frac{G}{2}\left[\hat{A}^*(z)\hat{E}e^{i\gamma} + \hat{A}(z)\hat{E}^* e^{-i\gamma}\right].$$

We express the left-hand part of this formula through the complex amplitudes and then equate expressions by the same exponential factors $\exp(\pm i\gamma)$. In view of Eq. (29) we find

$$\hat{A}(z) = \chi^*/(Gg)$$

Substituting this relation into Eq. (31), we finally derive unambiguous relation between the complex susceptibility and spectral function relevant to the Gross collision model[12]:

$$\chi^*(\omega) = \frac{gGzL(z)}{gx + iyL(z)} \tag{32}$$

[11]More information about the factor g is given in Section IV.G.1.

[12]In this collision model: (i) the velocity distribution function is assumed to be unperturbed by a.c. field; (ii) the employed induced distribution F in GT and in VIG was termed the "correlation-orientation" distribution, since in the work by Gross [37] the case $g = 1$ was originally considered. Here the term "Gross collision model" is ascribed to an arbitrary g-value, which is involved in Eq. (30).

This relation will be used further for calculation of the wideband spectra. Note again, that it is equally applicable for a rigid reorienting dipole and also for a nonrigid one formed by two oscillating charges.

It follows from Eq. (32) that the spectral function $L(z)$ actually determines the absorption coefficient. At high frequencies,[13] such that $x \gg y$, this coefficient is proportional to $x\,\mathrm{Im}[\chi^*(x)]$. In other limit, at low frequencies, one may neglect the frequency dependence $L(z)$ by setting $L(z) = L(iy)$. In this approximation, Eq. (32) yields the Debye-relaxation formula (VIG, p. 194):

$$\chi \cong \chi_\infty + \frac{\chi_s - \chi_\infty}{1 - ix/x_\mathrm{D}} \tag{33}$$

where $x_\mathrm{D} = \eta/\tau_\mathrm{D}$, τ_D is the Debye relaxation time, χ_∞ and χ_s are, respectively, susceptibilities at the high-frequency edge of the Debye relaxation band and at $x = 0$.

Let us now denote as χ''_D the maximum loss value χ''_max attained at $x = x_\mathrm{D}$. It is shown in VIG, Section 6, that for the Gross collision model our approximation $L(z) \approx L(iy)$ gives

$$\tau_\mathrm{D} = \frac{g\tau}{L(iy)} \tag{34a}$$

and

$$\chi''_\mathrm{D} = \frac{G}{2}[g - L(iy)] \tag{34b}$$

From Eq. (34a) we obtain the formula for the dimensionless frequency x_D of the loss maximum:

$$x_\mathrm{D} = yL(iy)/g \tag{34c}$$

B. Spectral Function of a Dipole Reorienting in a Local Axisymmetric Potential

1. *General Properties*

In subsequent sections we shall consider a few phenomenological axisymmetric potentials that determine the steady-state law of motion of a dipole. A polar fluid considered in most of our models is characterized by a local anisotropy—that is, by that in a *short-range* space scale. Correspondingly, we represent polar fluid as

[13]Details are given in Section II.C.

a medium comprising a continuum of identical and statistically independent subensembles differing by directions of their symmetry axes. As a consequence, the so-modeled polar fluid is *isotropic* as a whole. Due to the above-mentioned axial symmetry the projection **L** of a dipole's angular momentum about this axis C conserves during the lifetime of the potential well:

$$\mathbf{L} \equiv \mathbf{c} I \sin^2 \vartheta(t) \, \frac{d\psi(t)}{dt} = \text{const}(t) \tag{35}$$

where **c** is a unit vector along C, $\vartheta(t)$ is a current polar angle between the dipole-moment vector $\boldsymbol{\mu}$ and C, and $\psi(t)$ is a current turn of this vector about C during the time interval $[0, t]$. These angles are shown in Fig. 3. The function $\vartheta(t)$ is *periodic* due to axial symmetry of the potential. Let $\Im$ be the period of this function. In view of Eq. (35), the derivative $d\psi/dt$ is also periodic with the same period $\Im$:

$$\vartheta(t) = \vartheta(t + \Im), \qquad d\psi(t)/dt = d\psi(t + \Im)/dt \tag{36}$$

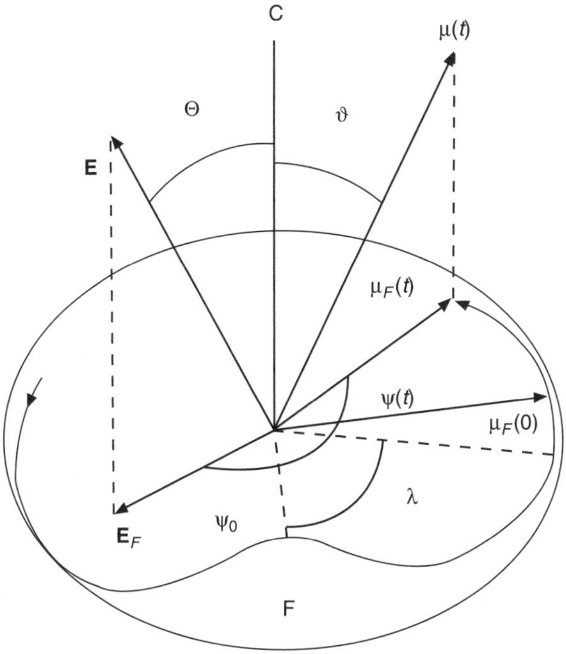

Figure 3. Locus of a dipole-moment projection onto the plane F perpendicular to the symmetry axis C: qualitative picture. Explanations are given in the text.

Let μ_F be the projection of the vector $\boldsymbol{\mu}$ on the plane F perpendicular to the axis C. The equalities (36) mean that the locus of μ_F comprises identical contours, and they are shown in Fig. 3 in a form of lobes. The identity property could be expressed, in addition to Eq. (36), as

$$\psi(J\Im/2) = J\lambda, \qquad \lambda \equiv \psi(\Im/2) \tag{37}$$

where J is integer. Functions $\cos[\psi(t)]$ and $\sin[\psi(t)]$ are nonperiodic since the angle λ is generally *noncommensurable* with π. Then the locus μ_F presents an open curve. Therefore, in view of Eqs. (36) and (37), the derivative $d\psi/dt$ comprises a constant component, equal to $2\lambda/\Im$, so that $\psi(t)$ could be represented as a sum of a linear and of a periodic functions:

$$\psi(t) = 2\lambda\frac{t}{\Im} + \tilde{\psi}(t) - \tilde{\psi}(0), \qquad \tilde{\psi}(t) = \tilde{\psi}(t+\Im) \tag{38}$$

It also follows from Eqs. (35) and (36) that

$$\psi(t, t_0) = \psi(t, t_0 + \Im) \tag{39}$$

We remind to the Reader that t_0 is an arbitrary constant, which in the case of a conservative system enters additively with t into the law of motion $\vartheta(t)$ of a particle.

Note that Eq. (24) holds due to the periodicity properties (36) and (39). This allows us to represent the spectral function in the form of the autocorrelator (27b) of the function $q(t)$. Taking into account (19) and Fig. 3, we represent q as

$$q = \cos\Theta\cos\vartheta + \sin\Theta\cos(\psi + \psi_0)\sin\vartheta \tag{40}$$

where:

- ψ_0 is the angle on the plane F between the projection $\mathbf{E}\sin\Theta$ of the a.c. field vector $\mathbf{E}$ and the initial (at $t = 0$) direction of the $\boldsymbol{\mu}_F$ component, with ψ being counted out of this direction.
- Θ is the angle between the symmetry axis C and the vector $\mathbf{E}$.

It is evident from Fig. 3 that the first and second summands in Eq. (40) are proportional, respectively, to the projection of the dipole moment on the axis C and on the plane F.

Let us introduce now the following dimensionless quantities:

$$\begin{aligned} &\varphi = t/\eta, \qquad \varphi_0 = t_0/\eta, \qquad \Phi = \Im/\eta, \\ &h = H/(k_B T), \qquad l = (2k_B T I)^{-1/2}|\mathbf{L}| \end{aligned} \tag{41}$$

The derivative over φ is denoted by a dot above the relevant quantity—for example, $\dot{\vartheta} \equiv d\vartheta/d\varphi$. Below we introduce the "effective" potential $w(\vartheta, l)$. Then the Hamiltonian of a rotating dipole with a pinned center of mass could be represented in terms (41) as follows:

$$h(\vartheta, \dot{\vartheta}, l) = \dot{\vartheta}^2 + w(\vartheta, l) \tag{42a}$$

$$w(\vartheta, l) \equiv \frac{U(\vartheta)}{k_B T} + \frac{l^2}{\sin^2 \vartheta} \tag{42b}$$

where $U(\vartheta)$ is axisymmetric potential. The notion of the effective potential is convenient, since in our problem we formally may consider a motion of a dipole as *one-dimensional* rotation stipulated by such a potential.

Formula (35) is equivalent to the following equation:

$$\dot{\psi} = l/\sin^2 \vartheta \tag{43}$$

A two-dimensional rotation of a dipole we may represent as a superposition of: (1) a deflection relative the symmetry axis and (2) a precession about this axis. The square of the polar velocity $\dot{\vartheta}$ presents in Eqs. (42a) and (42b) the "axial" component of a kinetic energy and $(l/\sin\vartheta)^2$—the precessional component. Below we consider two limiting cases.

Let $l = 0$. Then a dipole librates in a plane comprising the symmetry axis. Then in view of Eqs. (42a) and (42b) the effective potential w coincides with a "true" potential. It follows from Eq. (43) that precession is lacking,[14] and the angular velocity $\dot{\psi} = 0$.

Let now $l = l_{\max}(h)$. It follows from Eqs. (42a) and (42b) that l is maximum at $\dot{\vartheta} = 0$. In view of (43) the velocity $\dot{\psi}$ is then constant; the rotation of a dipole is reduced to a *regular precession* about the symmetry axis. Since the velocity $\dot{\vartheta}$ is *identically* equal to zero, the Hamilton equation written for the polar angle ϑ gives $\partial w(\vartheta, l)/\partial\vartheta = 0$. Hence, the precession angle corresponds to the minimum of an effective potential.

Thus, we conclude that

$$\begin{Bmatrix} \dot{\psi} \\ \dot{\vartheta} \end{Bmatrix} = 0, \quad \textit{viz.} \begin{Bmatrix} \text{libration in a plane comprising a symmetry axis} \\ \text{regular precession about the symmetry axis} \end{Bmatrix}$$

$$\text{at } l = \begin{Bmatrix} 0 \\ l_{\max}(h) \end{Bmatrix} \tag{44}$$

[14]A rigorous analyses shows that at $\varphi_0 = 0$ the velocity $\dot{\psi}$ is proportional to the sum of δ-functions with arguments $(\varphi - J\Phi/2)$. However, we may set $\dot{\psi} = 0$, since calculation of the contribution to the spectral function due to planar libration involves the averaging over time during a half-period.

2. *Ensemble Average*

The quantities $\{h, \varphi_0\}$ and $\{l, \psi_0\}$ constitute (to within a constant multiplier) two pairs of the canonically conjugate arbitrary constants. Therefore we may choose them as the phase variables while averaging over Γ in (27). We note that these quantities refer to a "local" phase space corresponding to any chosen direction of the symmetry axis. Hence, integration performed in the *overall* phase space $d\Gamma_{is}$ corresponding to an *isotropic fluid* should *additionally* include averaging over all possible inclinations Θ of the symmetry axis C to the a.c. field vector **E**. Thus,

$$d\Gamma_{is} = d\Gamma \sin\Theta \, d\Theta, \qquad \text{where} \quad d\Gamma = \frac{1}{2\pi} dh \, d\varphi_0 \, dl \, d\psi_0 \tag{45}$$

We note that the function averaged in Eq. (27b) rests the same, if l is changed for $-l$. This assertion is proved in GT, pp. 168 and 169 (see also VIG, pp. 98 and 99), starting from the general formula for the spectral function of the type (27) and from Eqs. (42a) and (42b). Therefore, averaging over l could be carried down in the interval $[0, l_{max}(h)]$. In view of Eqs. (27), (40), and (45), we then have

$$L = \frac{1}{3}(K_{\|} + 2K_{\perp}) \tag{46a}$$

$$\left\{ \begin{matrix} K_{\|}(z) \\ K_{\perp}(z) \end{matrix} \right\} \equiv 3iz \int_0^{\infty} \left\langle \left\{ \begin{matrix} \cos\vartheta(0)[\cos\vartheta(\varphi) - \cos\vartheta(0)] \\ \sin\vartheta(0)\cos\psi_0[\sin\vartheta(\varphi)\cos(\psi + \psi_0) \\ -\sin\vartheta(0)\cos\psi_0] \end{matrix} \right\} \right\rangle e^{iz\varphi} d\varphi \tag{46b}$$

Here and further the corner bracket $\langle\ \rangle$ means averaging only over a local-order phase space, namely,

$$\langle f \rangle \equiv \int_{\Gamma} f W(h) \, d\Gamma \tag{47}$$

Thus the spectral function $L(z)$ of an isotropic medium is represented as a linear combination of *two* spectral functions determined for an *anisotropic* medium pertinent to longitudinal ($K_{\|}$) and transverse ($K_{\perp}$) orientations of the symmetry axis with respect to the a.c. field vector **E**. It is shown in GT, Section V, that these spectral functions are proportional to the main components of the dielectric-susceptibility tensor.

3. *Representation of Spectral Function in Terms of Fourier Amplitudes*

In view of Eqs. (35) and (38), the functions

$$\cos\vartheta(\varphi), \qquad \sin\vartheta(\varphi)\cos\tilde{\psi}(\varphi), \qquad \text{and} \qquad \sin\vartheta(\varphi)\sin\tilde{\psi}(\varphi)$$

where $\tilde{\psi}$ is a periodical component of the function $\psi(\varphi)$ [see Eq. (38)], could be represented as a series with a fundamental frequency $2\pi/\Phi$:

$$\begin{Bmatrix} \cos\ \vartheta \\ \sin\ \vartheta \cos\ \tilde{\psi} \\ \sin\ \vartheta\ \sin\tilde{\psi} \end{Bmatrix} = \begin{Bmatrix} a_0/2 \\ c_0/2 \\ 0 \end{Bmatrix} + \sum_{n=1}^{\infty} \begin{Bmatrix} a_n \cos\left(2\pi n \dfrac{\varphi + \varphi_0}{\Phi}\right) \\ c_n \cos\left(2\pi n \dfrac{\varphi + \varphi_0}{\Phi}\right) \\ s_n \sin\left(2\pi n \dfrac{\varphi + \varphi_0}{\Phi}\right) \end{Bmatrix} \tag{48}$$

It follows from here (details are given in GT, pp. 166–167) that

$$\int_{-\Phi/2}^{\Phi/2} \cos\ \vartheta(\varphi) \cos\ \vartheta(0)\ d\varphi_0 = \frac{\Phi}{2} \sum_{n=0}^{\infty} a_n^2 \cos\left(2\pi n \frac{\varphi}{\Phi}\right) \tag{49a}$$

$$\int_{-\Phi/2}^{\Phi/2} d\varphi_0 \int_0^{2\pi} \sin\vartheta(\varphi) \sin\vartheta(0) \cos(\psi + \psi_0) \cos\ \psi_0 \frac{d\psi_0}{2\pi}$$
$$= \frac{\Phi}{2} \sum_{n=-\infty}^{\infty} b_n^2 \cos\left[2(\pi n - \lambda) \frac{\varphi}{\Phi}\right] \tag{49b}$$

where

$$a_n = \frac{\{2, 1\}}{\Phi} \int_{-\Phi/2}^{\Phi/2} \cos\vartheta(\varphi) \cos\left(2\pi n \frac{\varphi}{\Phi}\right) d\varphi \quad \text{at } \varphi_0 = 0, \{n > 0, n = 0\} \tag{50a}$$

$$b_n \equiv \frac{c_n + s_n}{2} = \frac{1}{\Phi} \int_{-\Phi/2}^{\Phi/2} \sin\vartheta(\varphi) \cos\left[\psi(\varphi) + 2(\pi n - \lambda) \frac{\varphi}{\Phi}\right] d\varphi \text{ at } \varphi_0 = 0 \tag{50b}$$

Integration over time φ in Eqs. (46a) and (46b) with account of (48)–(50a,b) yields the formula

$$L(z) = 4 \int_{\Gamma'} \frac{d\Gamma'}{\Phi} W(h) \sum_{n=-\infty}^{\infty} \left[\frac{(\pi n a_n/2)^2}{\left(\frac{2\pi n}{\Phi}\right)^2 - z^2} + \frac{(\pi n - \lambda)^2 b_n^2}{\left(\frac{2}{\Phi}\right)^2 (\pi n - \lambda)^2 - z^2} \right] \tag{51}$$

for the spectral function $L(z)$, where

$$d\Gamma' \equiv dh\, dl, \qquad 0 \le h \le \infty, \qquad 0 \le l \le l_{\max}(h) \tag{52}$$

The expression (51) is equivalent to the sum of Eqs. (5.123a) and (5.123b) given in VIG. The first and second terms in the brackets $[\cdot]$ present, respectively, the contributions to the spectral function $L(z)$ of the longitudinal and transverse components, $K_{\|}(z)$ and $K_{\perp}(z)$.

Thus, to calculate the spectral function for a given microscopic model one has to

- Find the dependences $\Phi(h,l)$, $\lambda(h,l)$, $a_n(h,l)$, $b_n(h,l)$
- Determine configuration of the phase region Γ'
- Calculate the norm of the steady-state distribution $W(h)$

All these characteristics are, in their turn, determined by a model potential $U(\vartheta)$.

C. About Correlation Between Dielectric Spectra and Parameters of a Molecular Model

Dielectric properties of a polar medium are determined by the formulas (32) and (51). In Sections II.C.1–II.C.3 we consider the case of a *high* frequencies, while in the last Section II.c.4 the low-frequency susceptibility will be considered.

1. *Localization of a Dipolar Ensemble*

At high frequencies, such that $x \gg y$, the right-hand side of Eq. (32) reduces to GL(z). Then the frequency dependences $\chi'(\omega)$ and $\chi''(\omega)$ are close to those given by

$$L'(z) \equiv \mathrm{Re}[L(z)] \quad \text{and} \quad L''(z) \equiv \mathrm{Im}[L(z)]$$

At $x \gg y$ the frequency dependence of the normalized *absorption coefficient* can be represented as

$$A(x) = xL''(z)$$

Let us compare the contributions to $A(x)$ of the longitudinal and transverse spectral functions $K_{\parallel}$ and $K_{\perp}$. We shall consider here the case of a strong local anisotropy. An anisotropy itself can be characterized by an averaged maximum angular deflection $\beta \equiv \langle \vartheta_{\max} \rangle$ of a dipole-moment vector above the symmetry axis. We shall term this quantity[15] *localization of a dipolar ensemble.* This anisotropy is more emphasized, if β is small. Then in view of Eq. (46b) the longitudinal component $K_{\parallel}$ of the SF has the following order of magnitude:

$$\begin{aligned} &\langle [\cos\vartheta(0) - 1 + 1][\cos\vartheta(\varphi) - \cos\vartheta(0)] \rangle \\ &\quad = \langle [\cos\vartheta(0) - 1][\cos\vartheta(\varphi) - \cos\vartheta(0)] \rangle \sim \beta^4 \end{aligned}$$

We have employed here an evident equality $\langle \cos\vartheta(\varphi) - \cos\vartheta(0) \rangle = 0$. On the other hand, $K_{\perp} \sim \langle \sin^2\vartheta \rangle \sim \beta^2$. Thus taking into account Eq. (46a), we have

$$K_{\perp} \gg K_{\parallel} \quad \text{and} \quad L \cong (2/3)K_{\perp} \qquad \text{at } \beta \ll 1 \tag{53}$$

[15] It will be calculated in Section IV.F for an example of a finite-depth rectangular well (viz., for the hat-flat model), where also a more general definition of this quantity will be given.

One may explain this result as follows. If a deflection ϑ is small, the locus of the vector $\boldsymbol{\mu}$ approximately coincides with its projection on the plane F (see Fig. 3). Therefore, interaction of a dipole with a.c. field (we mean here the particle's change of energy) would be more effective if the field vector is placed in the same plane F perpendicular to the symmetry axis.

Using the rule (53), we in most cases neglect the longitudinal spectral function $K_{\|}$ in Sections IV–VII.

2. *Estimation of the Absorption-Peak Frequency*

Now we shall estimate the frequency x_m, at which the absorption $A(x)$ attains its maximum value. In accord with Eq. (53) at small β, the absorption $A(x)$ is mostly determined by a transverse component of the spectral function, to which the sum of terms with denominators $(2/\Phi)^2(\pi n - \lambda)^2 - z^2$ in the integrand of Eq. (51) corresponds. These terms actually present the set of the resonance *Lorentz lines*, whose center frequencies are given by

$$x = (2/\Phi)(\pi n - \lambda)$$

Since the peaks intensities, as a rule, decrease with an integer n, the absorption is contributed mainly by the lines with $n = 0$ and $n = 1$, their center frequencies thus being

$$x = 2\lambda/\Phi \quad \text{and} \quad x = (2/\Phi)(\pi - \lambda)$$

Let us consider a small β approximation in terms of the *cone confined rotator* (CCR) model (see VIG, pp. 293–295).[16] In this approximation two largest terms (with $n = 0$ and $n = 1$) are *equal.* Based on this result, we assume that at small β these terms are close at *any* model potential. Then we set x to be approximately equal to the median frequency:

$$x \simeq \frac{1}{2}(2/\Phi)(\pi - \lambda + \lambda) = \pi/\Phi$$

We may estimate x_m by taking the *mean value* of this quantity[17]:

$$x_m \approx \pi\xi; \quad \xi \equiv \langle \Phi^{-1} \rangle \qquad \text{at } \beta \ll 1 \tag{54}$$

[16] For arbitrary β this model is also considered in Section III, where it is termed the "protomodel."

[17] This formula gives good result in the case of the rectangular potential well (see Sections III and IV). However, it is not applicable for the hat-curved potential well, for which another way of estimation of x_m will be suggested.

For a given angular velocity of a dipole the period Φ of $\vartheta(\varphi)$ decreases, if β decreases. Then, from Eq. (54), the following holds:

The narrower the angular localization of dipole's ensemble, the higher the absorption-peak frequency (55)

We should note that a *two-humped* absorption are pertinent to *aqueous* media. In terms of a microscopic molecular model, such a behavior could, partially, be explained by a *finite depth* of a potential well. Indeed, dipoles with rather small energies constitute a subensemble of particles *localized* in the well, so their maximum deflection $\breve{\beta}$ is determined by the angular width of the well, while dipoles with sufficiently large energies overcome the potential barrier. These dipoles perform a complete rotation; such particles occupy the whole sphere, so that $\mathring{\beta} = \pi$. This reasoning leads us to a conclusion that generally two types of motion could characterize a given potential well, so that

a number of absorption peaks $\leq$ *the number of types of motion* (56)

$$\breve{x}_m \geq \mathring{x}_m \qquad \text{at } \breve{\beta} \leq \mathring{\beta} \tag{57}$$

In Eq. (56) the inequality symbol refers to the case, when a contribution of one of subensembles is negligible; the equality means that these contributions are commensurable.

3. About Absorption Bandwidth

Using Eq. (51) we formulate a few qualitative rules concerning the absorption bandwidth[18] Δx. It follows from consideration of the aforesaid resonance denominators that

The shorter the lifetime τ *(the greater the collision frequency* $y \propto \tau^{-1}$*), the wider the absorption band* (58)

This rule will be illustrated in Section VII—for example, for the harmonic oscillator model.

If strong collisions are rare (e.g., if $y \ll 1$), another important factor—an *anharmonicity* of a molecular ensemble—plays an important role. This notion, introduced in GT, Section VII with respect to a parabolic potential well (or the well close to the latter), means dependence of the period Φ on the phase variables h and l.

[18] For definiteness, we determine Δx as a a distance between the frequencies at which the absorption comprises a half of the maximum value.

In the case of a parabolic well the period Φ is independent on the phase variables, the anharmonicity vanishes, and the bandwidth is nonzero only due to strong collisions. The more a potential profile differs from the parabolic one, the larger the anharmonicity and the wider the absorption band. The intensity of the absorption peak should then decrease since in accord with the Gordon rules (see, e.g., GT, Section III.G; or see Section VIIA.4 in the present chapter) in an isotropic medium the integrated absorption does not depend on parameters of the model.

The effect of the anharmonicity on Δx follows from estimation (54) of the frequency x_m. The right-hand part of this formula is actually the result of "gathering" of the Lorentz lines in (51), each being determined by a relevant pair $\{h, l\}$ of arbitrary constants. Evidently, the steeper the dependence of the period Φ on these variables—that is, the greater the $\langle \partial\Phi^{-1}/\partial h \rangle$ and $\langle \partial\Phi^{-1}/\partial l \rangle$—the more the centers of these lines will be apart. Then the more concave the absorption curve, the wider the absorption band. Thus,

An absorption band becomes blurred with increase of anharmonicity (59a)

If the lifetime τ is sufficiently long, then the form of the absorption band depends mainly on the anharmonicity—that is, on the following average values:

$$\langle \partial\Phi^{-1}/\partial h \rangle \quad \text{and} \quad \langle \partial\Phi^{-1}/\partial l \rangle \tag{59b}$$

Both these rules are employed in Section III, where we also suggest a *quantitative* definition of the anharmonicity.

In the case of a *limiting line*, corresponding to extremely rare collisions—that is, to $\tau \to \infty$—the absorption $A(x)$ is *explicitly* expressed through the derivatives of a period Φ over energy h or l. Indeed, the Lorentz lines in the integrand of Eq. (51) are reduced in this limit to δ-functions with arguments $(2\pi n/\Phi - x)$ and $[2(\pi n - \lambda)/\Phi - x]$. Single integration over h (or l) yields the expression, in which the integrand of the remaining integral is proportional to $(\partial\Phi^{-1}/\partial h)^{-1}$ or $(\partial\Phi^{-1}/\partial l)^{-1}$ taken in the points of a resonance. The larger these quantities, the higher the absorption peak and therefore the narrower is the band.

4. *Characteristics of the Relaxation Spectrum*

In the low-frequency range (with $x \ll x_m$) the spectral function $L(z)$ depends weakly on frequency x. Then Eq. (32) comes to the Debye-relaxation spectrum given by Eq. (33). Its main characteristics, such as the dielectric-loss maximum χ''_D and its frequency x_D, are given by Eq. (34). A connection between these quantities and the model parameters becomes clear in an example of a very small collision frequency y. In this case, relations (34) come to

$$x_D = \frac{y}{g} L(0); \; \chi''_D = \frac{G}{2}[g - L(0)] \tag{60a}$$

On the other hand, from Eq. (51) in view of (46a), (46b), and (50a) follows the relation

$$L(0) = \langle \cos^2 \vartheta - a_0^2 + \sin^2 \vartheta \rangle = 1 - \langle a_0^2 \rangle \tag{60b}$$

a_0 presents a constant component of the function $\cos \vartheta(\varphi)$. Substitution of (60a) in (60b) yields

$$x_D = \frac{y}{g}\left(1 - \langle a_0^2 \rangle\right), \qquad \chi_D'' = \frac{G}{2}\left[g - 1 + \langle a_0^2 \rangle\right] \tag{60c}$$

The quantity a_0 characterizes, as well as the maximum deflection β, the mean localization of dipoles. The narrower this localization, the lower the value of $\cos^2 \vartheta$ and, consequently, $\langle a_0^2 \rangle$ differs from 1. So, in view of (60c),

If collisions are rare, then the relaxation loss peak shifts to lower frequencies, with the angular localization becoming narrower (61)

Note that we consider here a *mean* localization, for which it is important to account for existence of *two* subensembles, characterized by the parameters $\breve{\beta}$ and $\mathring{\beta}$. In the above-mentioned case of a finite-depth potential well, the deeper the well, the greater the proportion of the particles localized in it and the narrower *in average* the localization. On the basis of these considerations, we shall show in Section IV, where such a potential will be considered, that dependence of the position x_D on the well depth agrees with the rule (61). Note, the loss-peak intensity χ_D'' also obeys the dependence given by Eq. (60c). However, its increase with an augmentation of the well depth is rather small, while the dependence of the relaxation time on the mean localization is much more essential.

III. PROTOMODEL: DIPOLE IN RECTANGULAR POTENTIAL WELL

A. A Dipole in a Conical Cavity

1. *Geometry and Law of Motion*

The word "protomodel" means that the employed here potential profile—an infinitely deep rectangular well—is the simplest in the class of the *rectangular-like* potentials considered in this review. The protomodel was represented in our previous review GT, Chapter 6B, and in VIG, Section 8.5, where it was termed "the cone-confined rotator" (CCR) model. In the present chapter we regard this model as a basis of further studies; most relations described here will be used and/or generalized in Section IV. Moreover, in Section III we (i) for the first time obtain (in Section III.E) an analytic representation for infinite series, which are involved in the expression for the spectral function and (ii) suggest a planar libration–regular precession approximation allowing us to replace double integrals for simple ones.

Geometrical image of a potential considered in this chapter is a conical cavity (it will be shown below) with perfectly reflecting walls. Let 2β be an opening of a cone; β is thus a maximum angular deflection of a dipole from the symmetry axis of the cone. The angle β determines the reorientation pertinent to a displacement of the *symmetry axis* of a dipole. We should emphasize that 2β is less than the angle $2\beta_c$, which actually restricts space occupied by a polar molecule reorienting inside a local-order cage of a liquid. We see in Fig. 4 that the difference $\beta_c - \beta$ is related to the "thickness" of a dipole; β_c will be estimated in the end of Section IV.

Geometrical scheme of dipole's reorientations is given in Fig. 5. Because of perfect reflections of a dipole-moment vector $\boldsymbol{\mu}$ from conical walls, a surface covered by this vector consists of identical sectors with a central angle 2α. The sectors are inclined at the angle $(\pi/2 - \theta)$ to the cone axis C, where θ is the angle between this axis and the direction $\mathbf{n}$ of the angular-velocity vector of

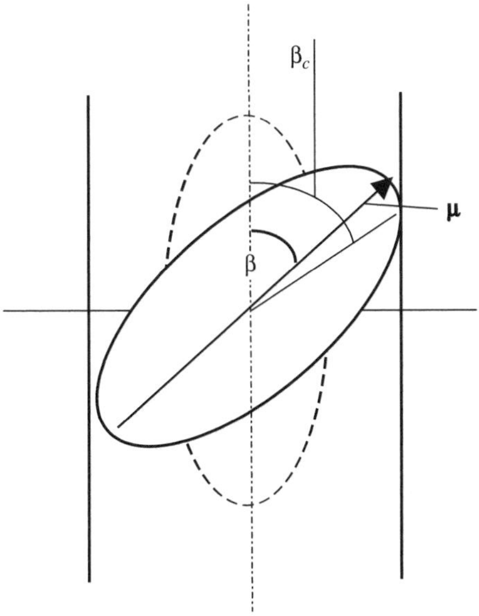

Figure 4. Physical (β_c) and model (β) angular dimensions of a local-order cage.

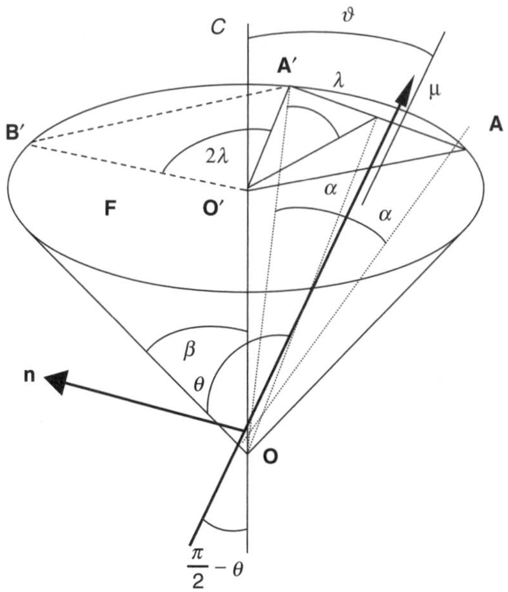

Figure 5. Geometric scheme of dipole's reorientation in a conical cavity. Explanations are given in the text.

a rotating dipole. The angle ϑ (φ) is, just as in Section II, the current deflection of vector $\boldsymbol{\mu}$ from the cone's axis C. The angle λ presents the projection of the angle α onto the plane F placed on the cone's base, so that $\lambda = (1/2)\angle \mathrm{AO'B}$. The triangle B′O′A′ illustrates the next (after reflection) position of the triangle AO′A′. We find from the figure the following relations connecting these angles:

$$\cos\alpha = \cos\beta / \sin\theta; \quad \sin\alpha = \sin\beta \sin\lambda \tag{62}$$

Let us express in terms of dimensionless units [Eq. (41)] the quantities referring to the law of motion of a dipole. The period Φ of the function ϑ (φ) is equal to the time interval between consecutive reflections of a dipole. During this interval it turns on the angle 2α, the square of its angular velocity being equal to a normalized energy h. Hence,

$$\Phi = \frac{2\alpha}{\sqrt{h}} \tag{63}$$

The projection of the angular velocity on the axis C is equal to the reduced angular momentum l above this axis:

$$l = \sqrt{h}\cos\theta \tag{64}$$

2. *Configuration of Phase Space*

Instead of h and l, we choose as phase variables the two components g and f of energy h given by

$$g \equiv \dot{\vartheta}^2(\vartheta = \beta) = h - f, \qquad f \equiv (l/\sin\beta)^2 \tag{65}$$

where g and f are, respectively, the "axial" (related to the change of ϑ) and precessional kinetic energies at the instants of reflections. Using Eqs. (62)–(65), we find

$$\alpha(g,f) = \arctan\left(\tan\beta\sqrt{\frac{g}{g+f}}\right), \qquad \lambda(g,f) = \arctan\left(\sec\beta\sqrt{\frac{g}{f}}\right) \tag{66}$$

In view of Eq. (65) the phase-variable differential (which we shall apply in Section III.B) is

$$d\Gamma' \equiv dh\,dl = \frac{1}{2}\sin\beta\,dg\,df/\sqrt{f} \tag{67a}$$

with

$$g \in [0,\infty], \qquad f \in [0,\infty] \tag{67b}$$

Now we shall find the norm C of the Boltzmann distribution $W(h)$. Instead of the phase variables g, f, and φ_0, it is convenient to take the angle ϑ, the angular velocity $\dot{\vartheta}$, and the projection l of the angular momentum. The angle ϑ and its time derivative $\dot{\vartheta}$ vary, respectively, in the intervals $[0,\beta]$ and $[-\infty,\infty]$; l varies in the interval $[0, \infty]$. Into the indicator of the Boltzmann exponent we should insert Eqs. (42a) and (42b), setting there $U = 0$. We note that the interval $\dot{\vartheta} \in [-\infty,\infty]$ corresponds to the interval $[0, \infty]$, in which g varies. As for the interval $\varphi_0 \in [0,\Phi]$, we should take into account the change of the angle ϑ in *two* directions (for seesaw motions). Therefore we should double the integral with the above-indicated limits of the phase variables ϑ and $\dot{\vartheta}$. Thus we have

$$1/C = 2\int_{-\infty}^{\infty} d\dot{\vartheta}\int_0^{\beta} d\vartheta \int_0^{\infty} \exp\left(-\dot{\vartheta}^2 - \frac{l^2}{\sin^2\vartheta}\right) dl = \pi(1-\cos\beta) \tag{68a}$$

Then the Boltzmann distribution is given by

$$W(h) = \frac{e^{-h}}{\pi(1 - \cos\beta)} \tag{68b}$$

B. Spectral Function (Rigorous Expression)

1. Representation by Series

The expression for the Fourier amplitudes a_n and b_n in Eqs. (50a) and (50b) obtained in terms of the CCR model [GT, Eqs. (6.31) and (6.62)] are given by

$$\begin{Bmatrix} a_n \\ b_n \end{Bmatrix} = (-1)^n \begin{Bmatrix} -2\sin\beta \\ \cos\beta \end{Bmatrix} \frac{\alpha\cos\beta\sin\lambda}{\cos\alpha}\left[\left(\pi n - \begin{Bmatrix} 0 \\ \lambda \end{Bmatrix}\right)^2 - \alpha^2\right]^{-1}, \qquad n \neq 0$$

Taking into account Eq. (66), we express them through the phase variables g and f. Then we have

$$\begin{Bmatrix} a_n \\ b_n \end{Bmatrix} = (-1)^n \begin{Bmatrix} -2\sin\beta \\ \cos\beta \end{Bmatrix} \alpha\sqrt{\frac{g}{g+f}}\left[\left(\pi n - \begin{Bmatrix} 0 \\ \lambda \end{Bmatrix}\right)^2 - \alpha^2\right]^{-1}, \qquad n \neq 0 \tag{69}$$

Substitution of Eqs. (62), (67a), (67b), (68a), (68b), and (69) into Eq. (51) yields

$$L(z) = \frac{\sin\beta}{\pi(1-\cos\beta)} \int_0^\infty e^{-g} g\, dg \int_0^\infty [\cos^2\beta S(\lambda) + \sin^2\beta S(0)]\alpha^3(g,f) \times \frac{e^{-f}\, df}{\sqrt{f(g+f)}} \tag{70a}$$

$$S(\lambda) \equiv \sum_{n=-\infty}^{\infty} [(\pi n - \lambda)^2 - \alpha^2]^{-2}[g + f - w_n^2(\lambda)]^{-1}, \qquad w_n(\lambda) \equiv \frac{z\alpha(g,f)}{\pi n - \lambda} \tag{70b}$$

the dependencies $\alpha(g,f)$ and $\lambda(g,f)$ are given by Eq. (66). This expression is equivalent to Eq. (6.41) in GT.

We shall transform now the expressions (70a) and (70b) to a form, which is more convenient for calculations. For this purpose we

1. Introduce new variables $q = \sqrt{g}$ and $p = \sqrt{f}$ instead of g and f.
2. Export the multipliers $\alpha^2/(\pi n - \lambda)^2$ from denominators in $S(\lambda)$.
3. Represent $S(0)$ as a series over positive numbers n.

Then we obtain

$$L(z) = \frac{4\pi \sin\beta}{1-\cos\beta}\int_0^\infty e^{-q^2} q^3\, dq \int_0^\infty [s(\lambda)\cos\beta + s(0)\sin\beta]\frac{\alpha e^{-p^2}dp}{\sqrt{q^2+p^2}} \tag{71a}$$

$$s(\{\lambda,0\} \equiv \{1,2\} \sum_{n=\{-\infty,1\}}^{\infty} \frac{(n-\{\lambda,0\}/\pi)^2}{\left[(\pi n-\{\lambda,0\})^2-\alpha^2\right]^2\left[(q^2+p^2)(\pi n-\{\lambda,0\})^2\alpha^{-2}-z^2\right]} \tag{71b}$$

where

$$\alpha(q,p) = \arctan\left(\frac{q\tan\beta}{\sqrt{q^2+p^2}}\right); \qquad \lambda(q,p) = \arctan\left(\frac{q}{p}\sec\beta\right)$$

2. *Analytical Representation*

We shall get from (70a) and (70b) another formula for the spectral function $L(z)$. At first, we express the series $S(\lambda)$ through elementary functions (the derivation is given in Section III.E):

$$S(\lambda)\cos^2\beta + S(0)\sin^2\beta = \frac{1}{(g+f-z^2)\alpha^2}\left[1+\frac{f}{g}-\frac{(z\sqrt{g+f}\cot\varsigma)(\sin^2\alpha-\sin^2)\varsigma}{\alpha(g+f-z^2)(\sin^2\lambda-\sin^2\varsigma)}\right] \tag{72}$$

where $\varsigma \equiv z\alpha(g+f)^{-1/2}$. Substituting Eqs. (71a) and (71b) into Eq. (70a), we finally have

$$L(z) = \frac{\sin\beta}{\pi(1-\cos\beta)}\int_0^\infty e^{-g}\,dg \int_0^\infty \left[\alpha\sqrt{\frac{g}{f}+1} - \cot\varsigma \right.$$
$$\left. \times \frac{\sin^2\alpha-\sin^2\varsigma}{\sin^2\lambda-\sin^2\varsigma}\frac{zg}{\sqrt{f}(g+f-z^2)}\right]\frac{e^{-f}\,df}{(g+f-z^2)} \tag{73a}$$

$$\varsigma(g,f) \equiv \alpha(g,f)z/\sqrt{g+f};\ \alpha(q,p) = \arctan\left(\sqrt{\frac{g}{g+f}}\tan\beta\right),$$
$$\lambda(g,f) = \arctan\left(\sqrt{\frac{g}{f}}\sec\beta\right) \tag{73b}$$

To interpret Eq. (73a), let us represent it as a difference:

$$L(z) = L_F(z) - L_R(z) \tag{74a}$$

$$L_F(z) = \frac{\sin\beta}{\pi(1-\cos\beta)}\int_0^\infty e^{-g}dg\int_0^\infty \sqrt{\frac{g}{f}+1}\,\frac{\alpha(g,f)e^{-f}}{g+f-z^2}\,df \tag{74b}$$

$$L_R(z) = \frac{z\sin\beta}{\pi(1-\cos\beta)}\int_0^\infty e^{-g}g\,dg\int_0^\infty \cot\varsigma\frac{\sin^2\alpha-\sin^2\varsigma}{\sin^2\lambda-\sin^2\varsigma}\frac{e^{-f}df}{\sqrt{f}(g+f-z^2)^2} \tag{74c}$$

Let us clarify the subscripts introduced here. If we use the variables h and $\cos\theta$, then in view of Eq. (64) we should write

$$d\Gamma' \equiv dh\,dl = \sqrt{h}\,dh\,d\cos\theta \tag{75a}$$

Comparing this with Eqs. (67a) and (67b), we replace $\sin\beta dg\,df/\sqrt{f}$ in Eq. (74a) by $2\sqrt{h}\,dh\,d\cos\theta$. Thus

$$L_F(z) = \frac{2}{\pi(1-\cos\beta)}\int_0^{\infty}\frac{e^{-h}h\,dh}{h-z^2}\int_0^{\sin\beta}\alpha(\theta)\;d\cos\theta \tag{75b}$$

Taking into account Eq. (62), we calculate the integral over θ:

$$\begin{aligned}\int_0^{\sin\beta}\alpha(\theta)\;d\cos\theta &= \alpha(\theta)\cos\theta\Big|_{\cos\theta=0}^{\cos\theta=\sin\beta} - \int_0^{\pi/2}\cos\theta\frac{d\alpha(\lambda)}{d\lambda}\,d\lambda\\ &= \int_0^{\pi/2}\left[\frac{d\alpha(\lambda)}{d\lambda}\right]^2 d\lambda = \sin^2\beta\int_0^{\pi/2}\frac{\cos^2\lambda\,d\lambda}{1-\sin^2\beta\sin^2\lambda}\\ &= \frac{\pi}{2}(1-\cos\beta)\end{aligned} \tag{76}$$

Substituting Eq. (76) into Eq. (75b), we express $L_F(z)$ through the integral exponential function $E_1(-z^2)$:

$$L_F(z) = \int_0^{\infty}\frac{e^{-h}h\;dh}{h-z^2} \equiv 1 + z^2e^{-z^2}E_1(-z^2) \tag{77}$$

We have obtained the expression given in GT, p. 225 for the spectral function of *free* rotors moving in a *homogeneous* potential in the interval between strong collisions; see also VIG, Eqs. (7.12) and (7.13). So, the subscript F means "free." The subscript R in Eq. (74c) is used as an initial letter of "restriction." Indeed, as it follows from the comparison of Eq. (77) with Eq. (74a), the second term of the last equation expresses the *steric-restriction* effect arising for free rotation due to a potential wall. If we set, for example, $\beta = \pi$, what corresponds to a complete rotation (without restriction) of a dipole-moment vector $\boldsymbol{\mu}$, then we find from Eqs. (74a)–(74c) that $L_R(z) = 0$ and $L(z) = L_F(z)$. This result confirms our statement about "restriction."

C. Planar Librations–Regular Precession (PL–RP) Approximation

To simplify the double integral in Eq. (70a) we assume that:

(i) Dipoles librate in a diametric section of a cone, if the precessional energy $f < g$.

(ii) They precess about the axis of a cone with a constant velocity, if $g < f < \infty$.

Then we divide the interval of integration over f by two above-mentioned intervals. We retain rigorous expression in exponential and power functions involved in the integrand in Eqs. (70a) and (70b), but in other co-factors we set $f = 0$ and $g = 0$ in the first and second intervals, respectively. The spectral function then is approximated by a sum

$$L(z) \approx L_{\mathrm{PL}}(z) + L_{\mathrm{RP}}(z) \tag{78}$$

1. *Calculation of the Librational Spectral Function $L_{\mathrm{PL}}(z)$*

We set the upper limit of the integral over f to be g (instead of ∞), $\alpha = \beta$, $\lambda = \pi/2$ and replace $g + f$ by g. Then Eqs. (70a) and (70b) yield

$$\begin{aligned}
&\cos^2\beta S(\lambda) + \sin^2\beta S(0) \\
&= \sum_{n=-\infty}^{\infty}\left\{\frac{\cos^2\beta}{[\pi(n-1/2)^2-\beta^2]^2\left[g-\frac{z^2\beta^2}{\pi^2(n-1/2)^2}\right]}+\frac{\sin^2\beta}{[(\pi n)^2-\beta^2]^2\left[g-\left(\frac{z\beta}{\pi n}\right)^2\right]}\right\} \\
&= 32\sum_{n=1}^{\infty}\frac{\sin^2(\pi n/2-\beta)}{[(\pi n)^2-(2\beta^2)]^2\left[g-\left(\frac{2z\beta}{\pi n}\right)^2\right]}
\end{aligned}$$

$$\begin{aligned}
L_{\mathrm{PL}}(z) &= \frac{\beta^3\sin\beta}{\pi(1-\cos\beta)}\int_0^\infty[\cos^2\beta S(\lambda)+\sin^2\beta S(0)]e^{-g}\sqrt{g}\,dg\int_0^g\frac{e^{-f}df}{\sqrt{f}} \\
&= \frac{32\beta^3\sin\beta}{\sqrt{\pi}(1-\cos\beta)}\int_0^\infty\sum_{n=1}^{\infty}\frac{\sin^2(\pi n/2-\beta)}{[(\pi n)^2-(2\beta)^2]^2\left[g-\left(\frac{2z\beta}{\pi n}\right)^2\right]}e^{-g}\mathrm{erf}(\sqrt{g})\sqrt{g}\,dg
\end{aligned} \tag{79}$$

2. *Calculation of the Precessional Spectral Function $L_{\mathrm{RP}}(z)$*

We set lower limit of the integral over f in Eq. (70a) equal to g and replace $g + f$ by f. Then Eq. (66) yield

$$\alpha \simeq \tan\beta\sqrt{g/f} \to 0; \lambda \simeq \sec\beta\sqrt{g/f} \to 0$$

so that terms of the series $S(0)$ in (70b) are reduced to

$$\left[(\pi n)^2 - \frac{g}{f}\tan^2\beta\right]^{-2}\left[f - \frac{zg\tan^2\beta}{(\pi n)^2 f}\right]^{-1} \to (\pi n)^{-4} f^{-1}$$

The composition of the letter expression and $g\alpha^3$ vanishes at $n \neq 0$. Since by definition the zero term of the series $S(0)$ equals zero, the composition $g\alpha^3 S(0)$

equals zero. As for the series with $\lambda \neq 0$, only the term with the subscript $n = 0$ gives a nonzero contribution at $g \to 0$ to the composition $g\alpha^3 S(\lambda)$:

$$\lim_{g\to 0}\left[g\alpha^3 S(\lambda)\right] = \frac{g(g/f)^{3/2}\tan^3\beta}{\left(\frac{g}{f\cos^2\beta} - \frac{g\sin^2\beta}{f\cos^2\beta}\right)^2 (f - z^2\sin^2\beta)} = \frac{\sqrt{gf}\sin\beta}{\cos^3\beta(f/\sin^2\beta - z^2)}$$

Taking this into account, we derive from Eqs. (70a) and (70b)

$$\begin{aligned} L_{\mathrm{RP}}(z) &= \frac{\sin\beta}{\pi(1-\cos\beta)}\int_0^\infty \frac{\sin\beta\cos^2\beta}{\cos^3\beta} e^{-g}\sqrt{g}\,dg \int_g^\infty \frac{\sqrt{f}}{f/\sin^2\beta - z^2}\frac{e^{-f}\,df}{\sqrt{f}} \\ &= \frac{\sin^2\beta}{\pi\cos\beta(1-\cos\beta)}\left[\left.\int_0^g e^{-g}\sqrt{g}\,dg \int_g^\infty \frac{e^{-f}\,df}{(f/\sin^2\beta) - z^2}\right|_0^\infty \right. \\ &\quad \left. + \int_0^\infty \frac{e^{-g}\,dg}{(g/\sin^2\beta) - z^2}\int_0^g e^{-g}\sqrt{g}\,dg\right] \end{aligned} \tag{80a}$$

Replacing g for f, we finally obtain

$$L_{\mathrm{RP}}(z) = \frac{\sin^2\beta}{2\sqrt{\pi}\cos\beta(1-\cos\beta)}\int_0^\infty \frac{\mathrm{erf}(\sqrt{f}) - 2\sqrt{f/\pi}e^{-f}}{f/\sin^2\beta - z^2}e^{-f}\,df \tag{80b}$$

Note, that "resonance" frequency here is the velocity of regular precession given by

$$\sqrt{f}/\sin\beta = l/\sin^2\beta = \dot{\psi}(\vartheta = \beta)$$

3. *Resulting Spectral Function. The Small β Approximation*

Summing up both terms in Eq. (78), we obtain from Eqs. (79) and (80b) the resulting spectral function:

$$\begin{aligned} L(z) \simeq{}& \frac{32\beta^3\sin\beta}{\sqrt{\pi}(1-\cos\beta)}\int_0^\infty \sum_{n=1}^\infty \frac{\sin^2(\pi n/2 - \beta)e^{-g}\mathrm{erf}(\sqrt{g})\sqrt{g}}{[(\pi n)^2 - (2\beta)^2]^2\left[g - \left(\frac{2z\beta}{\pi n}\right)^2\right]}dg \\ &+ \frac{\sin^2\beta}{2\sqrt{\pi}\cos\beta(1-\cos\beta)}\int_0^\infty \frac{\mathrm{erf}(\sqrt{f}) - 2\sqrt{f/\pi}e^{-f}}{f/\sin^2\beta - z^2}e^{-f}\,df \end{aligned} \tag{81}$$

At small angle β, this result can be reduced to Eq. (6.45) in GT:

$$L(z) \approx \frac{128\beta^2}{\pi^5}\int_0^\infty \frac{h\,e^{-h}\,dh}{h - (2\beta z/\pi)^2} = \frac{128\beta^2}{\pi^5}L_{\mathrm{F}}\left(\frac{2\beta}{\pi}z\right) \tag{82}$$

where $L_F(s)$ is given by Eq. (77). This approximation is based on the supposition that at small β the precessional component of the dipoles' trajectories could be neglected. Equation (81) is reduced to Eq. (82), if we omit in Eq. (81) the second, precessional, term and in the remaining, librational,

(i) Retain only the first term of the sum
(ii) Neglect β in comparison with π in the first multiplier in the denominator
(iii) Replace α and sin β for β; cos β for 1; and (1 − cos β) for $\beta^2/2$
(iv) Set $f = 0$ in $\exp(-f)$ in the integral over f; then integration gives $2\sqrt{g}$ instead of $\sqrt{\pi}\,\mathrm{erf}(\sqrt{g})$
(v) Replace designation g for h.

We remark that Eq. (81) gives a satisfactory value of the integral absorption (GT, p. 240) and presents a basis for the so-called "hybrid model"; also see Sections IV.5–IV.7 and VIG, Chapter 9.

4. *Absorption Frequency Dependence*

We show in Fig. 6a the dependence of the normalized absorption

$$A(x) = x\mathrm{Im}[L(z)]$$

on the normalized frequency x, calculated by using the rigorous formula (70a) and (70b) (solid curves) and the PL–RP approximation, Eq. (81) (dashed curves). We take two cone angles β: π/8 and π/4. The maximum-absorption frequency is larger for the lower β value; this property agrees with the rule (II.3.3). Next, for lower β the curve $A(x)$ is more concave. Both properties agree with the estimations given in the next section, where we shall also discuss Fig. 6b.

D. Statistical Parameters

In view of Eqs. (45), (47), (68), and (75), the mean value of any quantity s, which does not depend on the phase variables φ_0 and ψ_0, is defined as

$$\langle s\rangle = \pi^{-1}(1-\cos\beta)^{-1})\int_0^{\infty} e^{-h}\sqrt{h}\,dh\int_0^{\sin\beta} s\Phi\,d\cos\theta \tag{83}$$

Setting here $s = \Phi^{-1}$, we obtain the expression for a mean inverse period of the function $\vartheta(\varphi)$:

$$\xi \equiv \langle\Phi^{-1}\rangle = \frac{\sqrt{\pi}/2}{\pi(1-\cos\beta)}\sin\beta = \frac{\cot(\beta/2)}{2\sqrt{\pi}} \tag{84}$$

Estimation based on the rule (54) gives

$$x_m \approx \pi\xi = \frac{\sqrt{\pi}}{2}\cot(\beta/2) \tag{85}$$

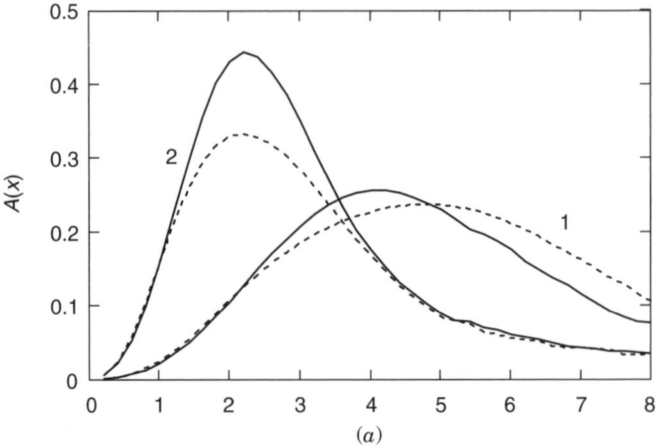

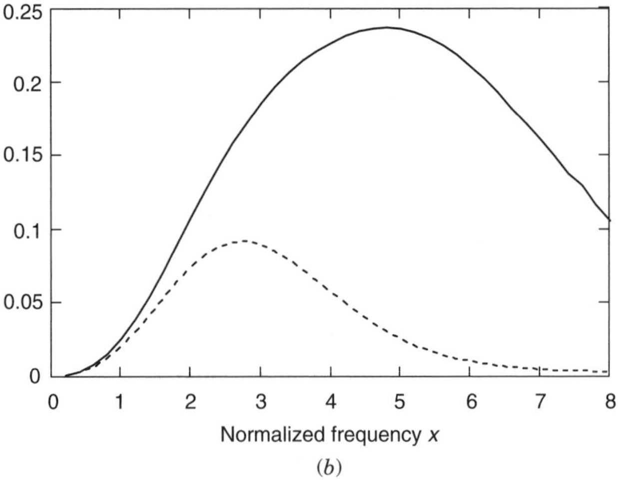

Figure 6. Frequency dependence of dimensionless absorption. Dimensionless collision frequency $y = 0.3$. (a) Calculation from rigorous formulas (70a) and (70b) (solid lines) and from the PL-RP approximation, Eqs. (78–80b) (dashed lines). Curves 1 refer to $\beta = \pi/8$ and curves 2 to $\beta = \pi/4$. Vertical lines mark the values of the absorption-peak frequencies estimated by Eqs. (85) and (86). (b) Comparison of the total absorption (solid line) with contribution of the precessional component (dashed line). Calculation for the PL–RP approximation, $\beta = \pi/8$.

The PL–RP approximation allows us to estimate this quantity in another fashion. It is seen in Fig. 6b that near the peak-absorption frequency the contribution of the "precessional" component L_{RP} is negligible, so that $L \cong L_{PL}$ at $x \cong x_m$. We shall estimate x_m by considering the limit $y \to 0$. The integrand in Eq. (79) comprises the Lorentz-like terms. In this limit their imaginary components transform to δ-functions, so we have

$$\frac{1}{\pi}\lim_{y\to 0}\mathrm{Im}\left[\int_0^\infty \frac{\mathrm{erf}\left(\sqrt{g}\right)e^{-g}\sqrt{g}\,dg}{g-\left(\frac{2z\beta}{\pi n}\right)^2}\right]$$

$$=\int_0^\infty \mathrm{erf}\left(\sqrt{g}\right)e^{-g}\sqrt{g}\delta[g-g_n(x)]\,dg=\mathrm{erf}\left(\sqrt{g_n(x)}\right)\sqrt{g_n(x)}e^{-g_n(x)}$$

where $g_n(x)=(2x\beta)^2/(\pi n)^2$ are the "resonance" values of the energy g. This expression is reduced to

$$\sqrt{g_1(x)}\,\mathrm{erf}\left[\sqrt{g_1(x)}\right]e^{-g_1(x)}$$

if we retain only the first term of the series in Eq. (79). Then absorption $A(x)$ becomes proportional to

$$x\sqrt{g_1(x)}\,\mathrm{erf}\left[\sqrt{g_1(x)}\right]e^{-g_1(x)}\propto g_1(x)\mathrm{erf}\left[\sqrt{g_1(x)}\right]e^{-g_1(x)}$$

This function attains maximum at $g_1\cong 1.2$. So, another estimation gives

$$x_m\approx\frac{\pi\sqrt{1.2}}{2\beta} \tag{86}$$

For the values of model parameters given in the capture to Fig. 6 we shall estimate now the following quantities:

(i) *Inaccuracy* $\Delta=\left|1-x_m^{\mathrm{est}}/x_m\right|$ concerning estimation of the absorption-maximum frequency. Here estimation x_m^{est} is given by Eq. (85) and x_m is obtained from the rigorous formula (70a). The results of estimations (85) and (86) are graphically indistinguishable and both are close to the position of $A_{\max}$ found in Eq. (70a). Hence, in the case of the protomodel the absorption-peak frequency is close to the mean "resonance" frequency.

(ii) The *anharmonicity* ANH. In accord with the rule (59b), we give the following *quantitative* definition of ANH:

$$\mathrm{ANH}\equiv\pi\left\langle\left(\frac{\partial\Phi^{-1}}{\partial\sqrt{h}}\right)_\lambda\right\rangle \tag{87}$$

In view of Eq. (63), $\frac{\partial\Phi^{-1}}{\partial\sqrt{h}}=(\Phi\sqrt{h})^{-1}$. Substituting this into Eq. (83) and comparing the result with Eqs. (84) and (85), we find that the dependence (87) differs from (85) only by a multiplier $2/\sqrt{\pi}$. Hence,

$$\mathrm{ANH}(\beta)=\cot(\beta/2) \tag{88}$$

TABLE I
Parameters of the Absorption Band Estimated for the Protomodel

β	$\Delta\%$	ANH/Δx	$\breve{m}$
$\pi/8$	6.1	1.09	7.1
$\pi/4$	2.7	1.01	3.4

(iii) The number $\breve{m}$ of reflections of a dipole from the conical surface performed during the mean lifetime τ. The frequency of such reflections is equal to a inverse period of the function $\vartheta(\varphi)$. Thus in accord with Eq. (84) we have

$$\breve{m} \equiv \xi\frac{\tau}{\eta} = \frac{\cot(\beta/2)}{2\sqrt{\pi y}} \tag{89}$$

This quantity characterizes some "figure of merit" of an elementary local-order cell in a liquid.

The results of these estimations are listed in Table I.
From Fig. 6 and this table we see that:

(1) The inaccuracy Δ is less[19] 10%.
(2) The absorption peak increases, if the angle β decreases, which agrees with Eqs. (85)–(86) and the rule (59a).
(3) In the third column of the table the ratio of quantity (87) to bandwidth Δx is given. We conclude that the anharmonicity parameter, introduced above, *is equal to the bandwidth* Δx.
(4) The value of Eq. (89) is $\gg 1$ and increases, if the cone angle β decreases, see last column of the table.

E. Summation of Series in Expression for the Spectral Function

We replace in Eq. (70a) the sum $g+f$ by h and introduce the variable $\varsigma = z\alpha/\sqrt{h}$. Using the formula 3 on p. 685 of Ref. 38, we get from Eq. (70b)

$$\begin{aligned} S(\lambda) &= \frac{1}{2h\alpha}\frac{\partial}{\partial\alpha}\sum_{n=-\infty}^{\infty}\frac{(\pi n-\lambda)^2}{\left[(\pi n-\lambda)^2-\alpha^2\right]\left[(\pi n-\lambda)^2-\varsigma^2\right]} \\ &= \frac{1}{2h\alpha}\frac{\partial}{\partial\alpha}\left\{\frac{1}{\alpha^2-\varsigma^2}\sum_{n=-\infty}^{\infty}\left[\frac{\alpha^2}{(\pi n-\lambda)^2-\alpha^2[(\pi n-\lambda)^2-\varsigma^2]}-\frac{\varsigma^2}{(\pi n-\lambda)^2-\varsigma^2}\right]\right\} \\ &= \frac{1}{2h\alpha}\frac{\partial}{\partial\alpha}\left[\frac{1}{\alpha^2-\varsigma^2}\left(\frac{\alpha\cos\alpha\sin\alpha}{\sin^2\lambda-\sin^2\alpha}-\frac{\varsigma\cos\varsigma\sin\varsigma}{\sin^2\lambda-\sin^2\varsigma}\right)\right] \\ &= (2h)^{-1}[\Xi_1(\lambda)+\Xi_2(\lambda)+\Xi_3(\lambda)] \end{aligned} \tag{90}$$

[19]It is true even for $\beta = \pi/4$, although the rule (54) is valid only for small β.

where

$$\Xi_1(\lambda) \equiv (\alpha^2 - \varsigma^2)^{-1} \frac{\partial}{\partial \alpha} \frac{\cos\alpha \sin\alpha}{\sin^2\lambda - \sin^2\alpha}$$

$$= (\alpha^2 - \varsigma^2)^{-1} \frac{\sin^2\lambda - 2\sin^2\lambda\sin^2\alpha - \sin^2\alpha + 2\sin^4\alpha + 2\cos^2\alpha\sin^2\alpha}{(\sin^2\lambda - \sin^2\alpha)^2} \tag{91a}$$

$$\Xi_2(\lambda) \equiv -\frac{\varsigma}{\alpha} \frac{\cos\varsigma\sin\varsigma}{\sin^2\lambda - \sin^2\varsigma} \frac{\partial}{\partial\alpha} \frac{1}{\alpha^2 - \varsigma^2} = \frac{2\varsigma}{(\alpha^2 - \varsigma^2)^2} \frac{\cos\varsigma\sin\varsigma}{\sin^2\lambda - \sin^2\varsigma} \tag{91b}$$

$$\Xi_3(\lambda) \equiv \frac{\cos\alpha\sin\alpha}{\sin^2\lambda - \sin^2\alpha} \frac{1}{\alpha} \frac{\partial}{\partial\alpha} \frac{\alpha}{\alpha^2 - \varsigma^2} = -\frac{1}{\alpha} \frac{\alpha^2 + \varsigma^2}{(\alpha^2 - \varsigma^2)^2} \tag{91c}$$

Then we sum termwise $\cos^2\beta\, \Xi_k(\lambda)$ and $\sin^2\beta\, \Xi_k(0)$. Accounting for Eq. (62), we find

$$\cos^2\beta\, \Xi_1(\lambda) + \sin^2\beta\, \Xi_1(0)$$

$$= (\alpha^2 - \varsigma^2)^{-1} \left[\frac{\cos^2\beta}{\sin^4\lambda\cos^4\beta} (\sin^2\lambda - 2\sin^4\lambda\sin^2\beta - \sin^2\lambda\sin^2\beta \right.$$

$$+ 2\sin^4\lambda\sin^4\beta + 2\cos^2\alpha\sin^2\lambda\sin^2\beta)$$

$$\left. + \frac{\sin^2\beta}{\sin^4\beta\sin^4\lambda} (-\sin^2\beta\sin^2\lambda + 2\sin^4\beta\sin^4\lambda + 2\cos^2\alpha\sin^2\beta\sin^2\lambda) \right]$$

$$= (\alpha^2 - \varsigma^2)^{-1} \sin^{-2}\lambda[\cos^{-2}\beta(1 - 2\sin^2\beta\sin^2\lambda - \sin^2\beta$$

$$+ 2\sin^4\beta\sin^2\lambda + 2\sin^2\beta\cos^2\alpha)$$

$$+ \sin^{-2}\beta(-\sin^2\beta + 2\sin^4\beta\sin^2\lambda + 2\cos^2\alpha\sin^2\beta)]$$

$$= (\alpha^2 - \varsigma^2)^{-1} \sin^{-2}\lambda(1 - 2\sin^2\beta\sin^2\lambda + 2\sin^2\beta/\sin^2\theta - 1$$

$$+ 2\sin^2\beta\sin^2\lambda + 2\cos^2\beta/\sin^2\theta)$$

$$= \frac{2}{(\alpha^2 - \varsigma^2)\sin^2\lambda\sin^2\theta} \tag{92a}$$

$$\cos^2\beta\, \Xi_2(\lambda) + \sin^2\beta\, \Xi_2(0) = \frac{2\varsigma\sin\varsigma\cos\varsigma}{(\alpha^2 - \varsigma^2)^2} \left(\frac{\cos^2\beta}{\sin^2\lambda - \sin^2\varsigma} - \frac{\sin^2\beta}{\sin^2\varsigma} \right)$$

$$= -\frac{2\varsigma}{(\alpha^2 - \varsigma^2)^2} \frac{\sin^2\alpha - \sin^2\varsigma}{\sin^2\lambda - \sin^2\varsigma} \cot\varsigma \tag{92b}$$

$$\cos^2\beta\, \Xi_3(\lambda) + \sin^2\beta\, \Xi_3(0) = -\frac{1}{\alpha} \frac{\alpha^2 + \varsigma^2}{(\alpha^2 - \varsigma^2)^2} \cos\alpha\sin\alpha$$

$$\times \left(\frac{\cos^2\beta}{\cos^2\beta\sin^2\lambda} - \frac{\sin^2\beta}{\sin^2\beta\sin^2\lambda} \right) = 0 \tag{92c}$$

Summing up Eqs. (92a) and (92b), we get with account of (91)

$$\begin{aligned}\cos^2\beta S(\lambda) + \sin^2\beta S(0) &= \frac{1}{h(\alpha^2-\varsigma^2)\sin^2\lambda\sin^2\theta} - \frac{\varsigma}{h(\alpha^2-\varsigma^2)^2}\frac{\sin^2\alpha-\sin^2\varsigma}{\sin^2\lambda-\sin^2\varsigma}\cot\varsigma \\ &= \frac{1}{(h-z^2)\alpha^2\sin^2\lambda\sin^2\theta} - \frac{z\sqrt{h}}{(h-z^2)^2\alpha^3}\frac{\sin^2\alpha-\sin^2\varsigma}{\sin^2\lambda-\sin^2\varsigma}\cot\varsigma \qquad (93)\end{aligned}$$

It follows from Eqs. (62) and (66) that

$$\sin^2\lambda\,\sin^2\theta = \frac{g}{g+f}$$

We get Eq. (72) after we express all quantities in Eq. (93) through g and f.

IV. HAT-FLAT MODEL

A. Classification of Trajectories

The depth of any reasonable potential well should of course be finite. Moreover, the recorded spectrum of such an important liquid as water comprises *two* absorption bands: One, rather narrow, is placed near the frequency 200 cm^{-1}, and another, wide and intense band, is situated around the frequency 500 or 700 cm^{-1}, for heavy or ordinary water, respectively. In view of the rules (56) and (57), such an effect can arise due to dipoles' reorientation of two types, each being characterized by its maximum angular deflection from the equilibrium orientation of a dipole moment.[20] The simplest geometrically model potential satisfying this condition is the rectangular potential with finite well depth, entitled "hat-flat" (HF), since its form resembles a hat. We shall demonstrate in Section VII that the HF model could be used for a *qualitative* description of wideband spectra recorded in water[21] and in a nonassociated liquid.

The profile of the model potential is described as follows:

$$U(\vartheta) = \begin{Bmatrix} 0 \\ U_0 \end{Bmatrix} \quad \text{at} \begin{Bmatrix} \vartheta \le \beta \\ \vartheta \ge \beta \end{Bmatrix} \qquad (94)$$

where U_0 is depth and 2β is an angular width of the potential well; ϑ is the current angle between dipole moment and symmetry axis of the potential (see

[20] We shall see further (in Sections VI, VII, and IX) that there exists also another important cause stipulating appearance of the second separate absorption band in the FIR spectral region. In terms of our approach, such a band originates due to vibration of the hydrogen-bonded molecules.

[21] In the case of water we actually shall employ two approximations of the hat-flat model.

Fig. 5). The potential considered in this section differs from that involved in the protomodel in two respects:

(a) The depth U_0 is *finite*. Therefore the scheme, shown in Fig. 5, refers only to dipoles, which do not overcome the potential barrier U_0 during their *regular motion inside the well*. Conventionally, we shall term such dipoles *librators* (a reservation could be made that "pure" librators are only those dipoles, which oscillate in a plane, sited in a diametric section of a hollow cone). The particles, which overcome the barrier, performing complete rotation around the cone axis, will conventionally be termed *rotators*.

(b) *Two* conical cavities are considered, which have a common apex and axis, one cavity being turned around another one at the angle π. Correspondingly with this configuration, besides the *axial* symmetry such potential is also characterized by *mirror* symmetry. For the sake of obviousness, the potential profile is shown in Fig. 7 in polar coordinates.

In Eq. (94) the angle ϑ is determined from the middle of "upper" well. If we replace ϑ by $\pi - \vartheta$, then we describe the potential in the "lower" well. For the configuration shown in Fig. 7 the subensemble of the librators comprises, in turn, two identical statistically independent subensembles localized inside the

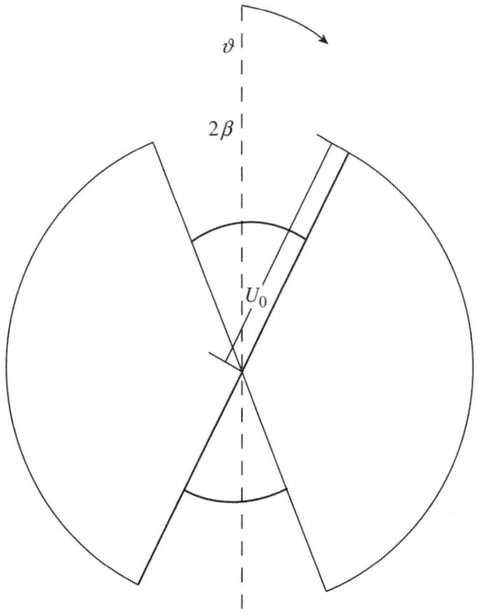

Figure 7. Profile of the model potential.

"upper" and "lower" cavities. It follows from Eq. (94) and from the Hamiltonian representation (42a) that the law of motion, represented in terms of ϑ, is the same as that for $\pi - \vartheta$. Then, in view of Eqs. (40) and (43), the expression for the autocorrelator given by Eq. (27a) does not change. Therefore, the contributions of both dipolar subensembles to the spectral function are equal. By the same cause the contributions to spectral function of the parts of the rotators' trajectories, which are placed in the "upper" and "lower" semispheres, are also equal. Thus, we may restrict our calculations by consideration of only one (any) well. For definiteness, we choose the upper well, where $\vartheta \leq \pi/2$. Then the dimensionless Hamiltonian is given by

$$h(\vartheta, \dot{\vartheta}, l) = \dot{\vartheta}^2 + w(\vartheta, l), \qquad w(\vartheta, l) \equiv w(\vartheta) = (l/\sin\vartheta)^2 + \begin{Bmatrix} 0 \\ u \end{Bmatrix} \text{ at } \begin{Bmatrix} \vartheta \leq \beta \\ \vartheta \geq \beta \end{Bmatrix} \tag{95}$$

Here $u = U_0/(k_B T)$, The quantity l is defined by Eqs. (35) and (41). The effective potential $w(\vartheta)$ was introduced in Section II, Eq. (42a), its profile is illustrated by Fig. 8. For the parameters chosen, the $w(\vartheta)$ dependence is emphasized in the left part of the figure and is weak in the right part. This distinction can be explained

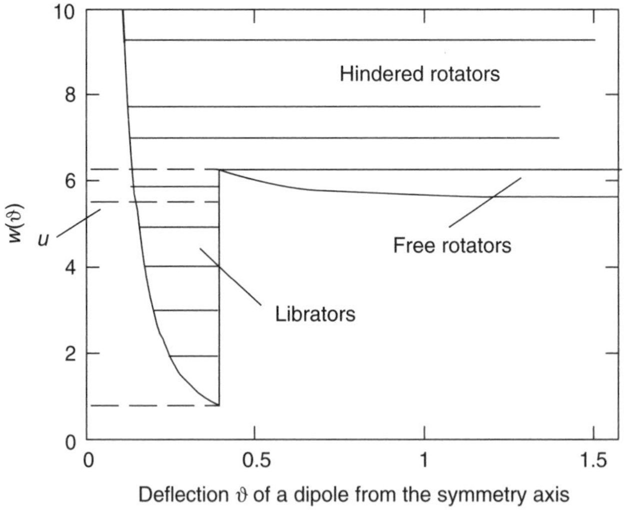

Figure 8. Profile of the effective potential $w(\vartheta)$. Dashed horizontal lines refer to the values $l^2/\sin^2\beta$, u, and $u + [l^2/\sin^2\beta]$ (from bottom to top). Calculation is made for $l = 0.3$, $\beta = \pi/8$, and $u = 5.5$. Solid horizontal lines mark various values of the total energy h of a dipole.

as follows. If the projection l of the angular momentum is small (in our example $l = 0.3$), the precession energy $(l/\sin\vartheta)^2$ is very small at $\vartheta \in [\beta, \pi/2]$, so the effective potential almost coincides with the homogeneous one (u). Note that in this interval the nonhomogeneity of $w(\vartheta)$ is neglected below in derivation of the interpolation formula (107).

Horizontal lines in Fig. 8 mark different values of the full energy h. One can see that the rotators' subensemble comprises the particles, with energies h less and larger than the threshold value $u_{\text{th}} = (l/\sin\beta)^2 + u$. The particles with $h > u_{\text{th}}$ occupy all the range of angles shown in Fig. 8, and the particles with $h < u_{\text{th}}$ occupy the interval $\vartheta \in [\beta, \pi/2]$. The distinction between two types of rotators is illustrated in Fig. 9. The plane R, tangential to the conical surface and inclined to the cone axis at the angle β, separates *hindered* and free *rotators*. The former (trajectory 1) at instants of intersection the conical surface change their rotation plane, since outside the cone they move in the potential U_0 and inside the cone they move in zero potential. All ranges of the angles shown in Fig. 8 correspond to them. The free rotators constantly exert action of a homogeneous potential U_0, since their trajectories (denoted 2) do not intersect

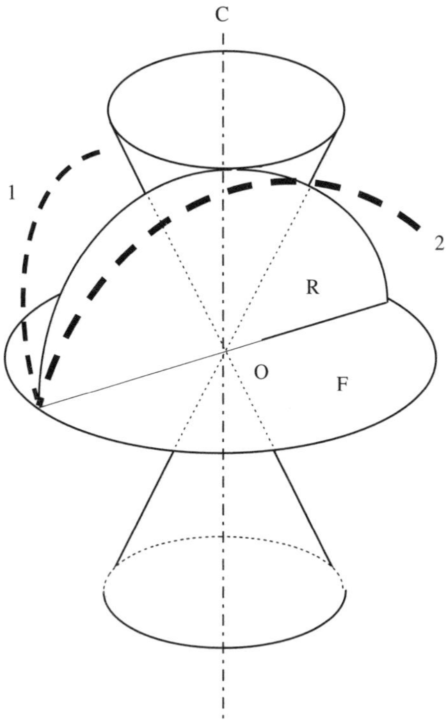

Figure 9. Schematic picture of trajectories of hindered (1) and free (2) rotators. Explanation is given in the text.

the conical surface. They occupy the range of the angles ϑ outside the well ($\vartheta > \beta$).

Thus, we should discriminate three subensembles. We mark the quantities referring to the librators by the superscript $\smile$, and we mark hindered and free rotators by the common superscript $\circ$. So, we represent the total spectral function (SF) as the sum

$$L(z) = \breve{L}(z) + \overset{\circ}{L}(z) \tag{96}$$

where, we repeat, $\overset{\circ}{L}(z)$ comprised the contribution to the SF of the rotators of both types. The terms on the right-hand side will be calculated below. In view of Fig. 8, membership of a dipole to any of these three subensembles is determined by the following conditions (here symbols $\frown$ and $\circ$ are used for hindered and free rotators, respectively):

$$\left(\breve{l}/\sin\beta\right)^2 \le \breve{h} \le u + \left(\breve{l}/\sin\beta\right)^2, \qquad 0 \le \breve{\vartheta} \le \beta \tag{97a}$$

$$u + \left(\overset{\frown}{l}/\sin\beta\right)^2 \le \overset{\frown}{h} \le \infty, \qquad 0 \le \overset{\frown}{\vartheta} \le \pi/2 \tag{97b}$$

$$u + \overset{\circ}{l}{}^2 \le \overset{\circ}{h} \le u + \left(\overset{\circ}{l}/\sin\beta\right)^2, \qquad \beta \le \overset{\circ}{\vartheta} \le \pi/2 \tag{97c}$$

Figure 10 demonstrates the phase regions occupied by these subensembles; here we show the cross sections at the angles ϑ_1 and ϑ_2, such that $\vartheta_1 \le \beta$ and $\vartheta_2 \ge \beta$. It follows from Eq. (97a) that the librators comprise some particles with energies exceeding the potential barrier u. It can be explained as follows. The u-value presents the threshold only with respect to the "axial" energy $\dot{\vartheta}^2$ but not

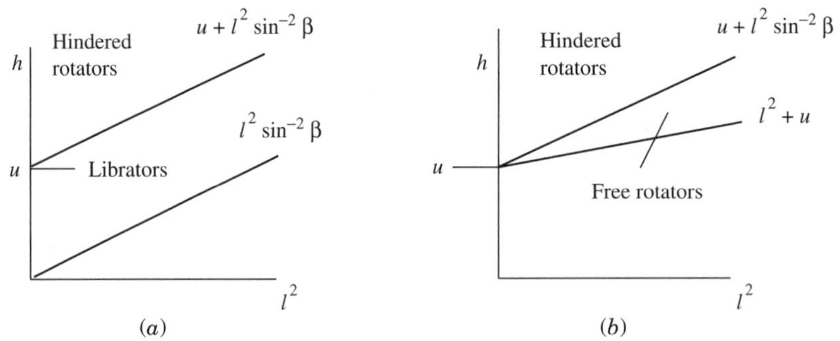

Figure 10. Phase regions occupied by subensembles of the dipoles at (a) $\vartheta < \beta$ and (b) $\vartheta > \beta$.

with respect to the total energy. If l is rather large, the former ($\dot{\vartheta}^2$) is less than the precession kinetic energy $(l/\sin\vartheta)^2$. In this case the kinetic energy is mostly pertinent to the precession of a dipole around the symmetry axis but is not pertinent to approaching toward (or to moving off) it.

Now we write down the expression for the norm C of the statistical distribution. By analogy with Eq. (68a) we have

$$1/C = 2\pi\left(\int_0^{\beta} \sin\vartheta \, d\vartheta + e^{-u}\int_{\beta}^{\pi/2} \sin\vartheta \, d\vartheta\right)$$
$$C = \frac{1}{2\pi E(\beta, u)} \tag{98a}$$

so that the steady-state distribution is expressed as

$$W(h) = \frac{e^{-h}}{2\pi E(\beta, u)}$$

where

$$E(\beta, u) \equiv 1 - (1 - e^{-u})\cos\beta \tag{98b}$$

B. Spectral Function of Librators

We shall apply the condition (97a), which in terms of the variables g and f, determined by Eq. (65), is given by

$$0 \le g \le u, \qquad 0 \le f \le \infty \tag{99}$$

1. *Representation in Form of a Series*

Comparing Eqs. (99) and (67b), we find that at the instant of reflection of the librator from the wall, when $\vartheta = \beta$, its "axial" energy $g \equiv \dot{\vartheta}^2$ is restricted by the value u, while in the case of the protomodel such energy tends to ∞. Therefore, in order to modify the spectral function (71a) of the protomodel with respect to librators, in the integral over $q \equiv \sqrt{g}$ we replace the infinite upper limit by $\sqrt{u}$.

Besides, the norm C of the distribution $W(h)$, Eq. (98a), differs from the norm given by Eq. (68a). To account for it, we introduce the multiplier $2C/C_0$, where C_0 denotes *now* the norm (68a) referring to the protomodel and the multiplier 2 is involved in view of the mirror symmetry of the potential (which comprises two wells).[22] As a result we have the following expression:

$$\breve{L}(z) = \frac{4\pi \sin\beta}{E(\beta,u)} \int_0^{\sqrt{u}} e^{-q^2} q^3\, dq \int_0^{\infty} \left[s(\lambda)\cos^2\beta + s(0)\sin^2\beta\right] \frac{\alpha e^{-p^2} dp}{\sqrt{q^2+p^2}} \tag{100a}$$

where

$$s(\{\lambda,0\}) \equiv \{1,2\} \times \sum_{n=\{-\infty,1\}}^{\infty} \frac{(n - \{\lambda,0\}/\pi)^2}{\left[(\pi n - \{\lambda,0\})^2 - \alpha^2\right]^2 \left[(q^2+p^2)(\pi n - \{\lambda,0\})^2 \alpha^{-2} - z^2\right]} \tag{100b}$$

$$\alpha(q,p) = \arctan\left(\frac{q \tan\beta}{\sqrt{q^2+p^2}}\right); \lambda(q,p) = \arctan\left(\frac{q}{p}\sec\beta\right) \tag{100c}$$

2. *Analytical Representation*

We modify Eq. (73a) in the fashion described above. Then we have

$$\breve{L}(z) = \frac{\sin\beta}{\pi E(\beta,u)} \int_0^{u} e^{-g}\, dg \int_0^{\infty} \left[\alpha\sqrt{\frac{g}{f}+1} - \cot\varsigma \frac{\sin^2\alpha - \sin^2\varsigma}{\sin^2\lambda - \sin^2\varsigma} \frac{zg}{\sqrt{f}(g+f-z^2)}\right] \frac{e^{-f} df}{(g+f-z^2)} \tag{101a}$$

$$\varsigma(g,f) \equiv \alpha(g,f)z/\sqrt{g+f}, \qquad \alpha(q,p) = \arctan\left(\sqrt{\frac{g}{g+f}}\tan\beta\right),$$

$$\lambda(g,f) = \arctan\left(\sqrt{\frac{g}{f}}\sec\beta\right) \tag{101b}$$

The results of calculation from Eqs. (101a) and (101b) coincide with graphical accuracy with those obtained from another representation (100a).

We shall show now that the integral in Eq. (101a) is pertinent to the first term in $[\cdot]$, namely

$$L_U(z) = \frac{\sin\beta}{\pi E(\beta,u)} \int_0^{u} e^{-g}\, dg \int_0^{\infty} \alpha\sqrt{\frac{g}{f}+1}\, \frac{e^{-f}\, df}{g+f-z^2} \tag{102a}$$

[22]Such a doubling will be employed further while calculating the other (besides the spectral function) additive quantities relating to the librators.

could be expressed through the spectral function $L_F(z)$ of "free rotors" introduced by Eq. (77). We shall transform this integral analogously with transformation of Eq. (74b) to the formula (77).

Furthermore, we replace the pair of variables, g and f, for $h = g + f$ and $\cos\theta = l/\sqrt{h}$. As is seen in Fig. 10a, the phase region of the librators can be represented as the total phase region without the region occupied by the hindered rotators. Taking it into account, we obtain in accord with (65) and (67)

$$L_U(z) = \frac{2}{\pi E(\beta, u)} \left(\int_0^\infty \frac{e^{-h} h\, dh}{h - z^2} \int_0^{\sin\beta} \alpha(\theta)\, d\cos\theta \right.$$
$$\left. - \int_u^\infty \frac{e^{-h} h\, dh}{h - z^2} \int_0^{\sqrt{1-u/h}\sin\beta} \alpha(\theta)\, d\cos\theta \right) \qquad (102b)$$

where $\cos[\alpha(\theta)] = \cos\beta/\sin\theta$. Earlier we have obtained Eq. (76) for the integral over cos θ in the first term of the difference in Eq. (102b); this term is expressed through the spectral function L_F (78). In the second term of the above-mentioned difference we change notation: $h \to h + u$. Then, finally, Eq. (102b) transforms to the sum

$$L_U(z) = \frac{1}{E(\beta, u)} \left[(1 - \cos\beta) L_F(z) \right.$$
$$\left. - \frac{2e^{-u}}{\pi} \int_0^\infty \frac{e^{-h} h\, dh}{h + u - z^2} \int_0^{\sqrt{\frac{h}{h+u}}\sin\beta} \alpha(\theta)\, d\cos\theta \right] \qquad (102c)$$

where the last integral is negligible, if u is large (in further calculations $u > 5$). Note that for calculations of the spectral function $\breve{L}(z)$, it is more convenient to use Eq. (73a), where the inner integrand comprises the difference $[\cdot]$.

3. *Planar Libration–Regular Precession (PL–RP) Approximation*

We modify Eqs. (79) and (80) in the fashion described in Section IV.B.1. Then we have the following expression for the librational spectral function:

$$\breve{L}(z) \simeq L_{PL}(z) + L_{RP}(z) \qquad (103)$$

$$L_{PL}(z) = \frac{32\beta^3 \sin\beta}{\sqrt{\pi} E(\beta, u)} \int_0^u \sum_{n=1}^\infty \frac{\sin^2(\pi n/2 - \beta)}{\left[(\pi n)^2 - (2\beta)^2 \right]^2 \left[g - \left(\frac{2z\beta}{\pi n} \right)^2 \right]} \operatorname{erf}(\sqrt{g}) e^{-g} \sqrt{g}\, dg$$

$$L_{RP}(z) = \frac{\sin^2\beta}{\pi E(\beta, u)\cos\beta} \int_0^u e^{-g} \sqrt{g}\, dg \int_g^\infty \frac{e^{-f} df}{f/\sin^2\beta - z^2} \qquad (104a)$$

The integral over f in the last formula differs from that in Eq. (80a). For our case of a finite u-value we have

$$\frac{\sqrt{\pi}}{2}\left[\mathrm{erf}(\sqrt{g}) - 2\sqrt{\frac{g}{\pi}}e^{-g}\right]\int_{g}^{\infty}\frac{e^{-f}\,df}{(f/\sin^2\beta) - z^2}\Bigg|_0^u$$
$$+\frac{\sqrt{\pi}}{2}\int_0^u\left[\mathrm{erf}(\sqrt{g}) - 2\sqrt{\frac{g}{\pi}}e^{-g}\right]\frac{e^{-g}\,dg}{(g/\sin^2\beta) - z^2}$$
$$=\frac{\sqrt{\pi}}{2}\left[\mathrm{erf}(\sqrt{u}) - 2\sqrt{\frac{u}{\pi}}e^{-u}\right]\int_{u}^{\infty}\frac{e^{-f}\,df}{(f/\sin^2\beta) - z^2}\Bigg|_0^u$$
$$+\frac{\sqrt{\pi}}{2}\int_0^u\left[\mathrm{erf}(\sqrt{g}) - 2\sqrt{\frac{g}{\pi}}e^{-g}\right]\frac{e^{-g}\,dg}{(g/\sin^2\beta) - z^2}$$

After further transformations we derive a more compact formula:

$$L_{\mathrm{RP}}(z) = \frac{\sin^2\beta}{2\sqrt{\pi}E(\beta, u)\cos\beta}\int_0^{\infty}\frac{\mathrm{erf}(\sqrt{s}) - 2\sqrt{\frac{s}{\pi}}e^{-s}}{(f/\sin^2\beta) - z^2}e^{-f}\,df, \quad s \equiv \begin{Bmatrix} f \\ u \end{Bmatrix} \quad \text{at} \begin{Bmatrix} f \le u \\ f \ge u \end{Bmatrix} \tag{104b}$$

C. Spectral Function of Rotators

We shall obtain here expression for the spectral function of the dipoles performing complete rotation by using an *interpolation* approximation, which allows us to ignore the distinction between hindered and free rotations. Both types of motion we mark by the same superscript $\circ$.

We come from two limiting cases.

1. *Small* β *Approximation*

If the angle β is much less than 1, then, in accord with Figs. 7 and 9, the most part of the rotators move freely under effect of a constant potential U_0, since their trajectories do not intersect the conical cavity. A small part of the rotators moves along a trajectory of the type 1 shown in Fig. 10. However, at $\vartheta \geq \beta$—that is, in the most part of such a trajectory—they are affected by the same constant potential U_0. Therefore, for this second group of the particles the law of motion is also rather close to the law of free rotation. For the latter the dielectric response is described by Eq. (77). We shall represent this formula as a particular case of the general expression (51), in which the contributions to the spectral function due to longitudinal $K_{\parallel}$ and transverse $K_{\perp}$ components are determined, respectively, by the first and second terms under summation sign. Free rotators present a medium *isotropic* in a local-order scale. Therefore, we set $K_{\parallel} = K_{\perp}$. Then the second term under the summation sign, in accord with Eq. (46a), is equal to the doubled first term. Thus we may write

$$L_F(z) = 3(2\pi)^2 C \int_0^{\infty} e^{-h}\, dh \int_0^{\sqrt{h}} \frac{a_1^2\, dl/\Phi}{(2\pi/\Phi)^2 - z^2} \tag{105}$$

where

$$\Phi = 2\pi/\sqrt{h}, \qquad a_1 = \sqrt{1 - \frac{l^2}{h}}, \qquad C = (2\pi)^{-1}$$

Here we have used the following formulas, given in GT:

- Eq. (7.1) for the normalized frequency $2\pi/\Phi$
- Eqs. (7.10) and (7.2) for the Fourier amplitude $a_n = a_1$
- Eq. (7.9) for the norm C and Eq. (7.5) for the phase-region boundary of the free rotors

The energy dependence of the period Φ and of the upper limit of the integral over l are determined as follows:

(a) In terms of reduced variables [Eq. (41)] the free-rotation frequency coincides with the angular velocity of a rotator.

(b) The squares of this frequency and of the maximum l-value are equal to the kinetic energy, which in our small-β case is equal to the total energy.

In the hat-flat model the kinetic energy of free rotators is equal to $h - u$. Therefore, we should replace h by $h - u$ in the above expressions for Φ, a_1, and $l_{\max}$. Besides, we replace the expression (68a) for the norm C by Eq. (98a) and, in view of Eq. (97c), the lower limit of the integral over h by u. Then we get for $\beta \ll 1$

$$\begin{aligned} \mathring{L}(z) &\simeq \frac{(2\pi)^2}{2\pi E(\beta, u)} \int_u^{\infty} e^{-h}\, dh \int_0^{\sqrt{h-u}} (h - u - z^2)^{-1} \frac{\sqrt{h-u}}{2\pi}\, dl \\ &= \frac{1}{E(\beta, u)} \int_u^{\infty} \frac{(h-u)e^{-h} dh}{h - u - z^2} = \frac{e^{-u}}{E(\beta, u)} \int_0^{\infty} \frac{e^{-h} h dh}{h - z^2} = \frac{e^{-u}}{E(\beta, u)} L_F(z) \end{aligned} \tag{106a}$$

2. *Rotation in Space Angle Close to Spherical Angle*

The trajectories of the major portion part of dipoles-rotators occupy in this case almost all the sphere (it is seen in Figs. 7 and 9). Consequently, a nonhomogeneity of the potential (94) could be neglected also in this case. Thus again we may consider free rotation of a dipole, but now in a *zero* potential, unlike the small-β case considered above. Hence, the kinetic energy of a dipole approximately coincides with its full energy. Therefore, for $\beta \approx \pi$ in Eq. (105) we replace the norm $C = (2\pi)^{-1}$ by Eq. (98a) and the lower limit of integral over h by u:

$$\overset{\circ}{L}(z) \simeq \frac{1}{E(\beta,u)}\int_u^\infty e^{-h}\,dh\int_0^{\sqrt{h}} \frac{\sqrt{h}dl}{h-z^2} = \frac{1}{E(\beta,u)}\int_u^\infty \frac{he^{-h}\,dh}{h-z^2}$$

$$= \frac{e^{-u}}{E(\beta,u)}\int_0^\infty \frac{e^{-h}(h+u)dh}{h+u-z^2} \tag{106b}$$

3. *Interpolation for Arbitrary Cone Angle* β

Let us replace $h+u$ in the nominator of the integrand of (106b) by $h + u\sin^2\beta$. The modified formula is valid at small β and at $\beta \approx \pi/2$, so it seems that this formula could be used as a satisfactory approximation for arbitrary β in the interval $[0, \pi/2]$. Thus we get a simple interpolation formula:

$$\overset{\circ}{L}(z) \simeq \frac{e^{-u}}{E(\beta,u)}\int_u^\infty \frac{h+u\sin^2\beta}{h+u\sin^2\beta - z^2}e^{-h}\,dh \qquad 107)$$

D. Statistical Parameters

1. *Proportions of Subensembles*

By definition, the proportion r of any subensemble is equal to the ratio of a partial statistical integral st to the total one, the latter being equal to $1/C$—that is, the inverse norm of the steady-state distribution. Here the statistical integral for rotators we represent as a sum of contributions of hindered and free rotators, st_{hin} and st_{F}. The corresponding proportions of the rotators we denote r_{hin} and r_{F}. Thus,

$$\{\breve{r}, r_{\text{hin}}, r_F\} \equiv C\{\breve{st}, st_{\text{hin}}, st_F\} \tag{108}$$

In accord with (45) and conditions (97a), (97c) the first and last statistical integrals are

$$\breve{st} = 2\int_0^\infty dl\int_{(l/\sin\beta)^2}^{u+(l/\sin\beta)^2} \breve{\Phi}(h,l)e^{-h}\,dh \tag{109a}$$

$$st_F = \int_0^\infty e^{-h}\,dh\int_{\sqrt{h-u}\sin\beta}^{\sqrt{h-u}} \overset{\circ}{\Phi}(h,l)dl \tag{109b}$$

The dependence $\breve{\Phi}(h,l)$ is given by Eqs. (62)–(64). The dimensionless kinetic energy $h-u$ of a free rotor and square of its angular velocity $2\pi/\overset{\circ}{\Phi}$ are equal to square of rotational frequency. Then

$$\overset{\circ}{\Phi}(h,l) = 2\pi/\sqrt{h-u} \tag{110}$$

To calculate the right-hand side of Eq. (109a), it is convenient to choose the variable $\cos\theta = l/\sqrt{h}$ instead of l. Then in accordance with (62) and (64) we have

$$\begin{aligned}\breve{st} &= 2\int_0^{\sin\beta} d\cos\theta \int_0^{u\frac{\sin^2\beta}{\sin^2\beta-\cos^2\theta}} e^{-h}\frac{2\alpha}{\sqrt{h}}\sqrt{h}\,dh \\ &= 4\int_0^{\sin\beta}\left[1-\exp\left(-u\frac{\sin^2\beta}{\sin^2\beta-\cos^2\theta}\right)\right]\alpha(\cos\theta)\,d\cos\theta\end{aligned}$$

For the part of this integral, which does not depend on u, Eq. (76) holds; the integrand of the other part is convenient to express through the angle λ. Since in view of Eq. (62) we have

$$\frac{\sin^2\beta}{\sin^2\beta-\cos^2\theta} = \frac{\sin^2\beta}{\sin^2\beta\left(1-\frac{\cos^2\lambda}{\cos^2\alpha}\right)} = \frac{\cos^2\alpha}{\cos^2\beta\sin^2\lambda};\quad \frac{\partial\cos\theta}{\partial\lambda} = -\frac{\sin\beta\cos^3\beta\sin\lambda}{\cos^3\alpha}$$

we obtain

$$\breve{st} = 2\pi(1-\cos\beta)[1-V(\beta,u)] \tag{111a}$$

$$\begin{aligned}V(\beta,u) &\equiv \frac{2}{\pi(1-\cos\beta)}\int_0^{\sin\beta}\exp\left(-u\frac{\sin^2\beta}{\sin^2\beta-\cos^2\theta}\right)\alpha(\cos\theta)\,d\cos\theta \\ &= \frac{2}{\pi}\cot\frac{\beta}{2}\cos^2\beta\int_0^{\pi/2}\alpha(\lambda)\exp\left(-u\frac{\cos^2\alpha}{\cos^2\beta\sin^2\lambda}\right)\frac{\sin\lambda\,d\lambda}{\cos^3\alpha(\lambda)}\end{aligned} \tag{111b}$$

From here and Eqs. (98a) and (108) we find

$$\breve{r} = \frac{1-\cos\beta}{E(\beta,u)}[1-V(\beta,u)] \tag{112}$$

The quantity st_F and free-rotators proportion r_F is found from Eqs. (98), (108), (109), (110), and condition (97c):

$$\begin{aligned}st_F &= \int_0^\infty e^{-h}dh\int_{\sqrt{h-u}\sin\beta}^{\sqrt{h-u}}\frac{2\pi}{\sqrt{h-u}}dl = 2\pi e^{-u}(1-\sin\beta), \\ \overset{\circ}{r} &= \frac{e^{-u}(1-\sin\beta)}{E(\beta,u)}\end{aligned} \tag{113a}$$

where $E(\beta, u)$ is determined by Eq. (98b).

The proportion of hindered rotators is found from evident equality:

$$\begin{aligned} r_{\text{hin}} &= 1 - \breve{r} - r_F = 1 - E^{-1}[(1 - \cos\beta)(1 - V) + e^{-u}(1 - \sin\beta)] \\ &= E^{-1}[1 - \cos\beta + {}^{-u}\cos\beta + (1 - \cos\beta)V - e^{-u}(1 - \sin\beta)] \\ &= E^{-1}(\beta, u)[(1 - \cos\beta)V(\beta, u) - e^{-u}(1 - \cos\beta - \sin\beta)] \end{aligned} \tag{113b}$$

Figure 11a shows the summary rotators' proportion

$$\mathring{r} = r_F + r_{\text{hin}} \tag{113c}$$

as a function of the potential well depth u. If u increases, this proportion decreases, since most of the particles move in the well (where they are ascribed to be librators). On the other hand, if the libration amplitude β increases, the summary proportion (113c) and that (r_{hin}) due to hindered rotators decrease for the same reason (see Fig. 11b, solid and dashed lines, respectively).

2. *Mean Localization*

To interpret the low-frequency loss spectrum, we have to consider the influence of the well depth on the *mean localization* σ, which is defined as a mean solid angle, occupied by the dipoles. As is seen in Fig. 7, for librators this angle is equal 4β, while rotators occupy the total sphere. Here we shall give the definition

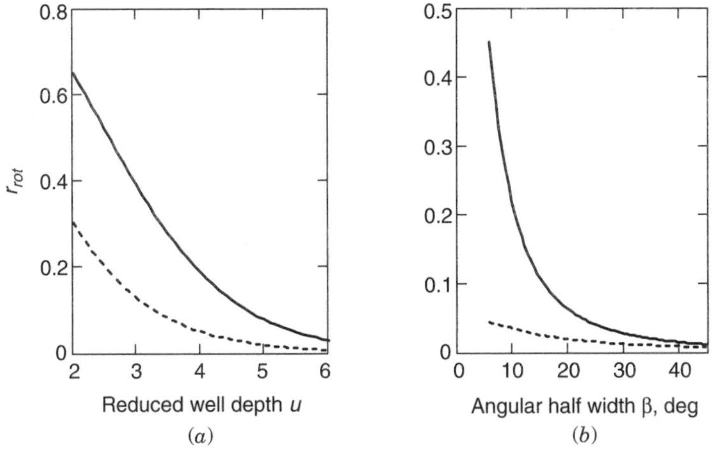

Figure 11. (a) Summary proportion of the rotators as a function of the well depth u at $\beta = \pi/8$ (solid line) and $\beta = \pi/4$ (dashed line); (b) proportions of the rotators versus angular half-width β at $u = 5.5$. Solid line refers to summary proportion and dashed line to proportion of the hindered rotators.

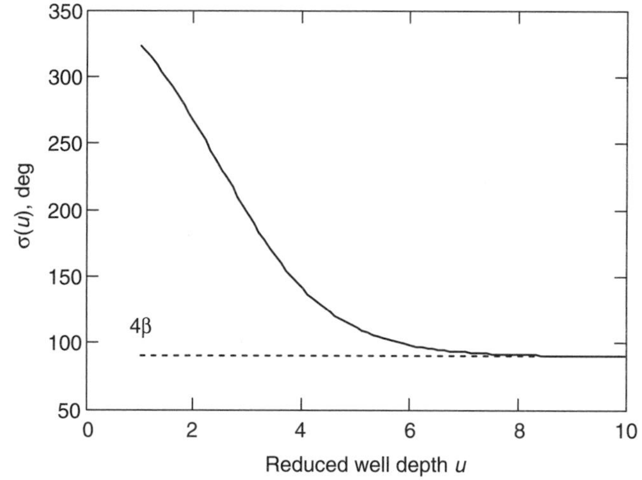

Figure 12. Dependenc of a mean localization σ on the well depth. $\beta = \pi/8$.

of σ, *taking into account of the proportions* corresponding to the librators and rotators; for the latter we again consider the summary quantity $\mathring{r}$ [Eq. (113c)], which is equal to $1 - \breve{r}$:

$$\sigma(\beta, u) = 2[2\beta\breve{r}(\beta, u) + \pi(1 - \breve{r}(\beta, u))] \tag{114}$$

The dependence of σ on the well depth u is illustrated in Fig. 12 for $\beta = \pi/8$. At large u this dependence becomes concave, since the swing between the two limiting σ values, (viz. 2π and 4β) decreases.

3. *Mean Number of Reflections. Estimation of Absorption-Peak Frequency*

It was shown in Section III that both these characteristics are determined by the mean inverse period of $\vartheta(\varphi)$ function, which we denote ξ. In our model, only librators perform reflections from conical surface. Correspondingly, the inverse period of $\breve{\vartheta}(\varphi)$ function is found by averaging *over librational* phase volume $\breve{\Gamma}$:

$$\xi \equiv \langle \breve{\Phi}^{-1} \rangle \equiv \frac{1}{\breve{r}} \int_{\breve{\Gamma}} \frac{d\Gamma}{\breve{\Phi}} \tag{115}$$

In view of (109a),

$$\int_{\breve{\Gamma}} \frac{W(h)\, d\Gamma}{\breve{\Phi}(h, l)} = 2 \int_0^{\infty} dl \int_{(l/\sin\beta)^2}^{u+(l/\sin\beta)^2} Ce^{-h} dh \tag{116}$$

Then we obtain from Eqs. (108), (111a), and (116):

$$\xi = \frac{2(1 - e^{-u})}{\breve{s}t} \int_0^\infty \exp(-l^2/\sin^2\beta)\, dl = \frac{\sqrt{\pi}(1 - e^{-u}) \sin\beta}{2\pi(1 - \cos\beta)[1 - V(\beta, u)]}$$
$$= \frac{1 - e^{-u}}{2\sqrt{\pi}[1 - V(\beta, u)]} \cot\frac{\beta}{2} \tag{117}$$

At $u \to \infty$ this formula reduces to (84), as it should be.

By analogy with (89), we find the mean number $\breve{m}$ of reflections during the lifetime as

$$\breve{m} = \frac{\xi}{y} = \frac{1 - e^{-u}}{2y\sqrt{\pi}[1 - V(\beta, u)]} \cot\frac{\beta}{2} \tag{118}$$

We remind the Reader that we use the hat-flat model in order to describe a *two-humped* absorption spectrum characteristic for water. The larger peak frequency ($\breve{x}$) we refer to the librators and the smaller one ($\mathring{x}$) to the *rotators of both types*. We estimate $\breve{x}$ by the same way as in Eq. (85):

$$\breve{x} \approx \pi\xi = \frac{\sqrt{\pi}}{2} \frac{1 - e^{-u}}{1 - V(\beta, u)} \cot\frac{\beta}{2} \tag{119a}$$

The alternative estimation

$$\breve{x} \approx \pi \frac{\sqrt{1.2}}{2\beta} \tag{119b}$$

given in Section III, Eq. (86), is applicable also for our case of finite u, since in this estimation the u-value is not at all involved. The results of both estimations will be compared in Section VI.

As for the rotational-peak frequency $\mathring{x}$, we estimate it in the same fashion as was used for derivation of Eq. (86). Namely, we find $\mathring{x}$ as the frequency, at which the rotational-absorption component $\mathring{A}(x)$ exhibits maximum in the limit $y \to 0$. It follows from Eq. (107) that in this limit

$$\mathring{A}(x) = x\mathring{L}(x) \propto \pi x^3 e^{-h_r} = \pi\left(h_r + u \sin^2\beta\right)^{3/2} e^{-h_r}$$

where $h_r \equiv \sqrt{x^2 - u \sin^2\beta}$ is the resonance value of the energy h. We find from these relations that the absorption component $\mathring{A}(x)$ exhibits maximum, if $\mathring{x}^2 = h_r + u \sin^2\beta = 3/2$. Then $\mathring{x}$ does not depend on the model parameters, and we have

$$\mathring{x} \simeq \sqrt{3/2} \tag{120}$$

TABLE II
Estimation of the Statistical Averages

u	$\overset{\circ}{r}\%$	$r_{\text{hin}}\%$	$\breve{x}$	$\overset{\circ}{x}$	$\breve{m}$
3.5	28	10	4.4	1.22	7
5.5	4.9	1.7	4.45	1.22	7.1

In Table II the results of our estimations are given for $\beta = \pi/8$, $y = 0.2$ and for two u-values. We see that the peak frequencies and the number of reflections m, unlike proportions, weakly depend on the well depth u.

E. Hybrid Model

This model was first introduced in VIG, Chapter 9, and in Refs. 32 and 39. The term *hybrid* was there used since *two types* of motion in a rectangular potential well were considered in a small β approximation: dipoles with energies h smaller than well depth u librate only in a *diametric* section of a cone, and those with energies exceeding this value perform complete rotation in a *homogeneous* potential equal to u. Because of its simplicity (the spectral function is determined by simple integrals) the hybrid model is widely used from 1997 for description of wideband dielectric relaxation in various polar fluids, while the first application of the hat model was given only in 2000 [3]. The precessional component of the dipole's motion is neglected in the hybrid model. One can consider the latter as a generalization of the *planar libration* approximation, given by Eq. (82) with account of the finiteness of the potential well depth.

The spectral function of the hybrid model can be found, if we

(a) Replace by u the infinite upper limit of the integral (82)
(b) Introduce the renormalizing multiplier $\frac{1-\cos\beta}{E(\beta,u)} \simeq \frac{\beta^2}{\beta^2+2e^{-u}}$
(c) Add to the result the rotational spectral function given by Eq. (106a) replacing there $E(\beta, u)$ dependence by its small β approximation—that is, by $\exp(-u) + \beta^2/2$

Then we have

$$L(z) = \breve{L}(z) + \overset{\circ}{L}(z) \tag{121}$$

$$\breve{L}(z) \approx \frac{128\beta^4}{\pi^5(\beta^2 + 2e^{-u})} \int_0^\infty \frac{e^{-h} h\, dh}{h - (2z\beta/\pi)^2} \tag{122a}$$

$$\overset{\circ}{L}(z) \approx \frac{2e^{-u}}{\beta^2 + 2e^{-u}} L_F(z) = \frac{2e^{-u}}{\beta^2 + 2e^{-u}} \int_0^\infty \frac{h\, e^{-h}\, dh}{h - z^2} \tag{122b}$$

We also give here the formulas for the proportion $\overset{\circ}{r}$ of the rotators, for estimation of the absorption-peak frequency $\breve{x}$ mentioned in Section IV.D.3, and for the mean number $\breve{m}$ of the reflections performed during the lifetime.

The proportion $\overset{\circ}{r}$ is found from VIG, Eq. (9.22a), in our approximation $\beta \ll 1$ as

$$\overset{\circ}{r} \simeq \frac{2e^{-u}}{\beta^2 + 2e^{-u}} \tag{123}$$

To find $\breve{x}$ we employ in the same approximation VIG, Eq. (9.25), where we change notation $\langle \breve{p} \rangle$ for $\breve{x}$:

$$\breve{x} \approx \frac{\sqrt{\pi}\,\mathrm{erf}(\sqrt{u}) - 2\sqrt{u}e^{-u}}{\beta(1 - e^{-u})} \tag{124}$$

Expression for $\breve{m}$ differs from Eq. (124) by multiplier $1/(\pi y)$, just as Eqs. (118) and (119a). Thus,

$$\breve{m} \simeq \frac{\sqrt{\pi}\,\mathrm{erf}(\sqrt{u}) - 2\sqrt{u}e^{-u}}{\pi y\beta(1 - e^{-u})} \tag{125}$$

We shall show below that the hybrid model gives a satisfactory description of the "usual" Debye relaxation at the microwave region and explains a quasi-resonance absorption band in the FIR region.

We should note that in many articles and in the book VIG other formulas for the summands of spectral function (121) of the hybrid model are employed:

$$\breve{L}(z) = \frac{4\beta^4}{\pi^5 E(\beta, u)} \left[\sin^2\beta \int_0^u \frac{h \exp(-h)\, dh}{h - (\beta z/\pi)^2} + 16\cos^2\beta \int_0^u \frac{h \exp(-h)\, dh}{h - (2\beta z/\pi)^2} \right] \tag{126a}$$

$$\overset{\circ}{L} = \frac{\exp(-u)}{E(\beta, u)} \int_0^\infty \frac{h \exp(-h)}{h - z^2}\, dh \tag{126b}$$

These formulas were previously used for all possible ranges of the cone angles—that is, for $\beta \in [0, \pi/2]$, but *without theoretical justification.* Because of the simplicity of Eq. (126a) in comparison with the formulas for the spectral function pertinent to the rigorous hat-plane model (see Section IV.C.3), the hybrid model was often applied for calculation of dielectric properties of various polar fluids.

Now we write down the formulas (9.22a) and (9.25), given in VIG for the above-mentioned parameter $\overset{\circ}{r}$, which is valid in terms of the hybrid model for any β value:

$$\overset{\circ}{r} = \frac{\exp(-u)}{1 - [1 - \exp(-u)] \cos \beta} \tag{127}$$

We obtain from Eqs. (9.30a) and (9.31) given in VIG the formula for a mean number[23] of the reorientation cycles:

$$\breve{m} = \frac{1}{2y\pi E(u, \beta)} \left\{ \sin \beta \left[\frac{\sqrt{\pi}}{2} \operatorname{erf}(\sqrt{u}) - \sqrt{u} \exp(-u) \right] + \frac{\sqrt{\pi}}{2} \exp(-u) \right\} \tag{128}$$

Then the absorption-peak frequency ν_{lib} (in cm^{-1}) could be estimated as a frequency of a periodic motion averaged over ensemble of the dipoles:

$$\langle \nu \rangle = \breve{m} / (\tau c) \tag{129}$$

The comparison of the spectra calculated for rigorous hat-flat model with various simplified variants will be given below.

F. Qualitative Spectral Dependences

First, we shall summarize the results of above calculations of the resulting spectral function (96).

For representation in the form of *infinite series* we have

$$L(z) = \breve{L}(z)|_{\text{from (100a)}} + \overset{\circ}{L}(z)|_{\text{from (107)}} \tag{130}$$

The spectral function found in the form of *analytic formula*

$$L(z) = \breve{L}(z)|_{\text{from (101a)}} + \overset{\circ}{L}(z)|_{\text{from (107)}} \tag{131}$$

For *the planar libration–regular precession* approximation

$$L(z) = L_{PL}(z)|_{\text{from (104a)}} + L_{RP}(z)|_{\text{from (104b)}} + \overset{\circ}{L}(z)|_{\text{from (107)}} \tag{132}$$

Finally, for *the hybrid model*

$$L(z) = \breve{L}(z)|_{\text{from (126a)}} + \overset{\circ}{L}(z)|_{\text{from (126b)}} \tag{133}$$

[23]Unlike Eq. (125) the quantity (128), which has the same notation, has slightly different sense: it roughly accounts for the effect of both subensembles.

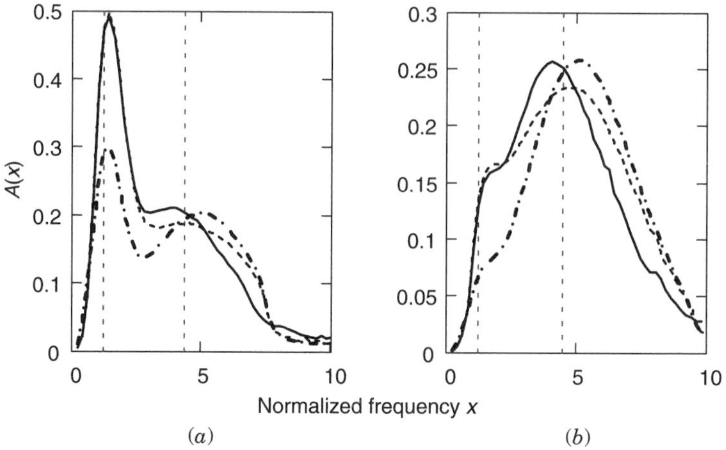

Figure 13. Dimensionless absorption versus normalized frequency calculated rigorously (solid lines), from the PL–RP approximation (dashed lines), and for the hybrid model (dashed-and-dotted lines). The cone angle $\beta = \pi/8$ and the reduced collision frequency $y = 0.2$. The reduced well depth $u = 3.5$ (a) and 5.5 (b). Left and righ vertical lines mark the frequency peaks estimated, respectively, in the "rotational" and "librational" ranges.

In Fig. 13 we juxtapose the frequency dependencies of the dimensionless absorption

$$A(x) = x\,\mathrm{Im}[L(x + iy)] \equiv x\,L(x + iy) \tag{134}$$

We choose the following set of the parameters: the cone angle $\beta = \pi/8$, collision frequency $y = 0.2$, and two u-values: 3.5 and 5.5. The results of rigorous and both approximate calculations agree *qualitatively.* We see a two-humped absorption curve. Its right part is stipulated by "librators" and the left part is stipulated by "rotators." At larger u-value all these curves have a distinct maximum in the "librational" range (at $x \approx 5$) and a shoulder in the "rotational" part (at $x \approx 1$). At a rather small u-value (see Fig. 13a), the "rotational" peak is sharply expressed. This form of the absorption curve, unlike that in Fig. 13b, is not typical for polar fluids. So, the potential barrier should be rather high ($u > 5$). The vertical lines mark the $\breve{x}$ values estimated from formulas (119a) and (119b) and values of $\mathring{x}$ found from (120). Two first estimations coincide with graphical accuracy. Note that such coincidence was also obtained in the case of the protomodel [see Eqs. (85) and (86)]. One can also see that dependence on u of $\breve{x}$ and $\mathring{x}$ values is weak.

In Fig. 14 the low-frequency loss spectrum $\chi''(x)$ is shown. It is calculated by using Eq. (32), in which the frequency dependence of the spectral function is neglected:

$$L(z) \equiv L(x + iy) \rightarrow L(iy) \tag{135}$$

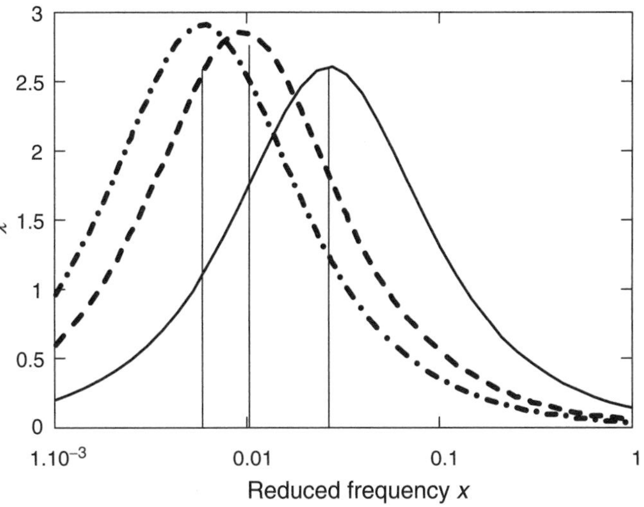

Figure 14. The low-frequency loss spectrum calculated from rigorous formulas (135) for the spectral function. $G = 2.4, g = 2.5, y = 0.2, \beta = \pi/8; u = 3.5$ (solid line), 5.5 (dashed line), and 10 (dash-and-dotted line). Vertical lines mark estimated x_D values, given by Eq. (34c).

where $L(iy)$ is found from rigorous formula (135). We set $G = 2.4$, $g = 2.5$, and the same values of y and β as were used above ($y = 0.2$, $\beta = \pi/8$). These values of the parameters are more or less typical with regard to the dielectric/FIR spectra of strongly polar fluids. Three curves differ by the u values, which are chosen equal to 3.5, 5.5, and 10. The last u value is so large that it actually corresponds to the protomodel (i.e., to infinite welldepth). The larger the value of u, the less the loss-peak frequency x_D. In agreement with the rule (61) and Fig. 12, this tendency takes place due to narrowing of the mean localization σ. The u value about 5.5 gives a reasonable description of the wideband spectra of water. In view of Fig. 12 at such large u-values the σ-value is close to the saturation level equal to 4β. We remark that the loss-maximum position x_D depends on the well depth u much strongly than the localization does. Namely, if u changes from 10 to 5.5, then x_D increases by 60%, while σ only by 15%.

We conclude. Restriction of the well depth u undertaken in the hat model is justified by at least two above-mentioned results.

(a) A shoulder (or a maximum) on an absorption curve arises at the normalized frequency $x \sim 1$ (see Fig. 13) due to subensemble of rotators, while in the case of the protomodel the curve $A(x)$ exhibits the *only* maximum (compare with Fig. 6). The feature (shoulder) shown in the left part of Fig. 13b qualitatively agrees with the experiment (see Section IV.G.2).

(b) Estimation gives a reasonable value of the loss-peak frequency x_D, while at $u \to \infty$ the estimated x_D value occurs to be too low (cf. curves in Fig. 14 referring to $u = 5.5$ and $u = 10$).

G. Application to Polar Fluids

In this section we calculate the complex permittivity $\varepsilon^*(\nu)$ and the absorption coefficient $\alpha(\nu)$ of ordinary (H_2O) water and of fluoromethane CH_3F over a wide range of frequencies. We shall first write down the list of the formulas useful for further calculations.

1. *General Expressions for Spectral Characteristics*

These quantities are related to one another as follows:

$$\alpha(\nu) = 2\pi\nu\varepsilon''(\nu)/n(\nu) \tag{136}$$

where the refraction index

$$n(\nu) \equiv \mathrm{Re}\left\{\sqrt{\varepsilon[x(\nu)]}\right\} \tag{137}$$

and $\nu \equiv \omega/2\pi c$ is the wavenumber, further called "frequency." We deal with the complex-conjugated permittivity $\varepsilon^* \equiv \varepsilon' + i\varepsilon''$ and susceptibility $\chi^* \equiv \chi' + \mathrm{i}\chi''$; thus the loss factor ε'' is defined as positive imaginary component of ε^*.

To relate the complex permittivity ε^* of a polar medium with the complex susceptibility χ^* provided by motions of the dipoles, we suggest that a polar medium under study is influenced by the external macroscopic time-varying electric field $\mathbf{E}_e(t) = \mathrm{Re}[\hat{\mathbf{E}}_m \exp(i\omega t)]$, where $\hat{\mathbf{E}}_m$ is the complex amplitude. This field induces some local field $\mathbf{E}_{ind}(t) = \mathrm{Re}[\hat{E}_i \exp(\mathrm{i}\omega t)]$ in a cavity surrounding each polar molecule. A given molecule directly experiences the latter field.

If we neglect the difference between two complex amplitudes, $\hat{\mathbf{E}}_m$ and $\hat{\mathbf{E}}_i$, then the complex permittivity ε^* of a polar medium and the complex susceptibility χ^* provided by motions of the dipoles would be related as follows:

$$\varepsilon^* - n_\infty^2 = 4\pi\chi^* \tag{138}$$

Here n_∞^2 is the permittivity (n_∞ is the refraction index near the HF edge of the orientational band; in the literature, n_∞^2 is usually denoted as ε_∞).

A more rigorous theory [40, 41] accounting for an internal field correction yields the following ratio of two field complex amplitudes:

$$\frac{\hat{E}_i}{\hat{E}_m} = \frac{3\varepsilon^*}{2\varepsilon^* + n_\infty^2}$$

Then Eq. (138) transforms to a relation (see GT, p. 200 or VIG, p. 7):

$$\chi^*(\varepsilon^*) = \frac{\varepsilon^* - n_\infty^2}{4\pi}\frac{2\varepsilon^* + n_\infty^2}{3\varepsilon^*} \tag{139}$$

which at zero frequency is reduced to the following relation between the static values ε_s and χ_s:

$$\chi_s = \frac{\varepsilon_s - n_\infty^2}{4\pi}\frac{2\varepsilon_s + n_\infty^2}{3\varepsilon_s} \tag{140}$$

The dependence inverse to Eq. (139) is given by

$$\varepsilon^*(\chi^*) = \frac{1}{4}\left[12\pi\chi^* + n_\infty^2 + \sqrt{\left(12\pi\chi^* + n_\infty^2\right)^2 + 8n_\infty^4}\right] \tag{141}$$

The relationship (139), of course, is not rigorous, but it is based on an elementary macroscopic consideration [41] of the internal-field correction. Being widely used (GT, VIG), such a relationship is sufficient for an accurate description of the low-frequency dielectric response of strongly polar fluids. We return to this problem later in this section.

The complex susceptibility in terms of the Gross collision model [(GT, p. 190; VIG, p. 188; see also Eq. (32)] is determined by the spectral function $L(z)$ as

$$\chi^*(x) = gGzL(z)\,[gx + iyL(z)]^{-1} \tag{142}$$

where

$$G = \mu^2 N (3k_\mathrm{B}T)^{-1} \tag{143}$$

where the number density N of molecules is expressed through the density of a liquid ρ as

$$N = N_\mathrm{A}\rho/M \tag{144}$$

M is the molecular mass, and N_A is Avogadro's number. A dipole moment μ of a molecule *in a liquid* is related to that of an *isolated* molecule μ_0 as

$$\mu = \frac{1}{3}\mu_0 k_\mu (n_\infty^2 + 2) \tag{145}$$

where k_μ is a fitting coefficient close to 1. The Kirkwood correlation factor

$$g = \frac{\varepsilon_s - n_\infty^2}{4\pi G}\frac{2\varepsilon_s + n_\infty^2}{3\varepsilon_s} \tag{146}$$

accounts for a spatial correlation[24] between the particles. It follows from Eq. (142) that the static susceptibility does not depend on the parameters of a model:

$$\chi_s = gG \tag{147}$$

The real and the imaginary components of the argument $z = x + iy$ of the spectral function L, see Eq. (142), are related to the frequency ν and the lifetime τ as

$$x = 2\pi c\eta\nu, \qquad y = \eta/\tau, \qquad \eta = \sqrt{\frac{I}{2k_B T}} \tag{148}$$

The spectral function $L(z)$ involved in Eq. (142) is determined by the profile of the model potential well (in this section it is the rectangular well). It follows from Eq. (148) that if we fix the dimensional quantities, such as frequency ν and temperature, then the spectral function $L(z)$ depends also on the lifetime τ and the moment of inertia of a molecule I. We consider a gas-like reorientation of a polar molecule determined by a dipole moment $\boldsymbol{\mu}$ of a molecule in a liquid.

Calculation of the moment of inertia I deserves special discussion.

(a) Although molecules H_2O and D_2O have the form of a flat asymmetric top, for simplicity we consider them to be linear molecules, with the moment of inertia I determined by the relation

$$\frac{1}{I} = \frac{1}{2}\left(\frac{1}{I_1} + \frac{1}{I_2}\right) \tag{149}$$

where I_1 and I_2 are two main moments of inertia[25] of an isolated molecule about the axes perpendicular to its dipole-moment vector $\boldsymbol{\mu}$.

This is a crude assumption. However, it appears that a quantum picture of discrete rotational lines, placed in the submillimeter wavelength range (ca. from few to $\sim$150 cm^{-1}), is essentially determined by a form of a molecule only for a *gas*. In the case of a liquid, discrete spectrum is *not* revealed, since separate rotational lines overlap due to strong intermolecular interactions, which become of primary importance. So, due to these interactions and the effect of a tight local-order cavity, in which molecules reorient, the maximum of the absorption band, situated in the case of vapor at $\sim$100 cm^{-1}, shifts in liquid water to

[24]As was mentioned in Section II.A.6, in the Gross collision model the angular velocities, unlike the orientations, are considered to be uncorrelated at an instant of a strong collision.

[25]The third principal moment of inertia I_3 is related to direction of a dipole moment and does not influence the dielectric response.

$\sim$700 cm^{-1}. Note that the contribution of "free rotors" to water spectrum, which reveals just at 100–200 cm^{-1}, still plays some role . But, as we shall see, this contribution is commensurable with the effect of specific interactions considered in Sections VI, VII, IX, and X.

Furthermore, we consider the moments of inertia I_1 and I_2 and thus I in Eq. (149) to be known, as well as the experimental position $\nu_{\rm lib}$. We remark that actually there is *no need* to know exactly the value of moment of inertia I. Indeed, a possible uncertainty in the employed I value given by Eq. (149), and therefore in determination of the center librational-band frequency $\nu_{\rm lib}$ from Eqs. (136), (137), (141), and (142), could easily be compensated by fitting the half-width β of the potential well.

(b) The same relation (149) and the like reasoning are valid also for a nonassociated liquid fluoromethane CH_3F, whose molecule has the form of a symmetric top, so that $I_1 = I_2$.

Although it is crude, our simplified approach permits us to investigate the main effect, namely, influence on wideband spectra of steric restrictions to reorientations of a strongly polar molecule.

We note that including the correlation factor g in Eq. (142) allows us to use the experimental static permittivity ε_s and optic permittivity n_∞^2 for the calculation of complex permittivity χ^*. This enables us to get the following from Eqs. (141) and (142): (a) the exact (experimental) value of the permittivity in the low-frequency limit $\omega \to 0$; (b) the correct value of the permittivity in the HF limit $\omega \to \infty$; (c) the reduction of the real (ε') and imaginary (ε'') parts of the complex permittivity to the Debye-relaxation formula at low frequencies, since at $\omega < \tau_{\rm D}^{-1}$ and $x \ll 1$ one may neglect the dependence of the spectral function L on x. Thus, in the low-frequency limit, setting in Eq. (142) $L_0 \equiv L(x+iy)_{x=0} = L(iy)$, we have the following estimation of the Debye relaxation time [see Eq. (34a)]:

$$\tau_{\rm D}^{\rm est} = g\eta/(yL_0) \tag{150}$$

Equating this to the experimental value, we may relate the parameters of the employed molecular model:

$$\tau_{\rm D}^{\rm est}(\text{free parameters of the model}) = \tau_{\rm D} \tag{151}$$

Finally, we remark that our microscopic approach is mainly aimed at the study of the resonance-interaction mechanisms, which are revealed in the FIR range. In this range, where $\omega \gg \tau_{\rm D}^{-1}$ and $x > 0.1$, inclusion of the internal field correction, which is accounted for in Eqs. (139) and (141), gives only a small effect.

2. *Application to Water H_2O*

In this section we have to calculate the complex permittivity $\varepsilon^*(\nu)$ and the absorption coefficient $\alpha(\nu)$ of ordinary (H_2O) water over a wide range of frequencies. It is rather difficult to apply rigorous formulas because the fluctuations of the calculated characteristics occur at a small reduced collision frequency y typical for water (in this work we employ for calculations the *standard* MathCAD program). Such fluctuations are seen in Fig. 13b (solid curve). Therefore the calculations will be undertaken for two simplified variants of the hat model. Namely, we shall employ the planar libration–regular precession (PL–RP) approximation and the hybrid model.[26]

a. Empirical Formula for the Complex Permittivity by Liebe et al. [17]. For comparison of our theory with the experiment we shall use the *recorded data* [42] and the *empirical* description of the permittivity $\varepsilon^*(\nu)$ over a wide range of frequencies, given in Ref. [17] using the double-Debye representation supplemented with two resonance Lorenz terms. This description, which is approximately valid for the range from 0 to 30 THz, actually presents a generalization of many experimental data and is very useful for estimation of the dielectric properties of pure water and of the low-concentrated aqueous solutions. The formula by Liebe et al. expresses the complex permittivity in terms of the frequency $f = \omega/(2\pi) = \nu c$ and of the function $\phi(T) = 1 - (300/T)$ of temperature:

$$\varepsilon^*(f) = \varepsilon_s - \left(\frac{\varepsilon_s - \varepsilon_1}{f + i\gamma_1} + \frac{\varepsilon_1 - \varepsilon_\infty}{f + i\gamma_2}\right) f + \sum_{j=1,2} \left(\frac{A_j}{f_j^2 - f^2} - \frac{A_j}{f_j^2}\right) \tag{152}$$

where the coefficients are given by

$$\varepsilon_s(T) = 77.16 - 103.3\phi + 7.52\phi \qquad \varepsilon_1(T) = 0.0671\varepsilon_s(T)$$

$$\varepsilon_\infty = 3.52 + 7.52\phi(T)$$

$$\gamma_1(T) = \{20.2 + 146.4\phi(T) + 316[\phi(T)]^2\} \times 10^9 \qquad \gamma_2(T) = 39.8\gamma_1(T)$$

$$f_1 = 5.11 \times 10^{12} \qquad \Gamma_1 = 4.46 \times 10^{12} \qquad A_1 = 25.03 \times 10^{24}$$

$$f_2 = 18.2 \times 10^{12} \qquad \Gamma_2 = 15.4 \times 10^{12} \qquad A_2 = 282.4 \times 10^{24}$$

$$\tau_D(T) = [2\pi\gamma_1(T)]^{-1}$$

$$\tau_2(T) = [2\pi\gamma_2(T)]^{-1}$$

b. Results of the Calculations. For definiteness we choose the temperature $T = 300$ K. Considering the frequency range up to 1000 cm^{-1}, we shall

[26]A *general* comparison of the spectra calculated by using rigorous formulas and above-mentioned approximations was given in Section IV.F.

TABLE III
Experimental Data and Parameters Fitted/Estimated for Water H_2O^a

A. Experimental Data [17, 42]								
ε_s	τ_D	ν_{tr}	α_R	$\nu(\alpha_{min})$	ν_R	$\Delta\nu_{lib}$	α_{lib}	ρ, $g \cdot cm^{-3}$
77.6	7.85	200	1445	260	670	530	3298	0.9986
Molecular constants			$n_\infty^2 = 1.7$; $\mu_0 = 1.84$ D; $I = 1.483 \times 10^{40}$ $g \cdot cm^2$, $M = 18$ g					

B. Free Parameters							
Planar Libration–Regular Precession Approximation				Hybrid Model			
β	τ	u	k_μ	β	τ	u	k_μ
19	0.3	6.2	1.15	19.9	0.34	5.5	1.12

C. Statistical Parameters					
Planar Libration–Regular Precession Approximation			Hybrid Model		
$\mathring{r}$, %	M	ν_{lib}	$\mathring{r}$, %	$\breve{m}$	ν_{lib}
3.3	5.8	644	6.4	6.1	595

[a]Frequency ν and absorption coefficient α are given in cm^{-1}, lifetime τ in ps, and the angle β in degrees. Temperature 27°C.

describe **(1)** the principal (Debye) relaxation region with the peak center at about 0.7 cm^{-1}, **(2)** the main (librational) absorption band with the absorption peak situated at $\nu_{lib} \approx 670$ cm^{-1}, and **(3)** the R-absorption band arising at a frequency $\nu_{tr} \sim 200$ cm^{-1}. Some useful experimental data are taken from Refs. 17 and 42. These are: the static permittivity ε_s, the Debye relaxation time τ_D, the positions ν_R and ν_{lib} and the values α_R and α_{lib} of the R- and the librational peaks, and the width $\Delta\nu_{lib}$ of the librational band. These data are given in Table IIIA together with value of the density ρ of water and its molecular mass M.

We employ the following equations: Eq. (142) for the complex susceptibility χ^*, Eq. (141) for the complex permittivity ε^*, and Eq. (136) for the absorption coefficient α. In (142) we substitute the spectral functions (132) for the PL–RP approximation and (133) for the hybrid model, respectively. In Table IIIB and IIIC the following fitted parameters and estimated quantities are listed: the proportion $\mathring{r}$ of rotators, Eqs. (112) and (127); the mean number $\breve{m}$ of reflections of a dipole from the walls of the rectangular well during its lifetime τ, Eqs. (118)

and (128)[27]; the estimated frequency of the librational absorption peak, Eqs. (119) and (148) for the PL–RP approximation and Eqs. (128) and (129) for the hybrid model.

It follows from Tables IIIB and IIIC that:

(a) The fitted parameters of the hybrid model are rather close to those fitted for the of PL–RP approximation. However, in the former the potential barrier u is lower; correspondingly, the proportion of rotators $\overset{\circ}{r}$ is greater and, as a result, the effect on spectra of the particles performing complete rotation is greater.
(b) The libration amplitude β (the cone angle) is rather small (about 20°).
(c) The lifetime τ, being equal to ~ 0.3 ps, is much shorter than the Debye relaxation time τ_D (7.85 ps).
(d) The normalized potential well is rather deep (comprises $\sim 6\ k_B T$).
(e) The correcting factor k_μ is close to unity.
(f) The proportion $\overset{\circ}{r}$ of rotators is small (only few per cent); interpretation of this result is given in Section IV.G.3.
(g) In view of our estimations we suggest that motion of dipoles in rectangular potential well is rather *regular*: The mean number of the reorientation cycles, performed by a dipole between strong collisions, substantially exceeds unity ($\bar{m} \approx 6$).
(h) Mean reorientation frequency calculated in terms of the hybrid model is indeed rather close to the recorded value of librational absorption-peak position (670 cm^{-1}): The difference comprises only 4%. This result confirms the idea that the librational band reflects resonance features of internal rotation in a liquid[28].

Figures 15a and 15b show the wideband absorption and loss spectra, respectively, calculated for the PL–RP version. Figures 15c and 15d demonstrate similar spectra pertinent to the hybrid model. It is seen that the spectra calculated for *both versions* of the rectangular-well model agree in their main features with the experimental dielectric/FIR spectra recorded in the region 0–1000 cm^{-1}.

The model used here has three main drawbacks.

(1) The calculated width of the librational band is wider than the recorded one, cf. solid and dashed lines in Figs. 15a and 15b. This disagreement is

[27]The last formula obtained by using the hybrid-model version refers to the quantity, slightly differing from this number, namely, to the number of reorientation cycles, averaged over the total ensemble.

[28]Note that the quantity (129) used for this estimation also roughly accounts for effect of the rotators, which is, however, rather small, since $\overset{\circ}{r} \ll 1$.

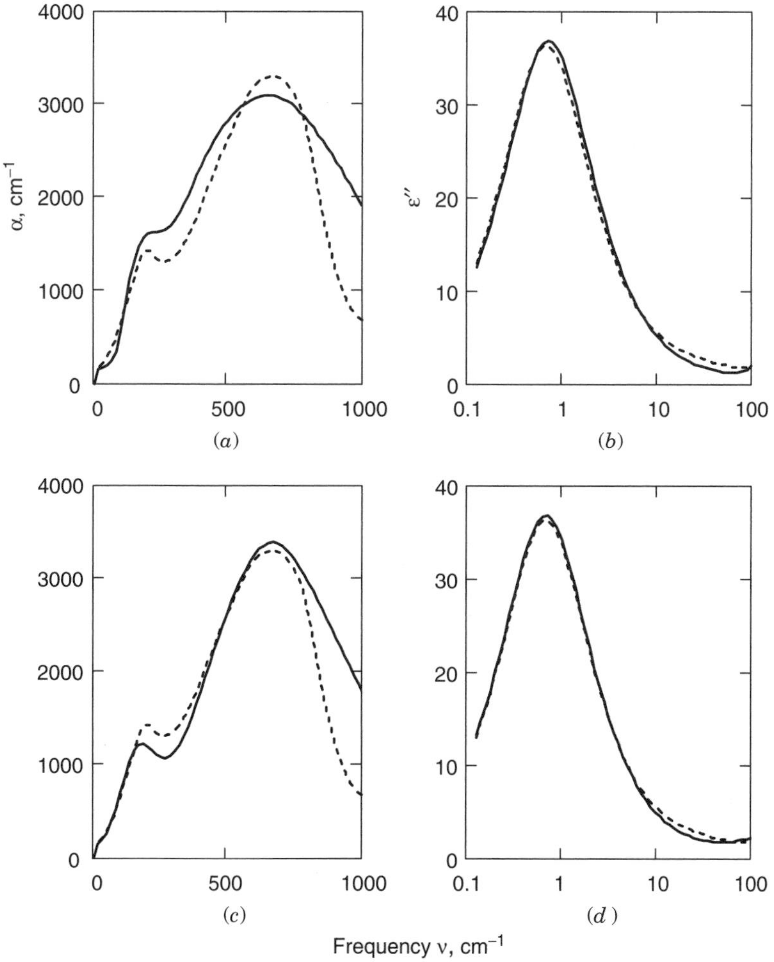

Figure 15. Frequency dependence of the absorption coefficient (a, c) and loss (b, d) calculated in PL–RP (a, b) and hybrid-model approximations (c, d). In Fig. (a,b): $\beta = 19°, u = 6.2, \tau = 0.3$ ps, $k_\mu = 1.15$; In Fig. (c, d): $\beta = 19.9°, u = 5.5, \tau = 0.34$ ps, $k_\mu = 1.12$. Dashed curves represent calculation from the empirical formulas of Liebe et al. [17].

due to the assumed (rectangular) profile of the intermolecular potential well. The alternative model presented in the next chapter will allow us to avoid this drawback.

(2) The hat-flat model *cannot* be applied to heavy (D_2O) water. Namely, for this liquid the *theoretical* center frequency of the R-absorption band is lower than for H_2O due to larger moment of inertia *I*, while the

experimentally recorded position $\nu_R(D_2O)$ is approximately the same as $\nu_{tr}(D_2O)$. In Section VI we shall improve the situation by considering (in a rough approximation) the effect on the R-band of *nonrigidity* of a reorienting polar molecule.

(3) There appears some disagreement of the calculated complex permittivity $\varepsilon^*(\nu)$ with the experimental data [17, 42] recorded in the *submillimeter* wavelength range—that is, from 10 to 100 cm^{-1}. It is evident from Fig. 15 and more clearly from Fig. 16 that a theoretical loss is less in this spectral interval than the experimental one. The reason of such a discrepancy can be explained as follows. Some *additional* mechanism of dielectric loss possibly exists in water. Such a mechanism will be studied in Sections VII, IX, and X, where we shall propose *composite* molecular models of water.

c. About Size of Intermolecular Cavity. Above, the model parameter β was estimated, which presents a maximum declination of a *librating dipole moment* from the symmetry axis of a potential. To account for the *size of a molecule*, we should replace β by the larger angle β_c (see Fig. 4), which restricts angular displacement of a molecule in a local-order cage. To simplify estimation of this angle, we represent the H_2O molecule and the intermolecular cavity as

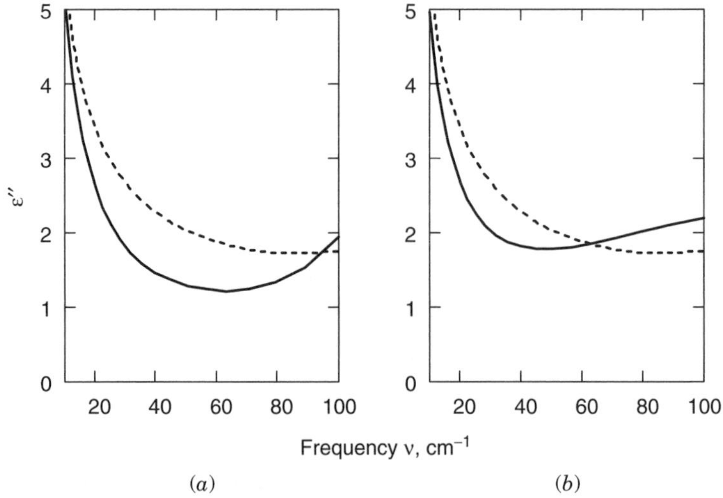

Figure 16. Solid curves show frequency dependence of the loss ε'' calculated for the PL–RP (a) and the hybrid-model (b) versions. In Fig. (a): $\beta = 19°, u = 6.2, \tau = 0.3$ ps, $k_\mu = 1.15$. In Fig. (b): $\beta = 19.9°, u = 5.5, \tau = 0.34$ ps, $k_\mu = 1.12$. Dashed curves represent calculation from the empirical formulas of Liebe et al. [17].

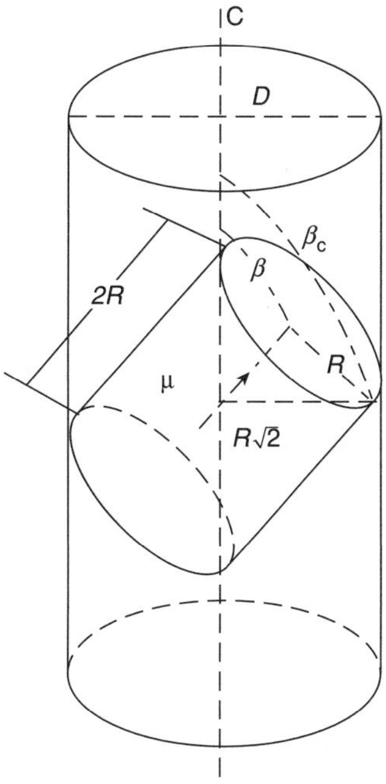

Figure 17. A scheme pertinent to displacement of a librating water molecule inside an intermolecular cavity.

cylinders, having, respectively, radius R and diameter D (see Fig. 17). The inner cylinder (water molecule) is supposed to have the height $2R$. It is seen that the following relations hold:

$$\tan(\beta_c - \beta) = 1 \quad \text{hence} \quad \beta_c = \frac{\pi}{4} + \beta \quad \text{and} \quad D = 2\sqrt{2}R \tag{153a}$$

If we set radius $R = R\ (H_2O) \approx 1.5$ Å and take from Table IIIB angle $\beta = 20°$, then we have

$$\beta_c \approx 65° \quad \text{and} \quad D \approx 4.24\,\text{Å} \tag{153b}$$

Thus, the range of angular displacements of a water molecule is large ($\approx$130°) even if the model potential well is rather narrow. One may regard the estimated D value as an *effective diameter* of a water molecule in a liquid. Note that this size is larger than the length $l \approx 3.29$ Å [35] between adjacent nodes in tetrahedron formed by five water molecules. In the above-cited work, this parameter was termed "hopping length"—that is, the length of jumps of H_2O molecule during the Debye relaxation time.

TABLE IV
Experimental Data Pertaining to Liquid Fluoromethane CH_3F^a

T (K)	ρ(g·cm^{-3})	ε_s	n_∞^2	τ_D(ps)	$I\times10^{40}$(g·cm^2)	M (g)	$\mu_0\times10^{18}$(D)	ν_{lib}	k_μ^{exp}
133	0.95	49	1.79	6.6	33	34	1.81	90	1.17
293	0.57	9	1.43	0.47				55	1.05

ν	10	15	20	25	30	35	40	45
α_{133}	176	190	203	224	241	264	288	325
α_{293}	115	190	251	292	325	346	380	388
ν	50	55	60	65	70	75	80	85
α_{133}	356	400	468	536	573	610	634	637
α_{293}	400	400	397	386	363	332	305	278

[a]Frequency ν and absorption coefficient α in cm^{-1}.

3. *Application to Strongly Absorbing Nonassociated Liquid*

We take, for example, liquid fluoromethane CH_3F. In Table IV we summarize the experimental data [43] for two temperatures.

In Table V the fitted free and estimated statistical parameters are presented. For calculation of the spectral function we use rigorous formulas (130) and Eqs. (132) for the hybrid model. For calculation of the susceptibility χ^*, complex permittivity ε^*, and absorption coefficient α we use the same formulas as those employed in Section IV.G.2 for water.[29]

TABLE V
Fitted and Estimated Parameters Pertaining to Liquid Fluoromethane CH_3F. (Frequency ν in cm^{-1})

T (K)	β (deg)	τ (ps)	u	k_μ	y	$\mathring{r}$	$\breve{m}$	ν_{lib}
			A. Rigorous Calculations					
133	17	0.4	6.45	1.25	0.75	0.035	1.22	102
293	45	0.15	2.5	0.9	2.5	0.23	0.4	48.5
			B. Hybrid Model					
133	21	0.37	6.25	1.25	0.81	0.028	0.91	82.6
293	43.4	0.1	3	1	2.1	0.16		49.6

[29]In particular, we use Eq. (142), derived for the Gross collision model in GT and VIG. Note that the application of the isothermal collision model [GT, p. 192; see also Eq. (370) in Section VII], gives for liquid CH_3F, where the dimensionless collision frequency y is large, unreasonable absorption spectrum.

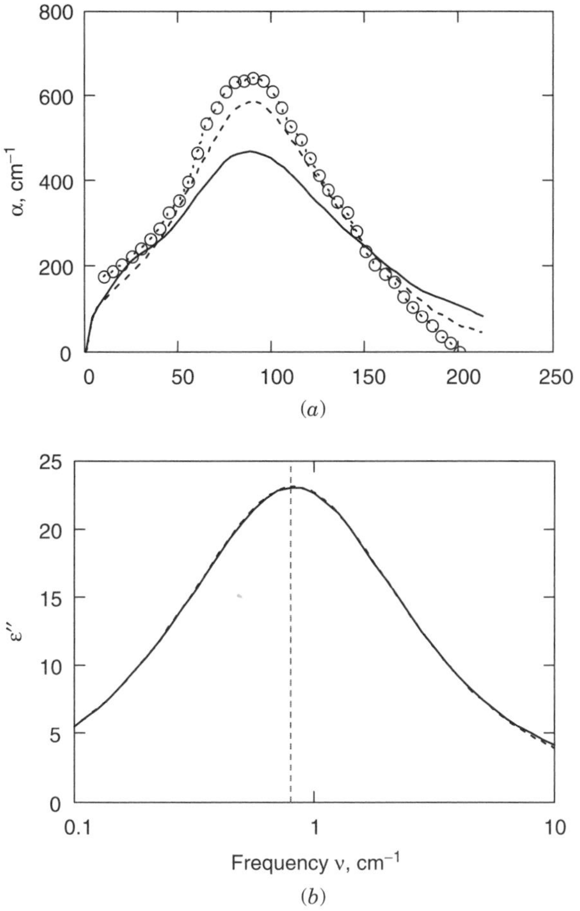

Figure 18. The absorption (a) and loss (b) spectrum of CH_3F at $T = 133$ K. Solid and dashed lines refer, respectively, to rigorous and the hybrid-model versions of the hat-flat model, circles do to experiment. The vertical line marks the frequency of relaxational loss peak.

The results of the calculations of the spectra are illustrated by Figs. 18 and 19. The first figure refers to the temperature $T = 133$ K, which is near the triple point (131 K for CH_3F). In this case the density ρ of a liquid, the maximum dielectric loss ε''_D in the Debye region, and the Debye relaxation time τ_D are substantially larger than those for $T = 293$ K (the latter is rather close to the critical temperature 318 K) to which Fig. 19 refers. The fitted parameters are such that the Kirkwood correlation factor is about 1 at $T = 293$ K.

It is apparent from both figures that the loss/absorption spectra, calculated by using rigorous formulas (130), qualitatively agree with the observed data [43] in

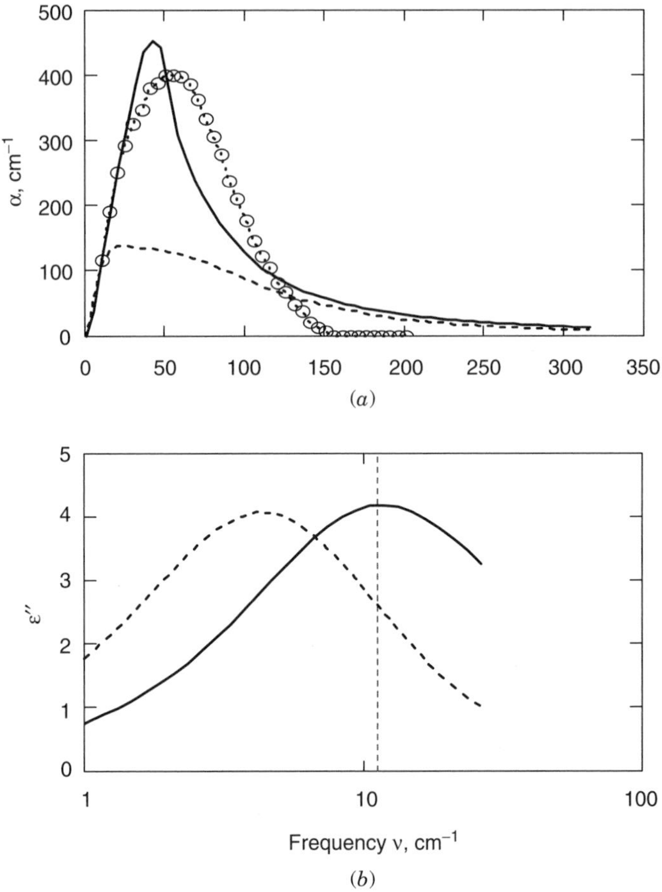

Figure 19. Same as in Fig. 18 but for $T = 293\,\mathrm{K}$.

the relaxation ($0.1 < \nu\ \mathrm{cm}^{-1} < 10$) and FIR ($10 < \nu\ \mathrm{cm}^{-1} < 150$) regions. However, the calculated absorption bandwidth exceeds the experimental one. The temperature dependence of the fitted/estimated parameters, demonstrated by Table V, corresponds to our understanding of molecular events occurring in polar nonassociated liquids (GT, p. 350; VIG, Chapter 14). Namely, if T increases, then:

(i) The cone angle β also increases because of increase of intermolecular distances.

(ii) The lifetime τ decreases, since strong collisions become more frequent.

(iii) As seen in Table V, the number $\breve{m}$ of reflections from the potential wall during the lifetime τ exceeds 1 near the triple point but is 3 times shorter near the critical temperature; in the last case, reorientations become overdamped.

(iv) The potential barrier considerably exceeds $k_B T$ even for high temperature (namely, for $u \approx 2.5$); these estimations suggest that the sites of *local order* unlike the gas state are strongly expressed in a nonassociated liquid just as in water at all possible temperatures.

(v) Correspondingly, the proportion $\mathring{r}$ of the rotators substantially *increases* with the rise of temperature.

(vi) The estimated absorption-peak frequency ν_{lib} is close to the recorded one, both decrease when the temperature rises due to increase of intermolecular distances.

Rigorous and simplified variants of the hat-flat model agree satisfactorily with the experiment at low temperature. The hybrid model yields worse result for the temperature $T = 293$ K near the critical one: The theoretical loss-peak position is noticeably shifted to lower frequencies in comparison with the experimental position. This important result shows that rigorous[30] consideration is preferable in the case of a nonassociated liquid.

Other distinctions between the two variants are as follows. *The rigorous consideration yields*

(a) *lower* proportion $\mathring{r}$ of over-barrier particles and lower effect on it of temperature

(b) *smaller* cone angle β at low temperature (note that its value is close to that obtained in the PL–PR version)

V. HAT-CURVED MODEL AND ITS APPLICATION FOR POLAR FLUIDS

Starting with this section we explore a possibility to reach a *quantitative* agreement of the theoretical and experimental spectra. (A part of our review from this section to the end could be looked through independently from Sections II–IV.)

This section presents central points of the review. (Another such point presents the structural–dynamical model descrbied in Section IX.) In Sections VI and VII the model described here will be further developed. The

[30]We remind the Reader that we have no theoretical justification (see the end of Section IV.E) for application of the hybrid model at high temperatures, at which the libration angle β is rather large.

following two sections V.A and V.B are introductory: Section V.A is devoted to discussion of the physical problem studied in this chapter, and Section V.B shortly reminds the Reader about the basic principles of our linear-response (dynamic) method. A brief list of the main formulas, which are the result of this method, was given in Section IV.G.1.

A. Problem of Unspecific Interactions

It is hardly possible to overestimate a fundamental importance of the studies of the wideband spectra of the complex permittivity $\varepsilon^*(\nu)$ and absorption coefficient $\alpha(\nu)$ of polar fluids. [The asterisk denotes the complex-conjugation; the wavenumber $\nu = \omega/(2\pi c)$ is termed "frequency," where ω is the angular frequency of electromagnetic radiation and c is the speed of light.] The spectra may cover a wide range from zero to $\sim$30 THz, corresponding to ν varying from zero to $\sim$1000 cm^{-1}. An *analytical molecular theory* of these spectra may supply one with spatial and time scales [GT, VIG] of molecular events, these scales being functions of the temperature T or of the structure of a liquid. Such a theory connects various physical and chemical disciplines; for example, the theory [14] of Brownian rotation of molecules is closely related to the structure of liquids, to various diffusion processes, and to reaction rate theory [44]. As a rule, an one-particle approximation is used in analytical linear response theories [44], in which interaction of a given particle (dipole) with all other particles of a liquid is accounted for *implicitly* as a motion of this particle in some intermolecular potential well. In terms of this *semiphenomenological* approach, the form of the well determines the trajectories of the particles and the specific spectral features of various fluids. However, the form of the well is chosen on an intuitive ground. Real progress in such spectroscopic studies, aimed at calculating wideband spectra, depends on a possibility of considering without insurmountable mathematical difficulty more or less nontrivial forms of the wells and/or of combinations of several such wells [VIG].

1. *Evolution of Rectangular-Like Models*

Intuitively, it appears that a polar molecule, moving in a liquid amongst other such molecules, should have a substantial rotational (as well as translational) mobility and should exert nearly elastic reflections from the nearby surroundings.

1. This idea has led long ago [20, 21, 45, 46] to a simple model termed the "confined rotator" model, where the dielectric response was found due to *free* planar librations performed *without friction* during the lifetime of a molecule in a site of near order. These librations occur between two reflective "walls," the

latter being perpendicular to the plane of rotation. The estimates have shown [GT, VIG, 20, 21] that the lifetime is very short (less than 1 ps). So, on average, only a small number[31] m_{lib} of the libration cycles are performed between the "strong collisions," which destroy the potential well in a near-order environment of a given dipole and change its velocity.

2. The next step was concerned with an approximate (VIG, p. 264; GT, p. 244) or more rigorous (VIG, p. 279; GT, p. 231) treatment of a dipole's rotation *in space*. The pair of reflecting "walls" was then replaced by a conical surface, restricting reorientation of a dipole. Between the instances of reflection from this surface, a *dipole rotates freely in a plane*; all such planes (relating to a given dipole) are inclined to the cone axis by the same angle. Thus, the spatial "cone confined rotator" model was suggested (VIG, p. 279; GT, p. 231). In this model, two components of molecular rotation arise: (i) librations corresponding to a time-varying deflection of a dipole moment from the cone axis and (ii) precession of a dipole moment about this axis.

Models 1 and 2 were applied to *simple nonassociated polar liquids.* Wideband {Debye relaxation + FIR} spectral dependencies of the permittivity $\varepsilon^*(\nu)$ and absorption $\alpha(\nu)$ were successfully described. Usually only one quasi-resonance absorption band (at ν between 10 cm^{-1} and 100–200 cm^{-1}) and one non-resonance Debye loss ε'' peak (at microwaves) arise in these fluids. Although a "spatial" model gives, unlike a planar one, a correct value of the integrated absorption $\int_0^\infty \nu\varepsilon''(\nu)d\nu$, the calculated spectral dependencies resemble those found for motion in a plane.

3. In spite of a considerable success and simplicity, the models 1 and 2 fail to explain the FIR absorption band in *water*, where beside the librational band (which is placed in the vicinity of 700 cm^{-1}) there arises also a less intensive R-band with its peak-absorption frequency at $\approx$200 cm^{-1}.

Models 1 and 2 were modified for description of wideband spectrum of H_2O. It seemed reasonable to distribute all dipoles over two groups, comprising respectively the "librators" and the "rotators." For the first group the models 1 and 2 could be applied. The dielectric response of the "rotators" was found by using the extended rotational diffusion model (VIG, p. 232; GT, p. 229) [22, 23]. In the latter a dipole rotates *freely* (as in a gas) between two successive strong collisions. Thus, no reflecting walls of the potential wall were introduced for the second group of molecules. Since a mean resonance frequency is lower for the "rotators" than for the "librators," free rotation may stipulate the

[31]This estimation concerns water. In nonassociated liquids, $m_{\text{lib}} \sim 1$ or still less (GT, VIG; see also Sections V.D.2 and V.D.3). This result indicates that in this case resonant processes are very damped in a liquid.

R-band characteristic for H_2O. Thus, it was suggested [47] that the "confined rotator–extended diffusion" model is capable of qualitatively describing the wideband spectra of water.

4. However, serious drawbacks of model 3 are that (i) the proportion $\overset{\circ}{r}$ of the rotators should be fitted; that is, it is not determined from physical considerations and (ii) the depth of the well, in which a polar particle moves, is considered to be infinite. Both drawbacks were removed in VIG (p. 305, 326, 465) and in Ref. 3, where it was assumed that: (a) The potential is zero on the bottom of the well: $U(\theta) \equiv 0$ at $[-\beta \leq \theta \leq \beta]$, where an angle θ is a deflection of a dipole from the symmetry axis of a cone. (b) Outside the well the depth of the rectangular well is assumed to be constant (and finite): $U(\theta) \equiv U_0$ at $[-\pi/2 \leq \theta \leq \pi/2]$. Actually, two such wells with oppositely directed symmetry axes were supposed to arise in the circle, so that the resulting dipole moment of a local-order region is equal to zero (as well as the total electric moment in any sample of an isotropic medium).

This model[32] was termed the "hat" model (VIG, pp. 326, 465) [62], since the potential profile resembles a hat. The low-energy dipoles move inside such a well, while the high-energy dipoles ($\equiv$ the "rotators") move over the potential barrier; their proportion $\overset{\circ}{r}$ could be calculated as a function of a reduced potential well depth $u = U_0/(k_B T)$ and the angle β of a cone (= angular half-width β of the well). The proportion $\overset{\circ}{r}$ decreases, if U_0 increases, and vanishes at $U_0 \to \infty$.

A simplified version of this model, termed the "hybrid model" (VIG, p. 305) [32–34, 39] (see also Section IV.E) was proposed for the case of a small cone angle β. In this model the rotators move freely over the barrier U_0 as if they do not "notice" the conical surface; the librators move in the diametric sections of a cone—that is, they librate. The hybrid model was widely used for investigation of dielectric relaxation in a number of nonassociated and associated liquids, including aqueous electrolyte solutions (VIG, p. 553) [53, 54]. The hat model was recently applied to a nonassociated liquid [3] and to water [7, 12c].

Models 1–4 have a fundamental drawback: The librational absorption band calculated for water appears to be too wide. This drawback at first glance could be overcome, if one employs the so-called "field" models, in which the static potential presents a smooth well (where a notion of a "collision of a dipole with a wall" actually has no physical sense). However, from the discussion given just below we shall see that this reasonable idea does not work properly with respect to calculating the wideband spectra in water.

[32]The hat (more precisely, the hat-flat) model is described in Section IV of this chapter.

2. *Evolution of "Field" Models*

Inside a rectangular well a dipole rotates freely until it suffers instantaneous collision with a wall of the well and then is *reflected*, while in the field models a continuously acting static force tends to decrease the deflection of a dipole from the symmetry axis of the potential. Therefore, if a dipole has a sufficiently low energy, it would start backward motion at such a point *inside* the well, where its kinetic energy vanishes. Irrespective of the nature of forces governing the motion of a dipole in a liquid, we may *formally* regard the parabolic, cosine, or cosine squared potential wells as the simplest potential profiles useful for our studies. The linear dielectric response was found for this model, for example, in VIG (p. 359) and GT (p. 249).

5. In our early work[33] [50] the constant field model was applied to liquid water, where the harmonic law of particles' motion, corresponding to a parabolic potential, was actually employed in the final calculations of the complex permittivity. In this work, qualitative description of only the libration band was obtained, while neither the R-band nor the low-frequency (Debye) relaxation band was described. Moreover, the fitted mean lifetime τ of the dipoles, moving in the potential well, is unreasonably short ($\approx$0.02 ps)—that is, about an order of magnitude less than in more accurate calculations, which will be made here.

6. More success was gained in the calculations [32] based on application of the cosine squared (CS) potential applied to nonassociated liquids (CH_3F and CHF_3). However, the results obtained by using the CS model are poor if compared with those given by the hybrid model, since the CS model yields (i) worse agreement with the low-frequency experimental spectra (the frequency ν_D of the maximum loss substantially differs from the experimental value) and (ii) unreasonably large ($>$25) value for the reduced well depth $u = U_0/(k_B T)$.

Thus, the field models could not give a satisfactory description of the *wideband* dielectric spectra in water and in typical nonassociated liquids, since *too narrow quasi-resonance and poor low-frequency spectra are characteristic of the field models.* Note, field models could be successfully employed for calculation of the R-band in water, see Sections VII.B, IX, and X.

3. *Hat-Curved Model as Symbiosis of Rectangular-Well and Parabolic-Well Models*

The problem considered now is connected with *unspecific interactions*, which in some measure are similar in water and nonassociated liquids. We shall present

[33]The paper [50] was written, when our general linear response-theory (ACF method) was still in progress, so the derivation of the spectral function described in Ref. 50 is more specialized than that given, for example, in VIG and GT.

the hat-curved model, which combines useful properties of models 4 and 6 based on the application of, respectively, the rectangular and parabolic potentials.

A general approach (VIG, GT) to a linear-response analytical theory, which is used in our work, is viewed briefly in Section V.B. In Section V.C we consider the main features of the hat-curved model and present the formulae for its dipolar autocorrelator—that is, for the spectral function (SF) $L(z)$. (Until Section V.E we avoid details of the derivation of this spectral function L). Being combined with the formulas, given in Section V.B, this correlator enables us to calculate the wideband spectra in liquids of interest. In Section V.D our theory is applied to polar fluids and the results obtained will be summarized and discussed.

Section VI will present a first step for description (again in terms of the hat-curved model) of the *collective (cooperative) effects* in water due to the H bonds (i.e., following Walrafen [16]), resulting from the specific interactions. The dielectric spectra of ordinary and heavy water will be calculated in this section. For this purpose we shall apply (with some changes) recent investigation [6, 8] based on the concept of a *nonrigid dipole*. Other applications of the hat-curved model to water will be described in Sections VII, IX, and X.

B. Basic Assumptions of a Linear Response Theory

The spectra $\varepsilon^*(\nu)$ will be described here in terms of a linear-response theory. We shall employ the specific form [GT, VIG] of this theory, called the ACF method, which previously was termed the "dynamic" method. The latter is based on the Maxwell equations and classical dynamics. A more detailed description of this method is given in Section II. Taking into attention the central role of the model suggested here, we, for the sake of completeness, give below a brief list of the main assumptions employed in our variant of the ACF method.

It is supposed that at some instant a strong collision interrupts the motion of a particle characterized by a fixed phase coordinate Γ, so that a new particle is produced with a coordinate Γ'. The main features of our response theory are as follows:

(1) A typical dipole moves in a potential well *without friction* during the lifetime t_v between two consecutive strong collisions. The well, the form of which should preferably be taken as simple as possible, is chosen on the basis of a phenomenological approach based on the experience gained in previous studies (VIG, GT). The time t_v is interpreted as a duration of local order in a liquid. The exponential distribution over t_v is used, where the *mean* lifetime τ (which later is referred to simply as "lifetime"), $\tau \equiv \langle t_v \rangle$, is a free parameter of the model.

(2) A *weak* transverse electromagnetic wave, which propagates through a polar medium, disturbs the coordinates and the momenta of the particles, as well

as the steady-state Boltzmann distribution $W = C\exp[-H/(k_B T)]$, where H is the full energy of a dipole. As a result, an *induced* distribution over the phase coordinates of the particles arises, which *explicitly* depends on the amplitude and initial phase of the a.c. field.

(3) The absorption coefficient *in the FIR spectral region* is proportional to the product $x\,\mathrm{Im}[L(z)]$. Here the *spectral function* (SF) $L(z)$ is the linear-response characteristic of the model under consideration, where the dimensionless complex frequency z is related to angular frequency ω of radiation and mean lifetime τ as follows:

$$z = x + iy, \qquad x = \eta\omega, \qquad y = \eta/\tau, \qquad \eta \equiv \sqrt{I/(k_B T))} \tag{154}$$

I is the effective moment of inertia of a dipole (we consider here a linear molecule), determined by the relation (149). The spectral function $L(z)$, calculated for *thermal equilibrium,* is linearly related to the *spectrum* $\tilde{C}^0$ of the dipolar autocorrelation function (ACF) $C^0(t)$ (VIG, p. 137; GT, p. 152) as

$$L(z) = 1 + (iz/\eta)\tilde{C}^0(z/\eta) \tag{155}$$

$L(z)$ is found [see Eqs. (170)–(183) below] for undamped reorientation of a dipole in a given potential field. Note that after considerable simplification we actually use for calculations the expressions (194), (171)–(173), and (179).

(4) The dipolar ACF and its spectrum are generally defined as follows:

$$C^0(t) = \langle \mu_E(0)\mu_E(t)\rangle \langle \mu_E(0)^2\rangle^{-1} \tag{156}$$

$$\tilde{C}^0(\omega) = \int_0^\infty C^0(t)\exp(i\omega t)\,dt \tag{157}$$

where $\langle\cdot\rangle$ denotes statistical average, μ_E is the projection of a dipole moment $\boldsymbol{\mu}$ onto the direction of the a.c. field. However, it is more convenient to find first the spectral function $L(z)$ of the model as the following ensemble average[34]:

$$L(z) = iz/(\mu^2\eta)\left\langle \mu_E(0)\int_0^\infty [\mu_E(t) - \mu_E(0)]\exp(izt/\eta)\,dt\right\rangle \tag{158}$$

Taking after that an inverse Fourier transform of $\tilde{C}^0(\omega)$, one may calculate (VIG, GT) from Eqs. (155) and (158), if necessary, the ACF itself, $C^0(t)$. (This ACF is not considered in our work.) In an isotropic medium, which we are interested in,

[34]Equation (158) is equivalent to Eq. (5.35) in VIG or to Eq. (3.18) in GT.

the average $\langle\cdot\rangle$ is taken over the canonical phase variables Γ *and* over all possible directions of the symmetry axes.

(5) The spectral function $L(z)$ determines the complex susceptibility χ^* of the medium,[35]

$$\chi^* = \chi' + i\chi'' \tag{159}$$

which is found from consideration of a *classical* law of motion of a dipole with a simplified account of the dissipation mechanism. The latter is described by a specific *model of collisions* (VIG, p. 203; GT, p. 191), in which the complex amplitude $\hat{P}$ of the a.c. polarization is expressed through its complex susceptibility as $\hat{P} = \chi\hat{E}$. Then $\hat{P}$ is equated to the dipole moment of a unit volume, which in turn depends on χ. As a result, the simple relation (142) between the susceptibility χ^* and the SF $L(z)$ is derived for the *Gross collision model* (VIG, p. 190; GT, p. 192; see also Section VII.C).

(6) A theory, accounting for an *internal field correction* [40, 41], gives the relationship $\chi^*(\varepsilon^*)$, Eq. (139), between the complex susceptibility and permittivity. For calculation of the wideband spectra it is more convenient to employ the reverse dependence $\varepsilon^*(\chi^*)$, Eq. (141).

C. Molecular Subensembles and Spectral Function

1. Form of Potential and Classification of Particles' Subensembles

Let θ be angular deflection of a dipole from the symmetry axis of the potential $U(\theta)$, let β be a small angular half-width of the well ($\beta \ll \pi/2$), and let U_0 be the well depth; its reduced value $u = U_0/(k_B T)$ is assumed to be $\gg 1$. Since in any microscopically small volume a dipole moment of a fluid is assumed to be zero, we consider that *two* such wells with oppositely directed symmetry axes arise in the interval $[0 \le \theta \le 2\pi]$. For brevity we consider now a quarter-arc of the circle. The bottom of the potential well is *flat* at $0 \le \theta \le f\beta$. In the remainder of the well we take the *parabolic* dependence U on θ. The form factor f is defined as the ratio of this flat-part width to the whole width of the well. Thus, the assumed potential profile is given by

$$\frac{U(\theta)}{k_B T} = \left\{ \begin{matrix} 0 \\ u\,(\theta - \beta f)^2(\beta - \beta f)^{-2} \\ u \end{matrix} \right\} \quad \text{at} \quad \left\{ \begin{matrix} 0 \le \theta \le \beta f \\ \beta f \le \theta \le \beta \\ \beta \le \theta \le \pi/2 \end{matrix} \right\} \tag{160}$$

[35] We consider the complex-conjugated quantities in order to calculate the loss factor χ'' or ε'' as determined by *positive* imaginary components of χ^* or ε^*.

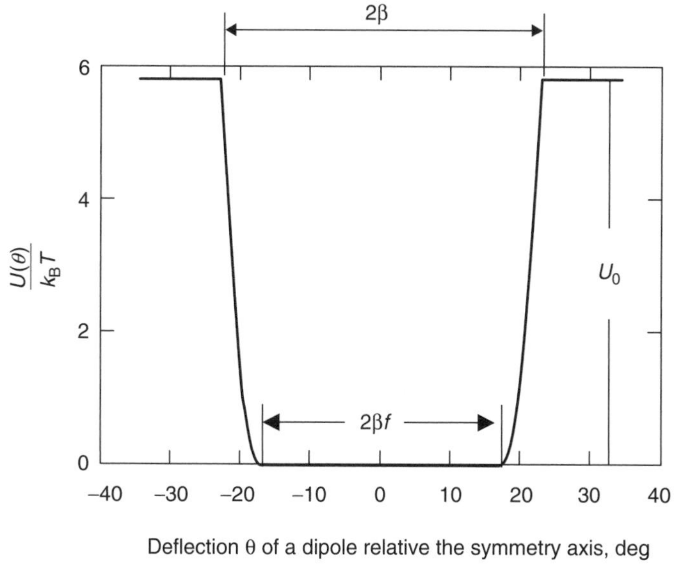

Figure 20. The form of the hat-curved potential well. The chosen parameters are typical for water; the half-width of the well, β, is 23°, the reduced well depth, u, is 5.8, and the form factor, f, is 0.75.

Introducing the wall's steepness S, we may rewrite the potential in the *parabolic part* of the well as

$$U(\theta)/(k_B T) = u(\theta - \beta f)^2(\beta - \beta f)^{-2}$$
$$= S^2(\theta - \beta f)^2 \qquad \text{so that } S \equiv \sqrt{u}\,(\beta - \beta f)^{-1} \tag{161}$$

The profile (160) is illustrated by Fig. 20 for the parameters typical for water; in this Figure the flat part $2\beta f$ and the total angular width 2β are shown. The rectangular profile, with which the dielectric properties of fluids were previously rather successfully studied (VIG) [3, 7], corresponds to $f = 1$ (see also Section IV).

In order to determine specific phase regions of a molecular ensemble, we express the full energy of a linear molecule as a function of a polar coordinate θ and of two angular velocities:

$$H\left(\theta, \frac{d\theta}{dt}, \frac{d\phi}{dt}\right) = \frac{I}{2}\left(\frac{d\theta}{dt}\right)^2 + \frac{I}{2}\left(\frac{d\phi}{dt}\right)^2 \sin^2\theta + U(\theta) \tag{162}$$

Polar angle θ denotes the angle between a dipole moment $\boldsymbol{\mu}$ and the symmetry axis of the potential, so the first term in Eq. (162) refers to the change of a polar angle. The second term in Eq. (162) describes precession of a dipole about the symmetry axis with a constant momentum P_ϕ, when azimuthal coordinate ϕ varies. Indeed, since H does not depend on ϕ, then $dP_\phi/dt = \partial H/\partial\phi = 0$, so that

$$P_\phi = \frac{\partial H}{\partial(d\phi/dt)} = I\frac{d\phi}{dt}\sin^2\theta = \text{const} \qquad \frac{I}{2}\left(\frac{d\phi}{dt}\right)^2 \sin^2\theta \equiv \frac{P_\phi^2}{2I\sin^2\theta} \tag{163}$$

We rewrite Eq. (162) for the energy H by taking into account Eqs. (160) and (163), dividing both sides of (162) by $k_B T$, replacing *inside the well* (where $\theta \le \beta \ll 1$) $\sin^2\theta$ by θ^2. Introducing the reduced full energy h and other quantities as follows,

$$\begin{aligned} h &= \frac{H}{k_B T}, \quad \varphi = \frac{t}{\eta}, \quad \dot{\theta} \equiv \frac{d\theta}{d\varphi}, \quad \dot{\phi} \equiv \frac{d\phi}{d\varphi} = \frac{l}{\sin^2\theta}, \\ l &= \frac{P_\phi}{\sqrt{2Ik_B T}}, \quad \eta = \sqrt{\frac{I}{2k_B T}} \end{aligned} \tag{164}$$

we have instead of (162)

$$V(\theta) = h - \dot{\theta}^2 = \left\{ \begin{array}{c} l^2/\theta^2 \\ l^2/\theta^2 + \dfrac{u(\theta - \beta f)^2}{(\beta - \beta f)^2} \\ l^2/\sin^2\theta + u \end{array} \right\} \quad \text{at} \quad \left\{ \begin{array}{c} 0 \le \theta \le \beta f \\ \beta f \le \theta \le \beta \\ \beta \le \theta \le \pi/2 \end{array} \right\} \tag{165}$$

We have introduced here an *effective potential* $V(\theta)$. Equation (165) formally coincides with the similar equation for a one-dimensional rotation of a dipole in such a potential. The potential (165) has its minimum at some angle $\bar{\theta}$, which is generally less than β, while in the case of rectangular well considered in Section IV, V is minimal at $\theta = \beta$.

The position $\bar{\theta}$ of the minimum potential, found from Eq. (165), obeys the equation

$$\bar{\theta}^3(\bar{\theta} - \beta f) = l^2(\beta - \beta f)^2 u^{-1} \tag{166}$$

The differential equation (165) determines the trajectory of any given dipole. Since $\dot{\theta} = \sqrt{h - V(\theta)}$ and $V(\theta) > 0$, it follows from Eq. (165) that its solution exists *only if* $h \ge V(\theta)$. Therefore the energy h of the particles *localised in the well* does not exceed $V(\beta)$. Their laws of motion $\{\theta(t), \phi(t)\}$ are such that the orientation θ of a dipole relative the symmetry axis *oscillates* between two values θ_{min} and θ_{max}, at which $\dot{\theta} = 0$, that is, $h = V(\theta)$.

To give an overview qualitatively of the main features of a dipole's motion, we consider two specific cases. In the first one, Fig. 21a, we set

$$V(\beta f) < V(\beta), \qquad \text{so that} \quad l^2[(\beta f)^{-2} - \beta^{-2}] < u \tag{167a}$$

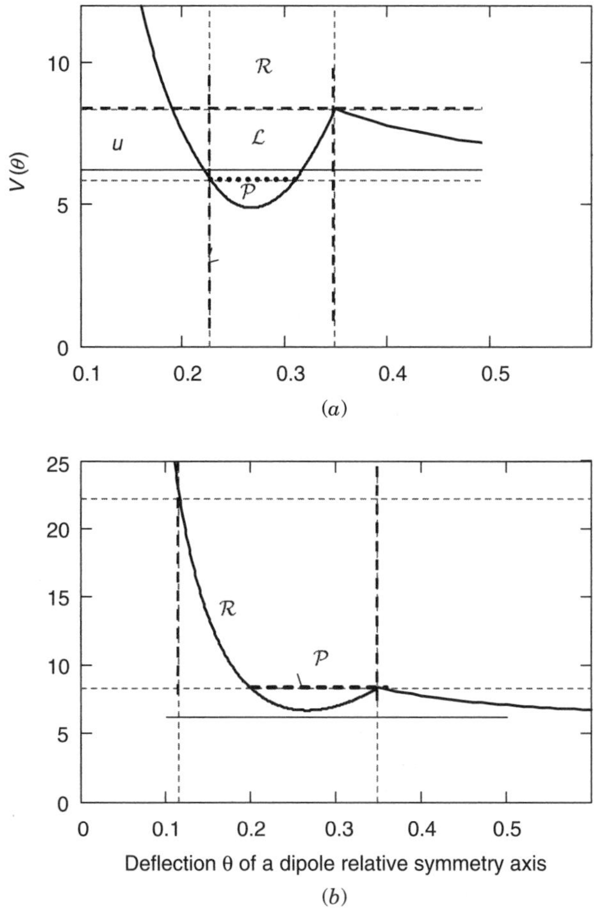

Figure 21. (a,b) The angular dependence $V(\theta)$; the V value on a boundary of the flat bottom is less (a) or greater (b) than at the edge of the well; $u = 6.2$, $\beta = \pi/9$. Solid horizontal lines refer to the chosen u value; left and right vertical lines refer to the values $\theta = \beta f$ and $\theta = \beta$, respectively, namely to the edges of the bottom's flat part. (c, d) Minimum value $V_{\min}$ of the effective potential V (c) and of the reduced precessional energy corresponding to this $V_{\min}$ (d) versus the reduced angular momentum l. Solid and dashed lines refer to the parameters typical for oridinary and heavy water, respectively. $\beta = 23°; u = 5.8, f = 0.75$ (solid lines); $u = 6.4, f = 0.65$ (dashed lines). Calculation from Eqs. (165), (166), and (168).

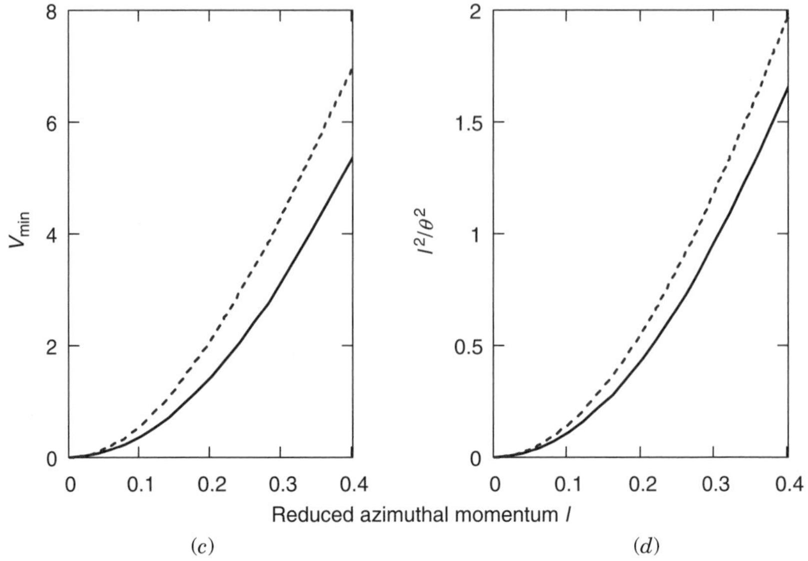

Figure 21 (*Continued*)

In the second we regard the inverse inequalities, Fig. 21b,

$$V(\beta f) < V(\beta), \qquad \text{so that} \quad l^2\left[(\beta f)^{-2} - \beta^{-2}\right] > u \tag{167b}$$

In view of Figs. 21a and 21b, we mark out three characteristic phase regions $\mathscr{L}$, and $\mathscr{P}$ and $\mathscr{R}$, to which three specific assemblies of the particles correspond; the quantities related to them are marked by the superscripts $\smile$, $\sim$, and $\circ$, respectively. The regions $\mathscr{L}$ and $\mathscr{P}$ are occupied by the particles localized in the well. The $\mathscr{L}$-particles, which exist only in the first case (167a), *penetrate into the flat part* of the well. We may conditionally call them "librators." The meaning of this term will become clear just below. The $\mathscr{P}$-particles are *concentrated in the parabolic part* of the well and are conditionally called "precessors." Indeed, if a dipole, which belongs to this subensemble, would have energy h close to that, for which the effective potential $V(\theta)$ undergoes its minimum,

$$V(\bar{\theta}) = [V(\theta)]_{\min} \tag{168}$$

the motion of such a dipole would resembles a regular precession around the symmetry axis.

In the region $\mathscr{R}$ each horizontal line in Figs. 21a, and 21b, corresponding to a constant h value, intersects the $V(\theta)$ curve *once*. Therefore, the condition $h \geq V(\theta)$ holds also outside the well, so trajectories of the $\mathscr{R}$-particles are not limited by the width of the well. They perform *complete rotation*, and we conditionally call them "*rotators*"; their spectral function is denoted $\overset{\circ}{L}$. We can, in turn, divide the region $\mathscr{R}$ into two subregions: $\mathscr{R}_{\rm hin}$ and $\mathscr{R}_{\rm free}$, where the energy of a rotator is, respectively, greater or less than $V(\beta)$. In $\mathscr{R}_{\rm hin}$ the dipoles are hindered rotators: They intersect the boundary of the well; therefore their angular velocities inside and outside the well are different. In $\mathscr{R}_{\rm free}$ the dipoles are outside the well; therefore they exert free rotation. If the angle β is small, the contribution of the hindered rotators to the SF $\overset{\circ}{L}$ is negligible. Thus the calculation of the spectral function $\overset{\circ}{L}$ due to rotators could be based on simple expressions derived before [GT, VIG] for *free rotators*—that is, for the extended diffusion model.

Figure 21c demonstrates the curve $V_{\rm min}(l)$, calculated from Eqs. (165) and (166) with u about 6 and f about 0.7. The potential minimum becomes deeper when l rises. Since $h > V_{\rm min}$ and the contribution of the dipoles to the absorption coefficient is very low for $h > u$, then, as seen in this figure, the potential minimum is low only for *small l values* (such that $l^2 < 0.1$). Therefore, the dipoles with a small axial momenta are main contributors to the spectra stipulated by the reorientation process.[36] (Note that at such l values the angle $\bar{\theta}$ in our example falls into the range 15–18 degrees; that is, $\bar{\theta}$ is rather close to the chosen β, which is equal to 23°.) The dependence on l of the reduced precessional kinetic energy $l^2/\bar{\theta}^2$ at the minimum of $V(\theta)$ also rises with l (see Fig. 21d). In view of Fig. 21c and 21d this precessional energy is at least 4 times lower than the full energy h (which exceeds $V_{\rm min}$).

To *simplify* the analytical expressions for the spectral function $L(z)$ we set:

(1) $l = 0$ for all the $\mathscr{L}$-particles; these are supposed to move in *planes* including the symmetry axis. Then the minimum $\bar{\theta}$ of the effective potential is placed in the middle of the well and Eq. (165) reduces to

$$V(\theta) = h - \dot{\theta}^2 = \begin{Bmatrix} 0 \\ S^2(\theta - \beta f)^2 \end{Bmatrix} \quad \text{at} \quad \begin{Bmatrix} 0 \le \theta \le \beta f \\ \beta f \le \theta \le \beta \end{Bmatrix} \tag{169}$$

(2) $\dot{\theta} \equiv 0$, $\theta \equiv \bar{\theta}$ for all the *$\mathscr{P}$-particles*; these *precess with a constant azimuthal velocity* $\dot{\phi}$ at the "frozen" polar angle $\theta \equiv \bar{\theta}$, corresponding to the minimum of the effective potential.

These assumptions constitute the *planar libration–regular precession* (PL–RP) approximation. Note that in Section IV the like one concerns the trajectories, which obey the *same* equation of motion. Here this approximation

[36]We may come to the same conclusion from consideration of the boundaries of the phase region, which are determined in the Section V.5.

means that *different* equations of motion correspond to two particular types of the trajectories. In other words, as compared with the hat-flat model, now the subensemble of "precessors" could be introduced in a more natural way since unlike the librators the precessors *do not* occupy the flat part of the hat-curved potential well.

The corresponding spectral functions, denoted $\breve{L}(z)$ and $\tilde{L}(z)$, are derived, as well as the SF $\mathring{L}$ for the rotators, in Section V.E in the form of simple integrals from elementary functions over a full energy of a dipole (or over some function of this energy). The total spectral function is thus represented as

$$L(z) = \breve{L}(z) + \tilde{L}(z) + \mathring{L}(z) \tag{170}$$

2. "Partial" Spectral Functions

We shall write down the formulas for the above-mentioned spectral functions pertinent to different dipolar subensembles. The spectral function of the *librators* is given by

$$\breve{L}(z) = \left(\frac{2}{\pi}\right)^5 \frac{4u(\beta f)^4}{D} \int_0^u S(h)[1+\sigma(h)]^7 \exp(-h)\, h\, dh \tag{171}$$

where

$$S(h) = \sum_{n=1}^{\infty} \frac{\sin^2\left[\frac{2n-1}{2}\frac{\pi}{1+\sigma(h)}\right]}{(2n-1)^2[(2n-1)^2\sigma^2(h) - (1+\sigma(h))^2]^2[(2n-1)^2 h - (2z\beta f/\pi)^2(1+\sigma(h)^2)]} \tag{172}$$

and

$$\sigma(h) \equiv \frac{\pi}{2}\frac{1-f}{f}\sqrt{\frac{h}{u}} \tag{173}$$

The spectral function of the *precessors* is

$$\tilde{L}(z) = \frac{4u^3\beta^2}{D(1-f)^{3/2}} \int_0^{\psi_{\max}} [(1-f)\psi + f]^{3/2} \sqrt{4(1-f)\psi + f} \times \frac{\Psi(\psi)\psi^{3/2} \exp[-h(\psi)]\, d\psi}{p^2(\psi) - z^2} \tag{174}$$

where ψ is a variable related to the reduced energy as

$$h(\psi) \equiv \frac{u}{1-f}\psi[2(1-f)\psi + f] \tag{175}$$

$$\Psi(\psi) \equiv [\psi(1-f) + f]^{3/2} - f\sqrt{2\psi(1-f) + f} \tag{176}$$

and

$$p(\psi) \equiv \beta^{-1}\sqrt{\frac{u\psi}{(1-f)[(1-f)\psi + f]}} \tag{177}$$

The upper limit in the integral over ψ is

$$\psi_{\max} = \frac{1}{4(1-f)}\left(\sqrt{f^2 + 8\frac{1-f}{1+f}} - f\right) \tag{178}$$

Finally, a simplified expression for the spectral function of the *rotators* is given by

$$\overset{\circ}{L}(z) = \frac{2u\exp(-u)}{D}\int_0^{\infty}\frac{\exp(-h)\,h\,dh}{h - z^2} \tag{179}$$

The coefficient D, being proportional to a normalizing coefficient C of the Maxwell–Boltzmann energy distribution $W = C\exp(-h)$, is determined by the parameters of the hat-curved model as

$$D = u/\pi C; \qquad C = \Sigma^{-1}, \qquad \Sigma \equiv \breve{\Sigma} + \tilde{\Sigma} + \overset{\circ}{\Sigma} \tag{180}$$

where

$$\breve{\Sigma}(u, \beta, f) = 4\beta^2 f\int_0^u \left[f + \frac{\pi}{2}(1-f)\sqrt{\frac{h}{u}}\right]\exp(-h)\,dh \tag{181a}$$

$$\tilde{\Sigma} = 2\int_0^{\psi_{\max}} \exp[-h(\psi)]\,\tilde{\Phi}(\psi)\left\{l_{\max}[h(\psi)] - \beta f\sqrt{h(\psi)}\right\} \times \frac{\partial h(\psi)}{\partial \psi}\,d\psi \tag{181b}$$

$$\overset{\circ}{\Sigma} = 2\pi\exp(-u) \tag{181c}$$

$h(\psi)$ and $\psi_{\max}$ are given by Eqs. (175) and (178) and

$$\frac{\partial h(\psi)}{\partial \psi} = \frac{u[4(1-f)\psi + f]}{1-f} \tag{182a}$$

$$\tilde{\Phi}(\psi) = \frac{2\pi\beta(1-f)}{\sqrt{u}} \sqrt{\frac{(1-f)\,\psi + f}{4(1-f)\,\psi + f}} \tag{182b}$$

and

$$l_{\max}(\psi) = \beta\sqrt{\frac{u\psi[(1-f)\psi + f]^3}{1-f}} \tag{183}$$

The *proportions* of the librators, precessors, and rotators are defined as the normalized weights:

$$\breve{r} = \frac{\breve{\Sigma}}{\Sigma}, \qquad \tilde{r} = \frac{\tilde{\Sigma}}{\Sigma}, \qquad \mathring{r} = \frac{\mathring{\Sigma}}{\Sigma} \tag{184}$$

In the case of the *rectangular well* (i.e., for $f = 1$), we have $\tilde{L} = 0$ and

$$\breve{L}(z) = \frac{128\,\beta^4}{\pi^5\left(\beta^2 + 2\exp(-u)\right)} \sum_{n=1}^{\infty} \int_0^u \frac{h\exp(-h)\,dh}{(2n-1)^2\left[(2n-1)^2 h - (2\beta z/\pi)^2\right]} \tag{185}$$

Letting $u \to \infty$ and retaining only the first term of the series, we get the formula in GT, Eq. (6.45), for the small-amplitude libration performed by a dipole in a diametric section of a cone.

In the case of the *parabolic* well (i.e., for $f = 0$), we have $\breve{L} = 0$ and

$$\tilde{L}(z) = \frac{\beta^2\left[1 - (1 + u + u^2/2)\exp(-u)\right]}{\left(\beta^2 + 2\,u\exp(-u)\right)\left(u/\beta^2 - z^2\right)} \tag{186}$$

At $u \gg 1$ this expression yields the Lorentz line [see VIG, Eq. (10.17) or GT, p. 258, or Section VII.C]:

$$L = \left(p^2 - z^2\right)^{-1}, \qquad p^2 = u/\beta^2 \tag{187}$$

3. *Qualitative Description of Absorption Spectrum*

Figures 22a and 22b illustrate a nondimensional absorption frequency dependence

$$\gamma(x) = x\,\mathrm{Im}[L(z)] \tag{188}$$

A two-humped absorption curve is generated by Eqs. (170) and (188). Curve 1 in Fig. 22a refers to the case of the rectangular well, $f = 1$, Eq. (185). Figure 22b refers to the case of a parabolic potential well (with $f = 0$), Eq. (186). When the form factor f decreases from 1 to 0, the high-frequency part of the absorption curve—that is, *the librational band* stipulated by reorientation of the dipoles in the well—changes from being rather broad to being very narrow. The *low-frequency* absorption band arises because the high-energy particles rotate *freely* over the potential barrier u; the peak frequency of this band weakly depends on the form factor f. Figures 22a and 22b demonstrate that the narrowing of the librational band is possible *by decreasing* the form factor f of the hat curved-potential.

It is important to compare the parts of the absorption band, generated by the librators, precessors, and rotators. In Fig. 22c and 22d we show evolution of these partial frequency dependencies, when the form factor f decreases from 0.8 to 0.65. These and other parameters ($u = 6$, $\beta = 23°$) are chosen to be typical for water. In the first case, Fig. 22c, the contribution of the precessors to the total absorption is negligible; in the second case, Fig. 22d, it is less than 5%. Therefore, for the set of parameters typical for water, *the absorption spectrum is actually determined by the librators and the rotators.*

Using this result, we may simplify calculation of the spectral function $L(z)$ by neglecting the precessional contribution to L. We shall estimate also in this approximation the peak frequencies x_{lib} and x_{rot} of the absorption bands determined by the librational and the rotational subensembles.

To *roughly* estimate the frequency of maximum absorption, we take the following steps:

(i) Considering the main (corresponding to $n = 1$) term in the denominator of Eq. (172) we write the resonance condition as

$$x_{\mathrm{res}}(h) = \frac{\pi\sqrt{h}}{2\beta f[1 + \sigma(h)]} \tag{189}$$

(ii) Starting from the *imaginary* part of Eq. (171), we transfer x under the sign of the integral, considering $x \approx x_{\mathrm{res}}(h)$. Then we find maximum

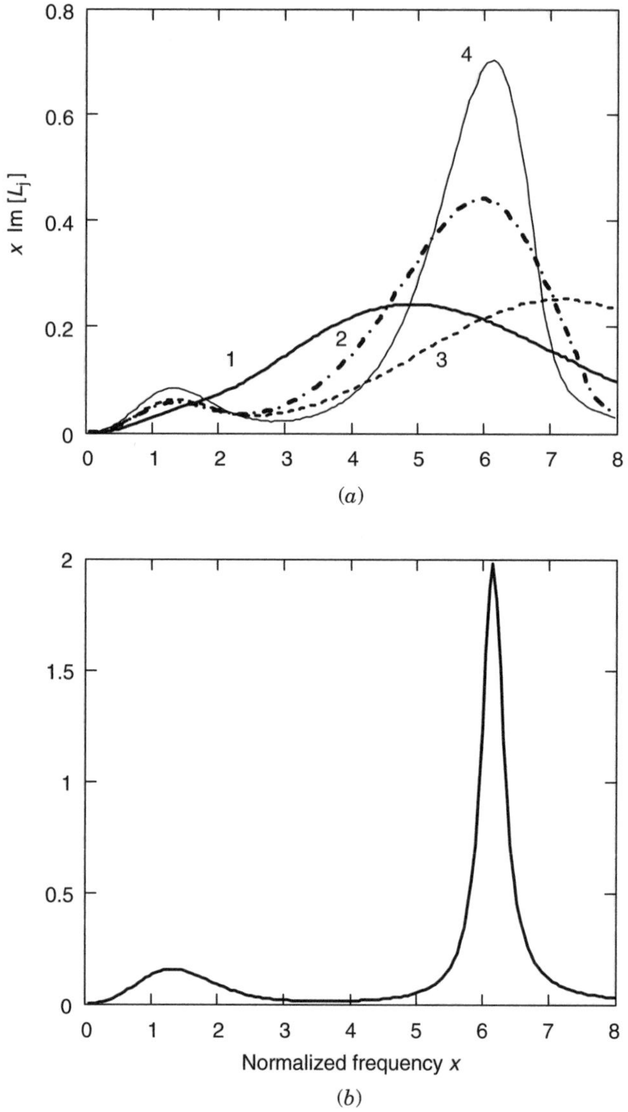

Figure 22. (a, b) Normalized absorption versus normalized frequency. Calculation for the hat-curved model, $u = 6$ and $\beta = 23°$. In Fig. (a) the form factor f is 1, 0.7, 0.5, and 0.3 for curves 1, 2, 3, 4, respectively. In Fig. (b), $f = 0$. Other parameters of the model are: $u = 6, \beta = 23°, y = 0.2$. (c, d) Frequency dependences of the reduced absorption due to the librators (2), precessors (3), and rotators (4). The total absorption is marked by solid curves (1). The form factor, f, is 0.8 in (c) and 0.65 in (d). Left and right vertical lines mark the reduced peak frequencies of the absorption bands arising due to the rotators, Eq. (191), and the librators, Eq. (190a).

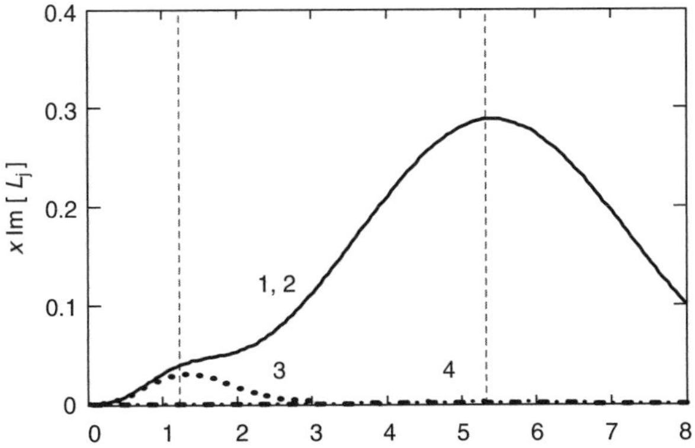

(c)

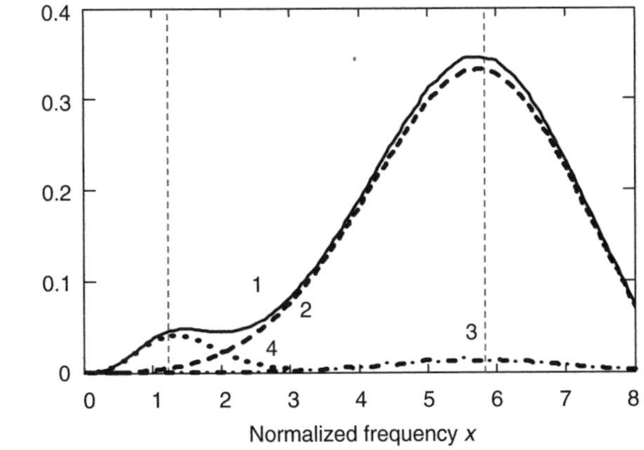

(d)

Figure 22 (*Continued*)

of the product $x_{\text{res}}(h)h\exp(-h) \propto F(h)$, where $F(h) = [1+\sigma(h)]^{-1} \times h\sqrt{h}\exp(-h) \approx h\sqrt{h}\exp(-h)$. The last equality is approximately valid, since $\sigma \ll 1$ at f close to 1. $F(h)$ attains maximum at $h \approx 3/2$.

(iii) Substituting this h value into $x_{\text{res}}(h)$, we assign the latter with the peak frequency of the absorption coefficient (in the librational band):

$$x_{\text{lib}} \approx x_{\text{res}}(3/2) = \frac{\pi\sqrt{3/2}}{2\beta}\left[f + \frac{\pi}{2}(1-f)\sqrt{\frac{3}{2u}}\right]^{-1} \tag{190a}$$

The period of librations $T_{\text{lib}} = 2\pi\eta x_{\text{lib}}^{-1}$ corresponds to the estimated librational frequency.

(iv) Taking into account that $x = 2\pi c\eta\nu$ and setting ν to be equal to the librational absorption-peak frequency ν_{lib}, we relate the form factor f to this frequency and to the well depth u as

$$f = \left(\frac{\sqrt{u}}{2\pi c\eta\nu_{\text{lib}}} - 1\right)\left(\frac{2}{\pi}\sqrt{\frac{2u}{3}} - 1\right)^{-1} \tag{190b}$$

For ensemble of the rotators we analogously derive:

$$x_{\text{rot}} \approx \sqrt{3/2} \tag{191}$$

Note Eq. (190a) reduces to (191) at $\beta = \pi/2$ and $f = 1$.

We shall also estimate the numbers m_{lib} and m_{rot} of the librational and rotational cycles performed on average by a librating or a rotating dipole during the lifetime τ.

$$m_{\text{lib}} = \frac{\tau}{T_{\text{lib}}} = \frac{\tau x_{\text{lib}}}{2\pi\eta} = \frac{x_{\text{lib}}}{(2\pi y)} \approx \frac{\sqrt{3/2}}{4\beta y}\left[f + \frac{\pi}{2}(1-f)\sqrt{\frac{3}{2u}}\right]^{-1} \tag{192a}$$

$$m_{\text{rot}} \approx \frac{\tau x_{\text{rot}}}{2\pi\eta} = \frac{\sqrt{3/2}}{2\pi y} \tag{192b}$$

The formulae (190a)–(192b) will be employed for interpretation of the results of further calculations.

The estimated positions of maximum absorption (see vertical lines in Figs. 22c and 22d are close to the peak frequencies found from Eqs. (171)–(173) and (179). The peak frequency (190a) is inversely proportional to β; it increases with the decrease in the form factor f or with the increase of the well depth u, see, respectively, Fig. 23a and 23b. The proportions of the librators, precessors, and rotators are calculated from Eq. (184) and presented in Fig. 23c and 23d by solid, dash–dotted, and dashed curves, respectively. Figure 23c shows relevant functions of the reduced well depth u and Fig. 23d—of the form factor f.

In an important case of a deep well and f close to 1, the main fraction is the *librators.* The proportion of the precessors is very small and their contribution to the FIR spectra is negligible as we have already demonstrated in Figs. 22c and 22d. The proportion of the rotators in this case is only a few percent. However, the rotators substantially affect the FIR and low-frequency spectra of water (it will be demonstrated in the Section V.D).

For the examples considered in Figs. 22c and 22d we have, respectively, $m_{\text{lib}} \approx 4$ and $m_{\text{rot}} \approx 1$. It appears, consequently, that *librational motion is more regular than free rotation* (both motions are assumed to be performed during the same lifetime).

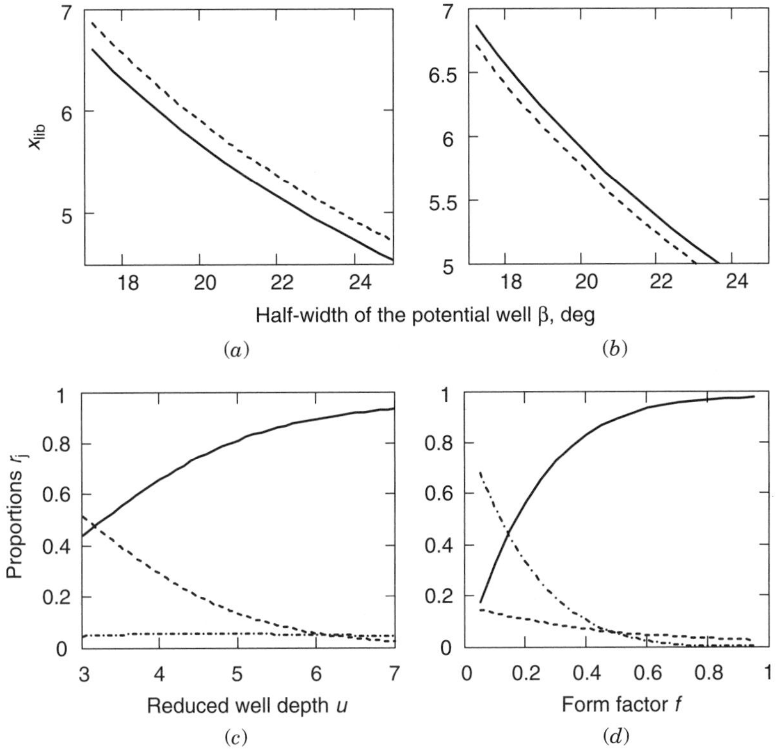

Figure 23. (a, b) Dependence of the estimated peak frequency of the librational absorption band on the angle β. (a) $f = 0.85$ (solid curve), $f = 0.65$ (dashed curve); $u = 5.6$; (b) $u = 5.6$ (solid curve), $u = 4.8$ (dashed curve); $f = 0.65$. The parameters of the hat-curved model are presented in Table VII. (c, d) The proportions of the librators (solid curves), rotators (dashed curves), and precessors (dashed-and-dotted curves) versus well depth (c) or form factor (d); $\beta = 23°$; $f = 0.5$ (in c) and $u = 6$ (in d).

Finally, neglecting the term in Eq. (184) pertinent to precessors, we can estimate the proportion $\overset{\circ}{r}$ of the rotators as

$$\overset{\circ}{r} \approx \overset{\circ}{\Sigma} \Big/ \left(\breve{\Sigma} + \overset{\circ}{\Sigma} \right) \tag{193}$$

where the quantities involved are determined by Eqs. (181a) and (181c). The proportion of other subensemble (of librators) is evidently equal to $1 - \overset{\circ}{r}$.

D. Calculated Spectra of Polar Fluids

In view of the above conclusion, if the form factor f is rather large (e.g., if $f > 0.65$), then we may retain in Eqs. (170) and (180) only two "partial" spectral functions:

$$L(z) \approx \breve{L}(z) + \mathring{L}(z) \tag{194a}$$

$$D = u/\pi C, \qquad C = \Sigma^{-1}, \qquad \Sigma \equiv \breve{\Sigma} + \mathring{\Sigma} \tag{194b}$$

where $\breve{L}(z)$ is given by Eqs. (171)–(173) and $\mathring{L}(z)$ by Eq. (179). Thus, the resulting spectral function of the hat-curved model is characterized by the four parameters

$$u, \ \beta, \ f, \ \tau \tag{195}$$

namely, by the reduced depth (u) and the angular width (β) of the potential well, by its form factor (f), and by the mean lifetime (τ). Note that τ is related to the reduced collision frequency y as η/τ. One can fit the parameters (195) by using the formulas given in Sections IV.G.1 and IV.C aided by comparison with the experimental frequency and/or temperature dependencies of the wideband spectra.

1. *Dielectric and FIR Spectra of Liquid H_2O*

Starting with the important example of *ordinary water*, we choose temperatures 22.2°C and 27°C. We compare our theory with the recorded FIR spectra [42, 56] of the complex permittivity/absorption. At *low frequencies* we use for this purpose an empirical formula [17] by Liebe et al.; these formulas were given also in Section IV.G.2.a. The values of the employed molecular constants are presented in Table VI and the fitted parameters in Table VII. The Reader may find more information about experimental data of liquid H_2O and D_2O in Appendix 3.

To facilitate parameterization of our model, we may employ Eqs. (150), (151), and Eq. (190b). The latter allows us to decrease the number of free parameters of the model, since the form factor f is thus expressed through u, β,

TABLE VI
Experimental Data Pertaining to Ordinary Water[a]

T °C	τ_D (ps)	ε_s	ρ (g $\times$ cm^{-3})	ν_{lib}	ν_R	α_{lib}	α_R	k_μ
27	7.85	77.6	0.9986	680	200	3298	1445	1.10
22.2	9.3	80.0	0.9986	690	201	3403	1208	1.02

$\mu_0 = 0.84\,D$, $n_\infty^2 = 1.7$, $I = 1.483 \times 10^{-43}$, $M = 18$.
[a]Frequency ν and absorption coefficient α in cm^{-1}.

TABLE VII
Parameters of the Hat-Curved Model Pertinent to Liquid Water

		Fitted Parameters				Estimated		
Liquid	T (K)	τ (ps)	u	β (deg)	f	m_{lib}	$\mathring{r}$ (%)	ν_L^{est} (cm^{-1})
H_2O	300.15	0.28	5.6	21	0.830	5.6	5.3	680
H_2O	295.35	0.28	5.7	21	0.724	5.5	5.8	690

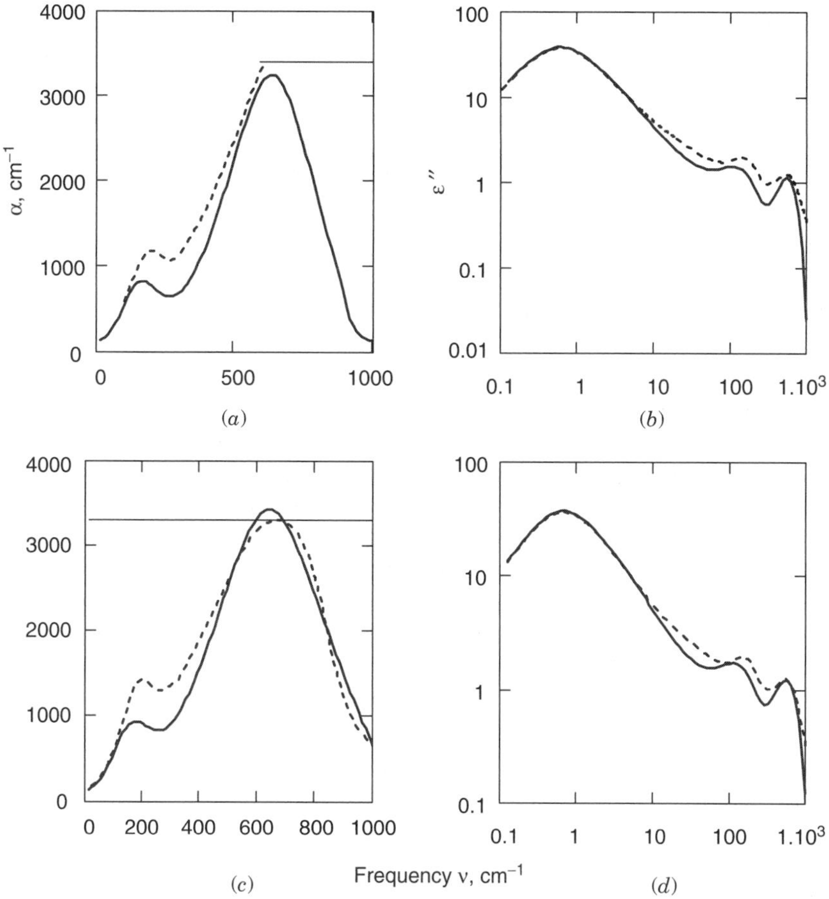

Figure 24. Absorption coefficient (a, c) and wideband diecltric loss (b, d) calculated for liquid H_2O water at 22.2°C (a, b) and 27°C (c, d) for the hat-curved model (solid lines). The experimental $\alpha(\nu)$ dependencies [17, 42, 56] are shown by dashed lines. The horizontal lines in Figs. (a) and (c) denote the maximum absorption recorded in the librational band.

τ, and experimental peak-absorption frequency ν_{lib}. On the other hand, using the former [Eqs. (150) and (151)], we may express one more parameter, say the lifetime τ, through u, β, and experimental value of the Debye relaxation time.

Figure 24 demonstrates by solid lines the results of the calculations; dashed lines show the experimental spectra. We comment on Fig. 24 as follows.

(a) A two-humped absorption curve $\alpha(\nu)$ typical for water appears in our theory, see Fig. 24a and 24c, if the rectangular-well depth U_0 is much greater than $k_B T$ (in our examples u is about 5.6). The greater u, the lower is frequency

$$\nu_D^{\text{est}} = \left(2\pi c \tau_D^{\text{est}}\right)^{-1} \tag{196}$$

of the principal Debye-loss peak ε'', estimated from Eq. (150). This peak is seen on the left-hand sides of Figs. 24b and 24d. Its position ν_D depends also on the lifetime: The lower the τ-value, the greater the frequency $\nu_{\text{D}}^{\text{est}}$, therefore the shorter the estimated Debye relaxation time.

(b) The condition (151) imposed on the Debye relaxation time holds, if the librational lifetime τ is much shorter than the Debye relaxation time τ_{D}—that is, if τ_{D}/τ is about 30 (at room temperature).

(c) The librational absorption peak ν_{lib} fits the experimental data, if the angle β is relatively small (about 21°). The lower the value of β, the greater the theoretical frequency [Eq. (190a)] of this peak. Its intensity fits approximately the experimental data, when the Fröhlich–Kirkwood expression [40] for a dipole moment approximately holds—that is, if the μ-correcting coefficient $k\mu$ is close to 1 [see Eq. (145)]. It is important that the formulas (190), derived from rough consideration, indeed work well.

(d) Two smaller FIR loss peaks ε'' seen on the right-hand sides of Figs. 24b and 24d correspond to the R- and librational absorption bands, respectively.

It follows from Table VII that when the temperature rises, the fitted lifetime τ and libration amplitude β practically do not alter; the reduced well depth u and the number m_{lib} of the librational cycles, performed during this time τ, increase only slightly; and the form factor f increases noticeably. These changes, stipulated by a *weakening of the water structure* occurring with an increase of T, lead to a specific property of a liquid state (as opposed to a gas state), namely a *decrease with T of the libration peak frequency* ν_L. Comparing the solid and dashed curves in Fig. 24, we ascertain that the theoretical and experimental spectra *qualitatively* agree.

In the R-band region (i.e., at 150–300 cm^{-1}) the disagreement between the calculated and experimental absorption spectra arises due to some fundamental

TABLE VIII
Fitted and Estimated Parameters of the Hat-Curved Model Pertaining to Liquid Fluoromethane CH_3F

	Fitted			Estimated	
β (deg)	τ (ps)	u	f	m_{lib}	$\mathring{r}$ (%)
22	0.41	6	0.96	1.0	2.5

reason. This drawback will be discussed (and removed) in Sections VI–X by using various approaches.

2. *Dielectric/FIR Spectra of Nonassociated Liquid (CH_3F)*

As a second example, we consider liquid fluoromethane CH_3F, which is a typical strongly absorbing nonassociated liquid. For our study we choose the temperature $T = 133$ K near the triple point, which is equal to 131 K. The relevant experimental data [43] were summarized in Table IV. As we see in Table VIII, which presents the fitted parameters of the model, the angle β is rather small. At this temperature the density ρ of the liquid, the maximum dielectric loss ε''_D, and the Debye relaxation time τ_D are substantially larger than they would be, for example, near the critical temperature (at 293 K). At such small β the theory given here for the hat-curved model holds. For calculation of the complex permittivity $\varepsilon^*(\nu)$ and absorption $\alpha(\nu)$, we use the same formulas as for water.

The calculated spectra are illustrated by Fig. 25. In Fig. 25a we see a quasi-resonance FIR absorption band, which, unlike water, exhibits only one maximum. Figure 25b demonstrates the calculated and experimental Debye-relaxation loss band situated at microwaves. Our theory satisfactorily agrees with the recorded $\alpha(\nu)$ and $\varepsilon''(\nu)$ frequency dependencies. Although the fitted form factor f is very close to 1 ($f \approx 0.96$), the hat-curved model gives better agreement with the experiment than does a model based on the rectangular potential well, where $f = 1$ (see Section IV.G.3).

3. *Concluding Remarks*

Evident progress in studies of liquids has been achieved up to now with the use of computer simulations and of the models based on analytical theory. These methods provide different information and are mutually complementary. The first method employs rather rigorous potential functions and yields usually a chaotic picture of the multiple-particle trajectories but has not been able to give, as far as we know, a satisfactory description of the wideband spectra. The analytical theory is based on a phenomenological consideration (which possibly gives more regular trajectories of the particles than arise in "reality") in terms of a potential well. It can be tractable only if the profile of such a well is rather

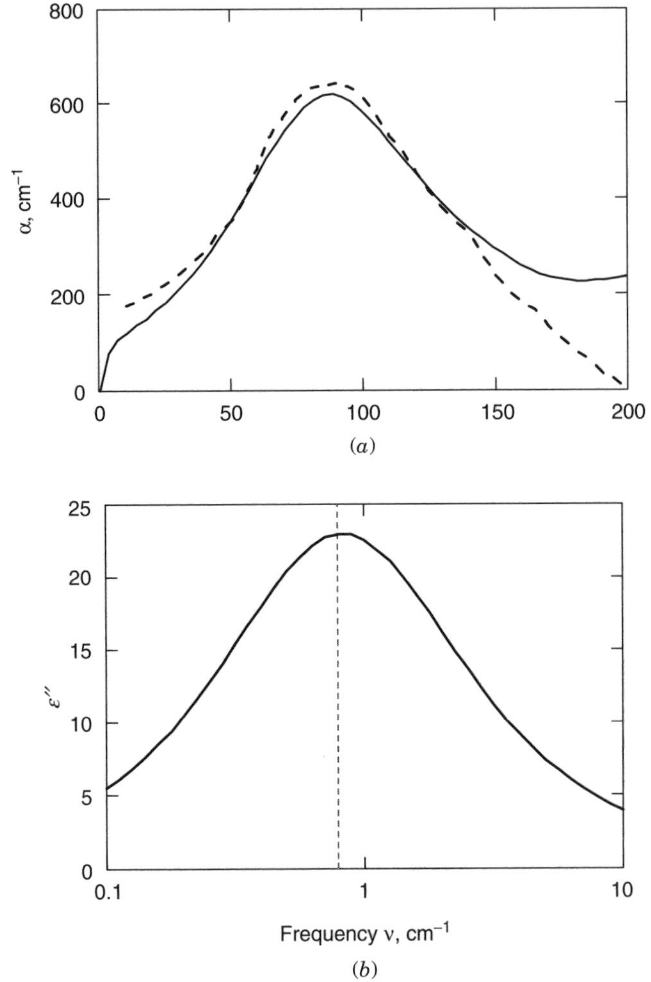

Figure 25. Frequency dependence of the absorption coefficient (a) and dielectric loss (b). Liquid fluoromethane CH_3F at 133 K calculated for that hat-curved model (solid lines). Dashed curve in Fig. (a) refers to the experimental [43] data, vertical line in Fig. (b) marks the experimental position of the maximum dielectric loss. The parameters of the hat-curved model are presented in Table VIII.

simple. In the famous Debye relaxation theory [52] a motion of a *single* dipole is actually considered. It is very surprising that such a primitive model is applicable to water. The reason for this is clearly explained by Agmon [35]. After a profound analysis of the Debye model, it was stated in Ref. 35 that the translational hops of a dipole from an occupied to an unoccupied site inside a tetrahedron of five water molecules are inherently related to a *single-molecule* reorientation, which occurs simultaneously with this translation. The hat-curved model widely extends the correctness of this viewpoint but on a more general physical basis, since rather short periods of librational motions of a molecule are superimposed on a slow Debye relaxation process. As a result, our approach is valid far away from the relaxation region—that is, also in the librational band (from 300 to 1000 cm^{-1}). The employed hat-curved potential allows us to decrease the width of the librational absorption band as compared with that found in previous studies [VIG, 49, 3, and others]; compare with the results described in Sections IV.G.2 and IV.G.3. We have obtained better agreement of theory and experiment, since the spectrum stipulated by to-and-fro motions of a dipole becomes narrower in such a well.

We recognize that it would be interesting to find a "self-consistent" form of the well starting from the near-order structure of a liquid. A preliminary attempt was undertaken recently: We considered *attractive* string-like interactions in an ensemble of the H-bonded molecules [12, 12a, 12b]; see Section IX. This way of modeling may open new perspective for modeling of dielectric relaxation in aqueous media.

The main results of this section are follows.

1. A semiphenomenological molecular model is described, in which a flat-bottom part of the potential well is followed by the parabolic part and then by a flat rim, with the depth and angular width of the well being, respectively, U_0 and β (Fig. 20). For a rather deep and narrow well (with $U_0 \gg k_B T$ and $\beta \ll 1$) a simple analytical linear-response theory is given. The essence of this theory is expressed by ordinary integrals over energy, Eq. (170), which determine a dipole autocorrelator (spectral function L) as a function of the free parameters (195) of the model. The spectral function (194a) accounts for the contributions $\breve{L}(z)$ and $\mathring{L}(z)$ to $L(z)$, respectively, of the "librating" and the "rotating" particles. This calculation scheme is valid for a rather large form factor ($f < 0.65$). If the well is rounded to a greater extent (i.e., if $f < 0.65$), then the third term (174), corresponding to a precession of the dipoles, possibly should also be accounted for.

The complex susceptibility χ^*, generated by the motions of polar molecules in an adopted static potential (160), is connected with the spectral function $L(z)$ by Eq. (142); the complex permittivity ε^* and absorption coefficient α of a liquid are determined by Eqs. (136), (137), (141), and (170).

2. Using this calculation scheme, we have found the frequency dependencies $\alpha(\nu)$ and $\varepsilon''(\nu)$ for ordinary (H_2O) water at two temperatures. Figure 24 demonstrates for water a *qualitative* agreement in the calculated and experimental spectra in a very wide frequency band, comprising the microwave Debye relaxation region and the FIR range. This advance in application of the

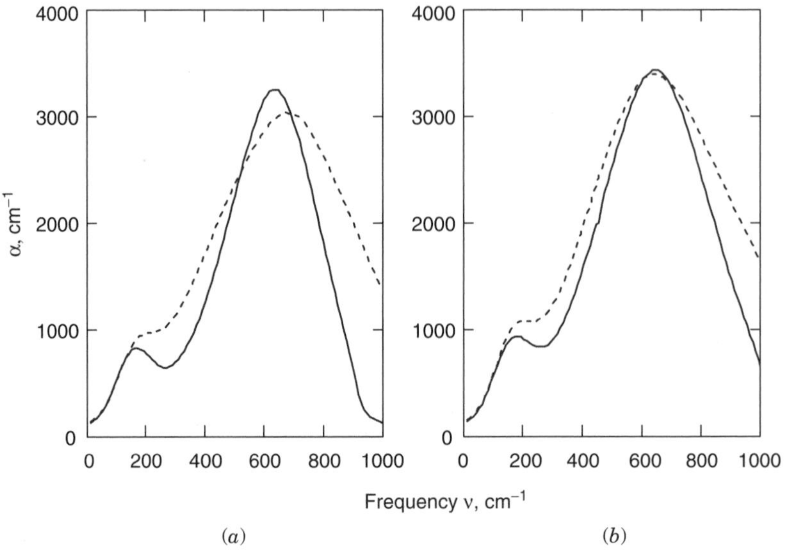

Figure 26. Comparison of the FIR absorption spectra of liquid water calculated for the hat-curved model (solid lines) and the rectangular-well model (dashed lines) with the same u and β values at 22.2°C (a) and 27°C (b). The parameters u, β, τ, f of the hat-curved model are given in Table VII.

rounded potential well has brought an important new result: a *narrowing of the FIR absorption band* as compared with application of the rectangular well. This statement becomes evident when we examine the $\alpha(\nu)$ water spectra (see Fig. 26) calculated with the same set of the parameters (195): (i) for the hat-curved model (solid lines) and (ii) for the rectangular-well potential model[37] (dashed lines). In Section VI we shall strengthen this statement by an example of another polar fluid (liquid D_2O).

3. The *hat-curved* model also gives a satisfactory description of the wideband dielectric/FIR spectra of a nonassociated polar fluid (CH_3F) (Fig. 25). It is worthwhile mentioning that only a poor description of the low-frequency (Debye) spectrum could be accomplished, if the *rectangular* potential were used for such a calculation [32]; see also Section IV.G.3. Unlike Fig. 25b, the estimated peak-loss frequency does not coincide[38] in this case with the experimental frequency ν_D.

[37]For the case $f = 1$ (to which the dashed lines in Fig. 26 correspond), the formulas for the SF $L(z)$ are given in Section IV.E.

[38]The estimated frequency ν_D is 1.22 cm^{-1}, the experimental [43] one is 0.8 cm^{-1}.

It is important that the form of the wideband dielectric/FIR spectrum in CH_3F resembles that in liquid water and thus could be described in terms of the same hat-curved potential. However, due to a difference between the structures of both fluids and the near-order lifetimes, characterizing them, two important distinctions arise:

(a) Only about one libration cycle is performed in CH_3F by a dipole during the lifetime τ of the HC potential. This could be compared with ≈ 6 cycles in the case of water (cf. Tables VII and VIII).
(b) The low-frequency R-absorption band does not arise in CH_3F unlike in water.
(c) An important reason for this difference is absence of the H bonds in liquid CH_3F.

4. Thus, evolution of semiphenomenological molecular models mentioned in Section V.A (items 1–6) have led to the hat-curved model as a model with a rounded potential well. This model combines useful properties of the rectangular potential well and those peculiar to the "field" models based on application of the parabolic, cosine, or cosine-squared potentials. Namely, the hat-curved model retains the main advantage of the rectangular-well model—its possibility to describe both the librational and the Debye-relaxation bands.

On the other hand, the hat model acquires a useful property of the field models—a possibility to describe also a rather *narrow* absorption band. The form of the latter depends on the form factor f. It appears that the $(1 - f)$ value characterizes the *spread* of the intermolecular interactions. In the case of an associated liquid (H_2O) characterized by rather long-range interactions, f is about 0.7–0.75. On the contrary, in the case of a nonassociated liquid, for which a short-range interactions are characteristic, f is close to 1 ($f \approx 0.96$).

5. In this chapter, only *four* parameters are used for the description of a very wide frequency band (from 0 to 1000 cm^{-1}). On the other hand, we show that this approach of rigid dipoles cannot describe some fine experimental features. The difference between the calculated and the measured spectra of water, seen in Figs. 24a and 24b in the range of the submillimeter wavelengths (viz. at [10–300 cm^{-1}]), appears to be of *principal importance*. This difference is hardly possible to eliminate by further correcting the form of intermolecular potential. We shall show in the next section that consideration of a single-molecule rotation of H_2O in the hat-curved potential well could be successfully added by taking approximately into account the *collective (cooperative) effects* in water. For such an improvement of our model a few new free model parameters will be introduced and the calculation scheme should be slightly modified. We shall show that the so-corrected hat-curved model could be also applied to heavy (D_2O) water.

E. Spectral Function (Derivation)

1. *Norm C of Boltzmann Distribution* $W = C\,exp[-h(\Gamma)]$

We define the norm C of this distribution by using the normalizing condition

$$\int_\Gamma W(h)\,d\Gamma = C\int_\Gamma \exp(-h)\,d\Gamma = 1 \tag{197a}$$

We represent the phase-volume element $d\Gamma$ in the form $d\Gamma = dh\,dl\,d\varphi_0$, where h and φ_0 are the pair of reduced canonically conjugated variables, φ_0 is an "initial" instant that enters in addition to time φ in the law of motion of a dipole. Another pair of canonical variables is l, ϕ_0; we omit differential $d\phi_0$ in $d\Gamma$, since the variables we use do not depend on the azimuthal coordinate ϕ_0.

Let $\breve{\Sigma}$, $\tilde{\Sigma}$, and $\mathring{\Sigma}$ denote the partial statistical integrals corresponding, respectively, to librational, precessional, and rotational subensembles, which are defined as integrals of the Boltzmann factor exp($-h$) over corresponding phase volumes:

$$\breve{\Sigma} \equiv \int_{\breve{\Gamma}} \exp(-h)\,dh\,dl\,d\varphi_0,$$

$$\tilde{\Sigma} \equiv \int_{\tilde{\Gamma}} \exp(-h)\,dh\,dl\,d\varphi_0, \quad \text{and} \quad \mathring{\Sigma} \equiv \int_{\mathring{\Gamma}} \exp(-h)\,dh\,dl\,d\varphi_0 \tag{197b}$$

Then it follows from (197a) and (197b) that the norm C is given by

$$C = \Sigma^{-1}, \qquad \text{where } \Sigma = \breve{\Sigma} + \tilde{\Sigma} + \mathring{\Sigma} \tag{197c}$$

The integrals (197b) will be calculated in Section V.E.7 in the planar libration–regular precession (PL–RP) approximation as functions of the parameters u, β, f of the hat-curved model. For further consideration it is convenient to introduce the coefficient

$$D = \frac{u}{\pi C} = \frac{u\Sigma}{\pi} = \frac{u}{\pi}\left(\breve{\Sigma} + \tilde{\Sigma} + \mathring{\Sigma}\right) \tag{197d}$$

2. *Phase Regions of Subensembles*

a. Boundaries for Librators. In view of Fig. 21a and Eq. (165) the phase region $\mathscr{L}$ is restricted by two straight lines AB and 0B; see Fig. 27a, where coordinates h, l^2 are employed. In this case, $V(\beta f) \le V(\beta)$ and

$$\frac{\breve{l}^2}{\beta^2 f^2} \le \breve{h} \le u + \frac{\breve{l}^2}{\beta^2}, \qquad 0 \le \breve{l}^2 \le u\left(\frac{1}{\beta^2 f^2} - \frac{1}{\beta^2}\right)^{-1} \tag{198}$$

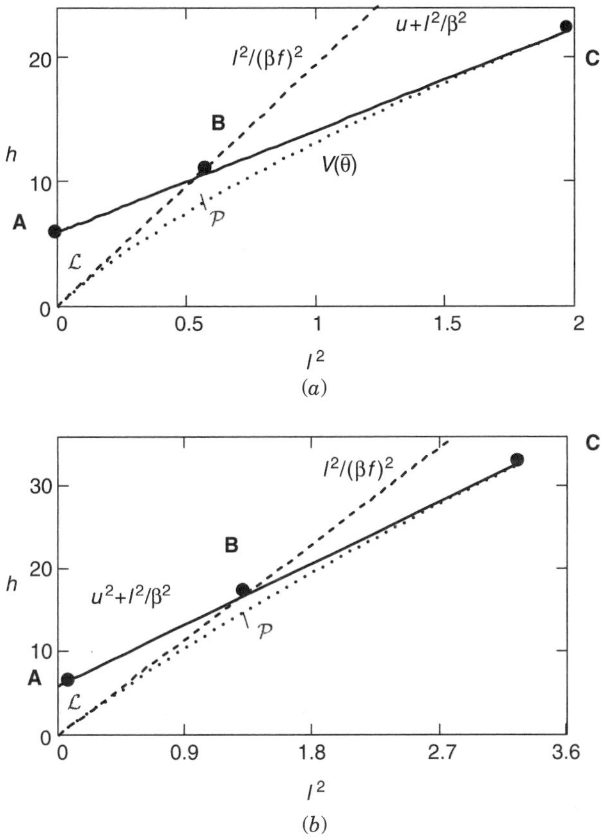

Figure 27. Evolution of the phase regions occupied by the librators ($\mathscr{L}$) and by the precessors ($\mathscr{P}$) with the increase of the form factor f from 0.65 (a) to 0.85 (b). The angular width of the hat-curved potential well, β, is $\pi/9$, and the reduced well depth, u, is 5.9.

Intersection of the above-mentioned lines at point B determines the maximum values of $\breve{h}, \breve{l}$, which we denote by $\breve{h}_m, \breve{l}_m$. Since at the intersection point $u + \breve{l}_m^2\beta^{-2} = \breve{l}_m^2\beta^{-2}$, we find

$$\breve{l}_m^2 = u\beta^2 f^2 \left(1 - f^2\right)^{-1} = \frac{u\beta^2 f^2}{1 - f^2}, \qquad \breve{h}_m = u + u\frac{\breve{l}_m^2}{\beta^2} = \frac{u}{1 - f^2} \tag{199}$$

b. Boundaries for Precessors. In view of Fig. 21 and Fig. 27 the precessors occupy the phase region placed between the straight lines 0B, BC and the curve

0C. The latter represents the potential V taken in the position $\theta = \bar{\theta}$ of its minimum. In cases (a) and (b) we have respectively $V_{\min} \le \tilde{h} \le V(\beta f)$ and $V_{\min} \le \tilde{h} \le V(\beta)$, so that the corresponding inequalities become

$$V(\bar{\theta}) \le \tilde{h} \le \frac{\tilde{l}^2}{f^2\beta^2}, \qquad 0 \le \tilde{l}^2 \le u\left(\frac{1}{\beta^2 f^2} - \frac{1}{\beta^2}\right)^{-1} \tag{200a}$$

$$V(\bar{\theta}) \le \tilde{h} \le u + \frac{\tilde{l}^2}{\beta^2}, \qquad \tilde{l}^2 \ge u\left(\frac{1}{\beta^2 f^2} - \frac{1}{\beta^2}\right)^{-1} \tag{200b}$$

The angle $\bar{\theta}$, at which the minimum of $V(\theta)$ occurs, is a function of the parameters l, β, u, f; $\bar{\theta}$ obeys Eq. (166) and varies between the boundaries $f\beta$ and β of the precession region. In view of Eqs. (165) and (166), we obtain

$$V(\bar{\theta}) = l^2\bar{\theta}^{-2} + u\frac{(\bar{\theta} - \beta f)^2}{(\beta - \beta f)^2} = u\frac{(2\bar{\theta} - \beta f)(\bar{\theta} - \beta f)}{\beta^2(1 - f)^2} \tag{201}$$

The point $\{\tilde{l}_m, \tilde{h}_m\}$ corresponds to a maximum value of the polar angle $\bar{\theta}$. Physical arguments, supported by Fig. 21, suggest that $\bar{\theta}_{\max} = \beta$. We can prove the validity of this relation if we take into account that this point in Fig. 27 is at the intersection of the straight line $h(\bar{\theta}) = u + (l/\beta)^2$ and the curve $h(\bar{\theta}) = h_{\min}[l(\bar{\theta})] = V(\bar{\theta})$. The first dependence is found, if we substitute l^2 from (166); the second dependence is found from (201). Thus we have

$$u\left[1 + \frac{\bar{\theta}^3(\bar{\theta} - \beta f)}{\beta^4(1 - f)^2}\right] = u\frac{(\bar{\theta} - \beta f)(2\bar{\theta} - \beta f)}{\beta^2(1 - f)^2}$$

This equation becomes an identity at $\bar{\theta} = \beta$; our suggestion is proven: $\bar{\theta}_{\max} = \beta$.

The phase regions occupied by the librators and precessors are depicted in Fig. 27 in coordinates h, l^2 for the parameters $u = 5.9$, $\beta = \pi/9$. We take two values of the form factor f: 0.65 in (a) and 0.85 in (b). When f increases, the $\mathscr{L}$ and $\mathscr{P}$ areas extend to the larger h and l values. The values of $V_{\min}$ are shown as functions of l in Fig. 21c. In this example (and in the calculations described in Section V.C) the potential well depth U_0 is much greater than $k_B T$, that is, $u \gg 1$. We see in Fig. 27 that for the $\mathscr{L}$ and $\mathscr{P}$ areas the boundary values for h are still greater than u. This property is used to simplify analytical expressions for the spectral functions.

3. General Representation of Spectral Functions

Using the ACF method (GT, VIG), we shall derive the analytical expressions for the spectral function (SF) $L(z)$ of an isotropic ensemble of polar molecules. We choose two pairs of the canonical variables as $\{h, \varphi_0\}$ and $\{l, \phi_0\}$. Here φ_0 is a constant of integration, which is additive to the reduced time variable φ in the law of motion $\theta(\varphi)$; l is the reduced projection of the angular moment on the symmetry axis; and ϕ_0 is an "initial" value of the variable (azimuthal coordinate) ϕ. Due to periodic-time dependencies of the coordinates θ and ϕ, we may generally represent the integrand of Eq. (158) for the correlator $L(z)$ in the form of a Fourier series. Averaging over the phase variables φ_0 and ϕ_0 can be accomplished in a standard fashion (GT, VIG); the results are denoted *here* by $\breve{Q}(h,l)$ and $\tilde{Q}(h,l)$. The expression for the spectral function can be found, if we take the integral over two other phase variables, h and l. Then we can represent the first two spectral functions in Eq. (170), for the *librators* and *precessors*, in the form

$$\breve{L}(z) = 2\int_{[\breve{h},\breve{l}]} \breve{Q}(h,l)W(h)\,dh\,dl, \qquad \tilde{L}(z) = 2\int_{[\tilde{l},\tilde{h}]} \tilde{Q}(l,h)W(h)\,dl\,dh \quad (202)$$

The multiplier 2 here accounts for existence of two potential wells with oppositely directed symmetry axes (the case of mirror symmetry is assumed). The derivation of the spectral function of the *rotators*, namely the last term in Eq. (170), will be given in Section V.E.6.

In accord with the planar libration (PL)–regular precession (RP) approximation, in the subensemble of the librators we set $l = 0$,

$$\breve{Q}(h,l) \cong \breve{Q}(h,0) \equiv \breve{Q}(h) \quad (203)$$

As applied to the subensemble of the precessors, for any given energy h we choose in $\breve{Q}(h,l)$ the maximum value $l = l_{\max}(h)$ of the reduced angular momentum. Indeed, as it follows from Eq. (165) that if we regard l as a function of the reduced energy h, then l attains its maximum at $\dot{\theta} = 0$. In our regular precession (RP) approximation $\dot{\theta}$ vanishes at a fixed angle $\theta \equiv \bar{\theta}$, to which a minimum energy $h = h_{\min} \equiv V(\bar{\theta})$ corresponds. Thus, in the law of motion of the precessors we set $\theta(\varphi, l, h) \approx \theta[\varphi, l_{\max}(h), h]$ and correspondingly

$$\tilde{Q}(l,h) \cong \tilde{Q}[l_{\max}(h),\, h] \equiv \tilde{Q}(h) \quad (204)$$

Consequently, we rewrite Eqs. (202) as

$$\breve{L}(z) = 2\int_{[\breve{h},\breve{l}]} \left[\breve{Q}(h)\right] W(h)\, dh\, dl \tag{205a}$$

$$\tilde{L}(z) = 2\int_{[\tilde{h},\tilde{l}]} \tilde{Q}(h)\, W(h)\, dl\, dh \tag{205b}$$

For the subensemble of rotators we use a similar representation

$$\mathring{L}(z) = \int_{[\mathring{h},\mathring{l}]} \mathring{Q}(h) W(h)\, dh\, dl \tag{205c}$$

the multiplier 2 is not needed here, since this subensemble embraces both potential wells.

Furthermore, we shall omit the symbols $\sim$, $\smile$, and $^\circ$, if this will not lead to a confusion.

4. *Spectral Function of Librators*

In accord with the planar libration approximation, we first come from representation of the spectral function for motion of a dipole *in a plane*, where integration over l is lacking by definition, so only integration over energy h is employed. We shall find in this way the function (203). As a next step we carry out integration over l, so that a rather simple expression (171) for the spectral function $\breve{L}(z)$ will be obtained.

For motion *in a plane* the spectral function $L_{\text{plan}}(z)$ of an isotropic ensemble of polar molecules is represented [GT, Eqs. (3.72), (3.36)] as a sum of two components, $K_{\parallel}(z)$ and $K_{\perp}(z)$, produced by a periodic time dependence of the projections $\mu_{\parallel}$ and $\mu_{\perp}$ of the dipole moment $\boldsymbol{\mu}$. Here $\mu_{\parallel}$ and $\mu_{\perp}$ are, respectively, the projections of $\boldsymbol{\mu}$ collinear and perpendicular to the symmetry axis of the potential. In a small β approximation the axial component $K_{\parallel}(z)$ could be neglected. It is convenient to represent the other component $K_{\perp}(z)$ in a form of a trigonometric series [VIG, Eq. (5.88); or GT, Eq. (3.36)]:

$$L_{\text{plan}}(z) \approx K_{\perp}(z) = \frac{\pi^2}{2}\int_{h_{\min}}^{h_{\max}} \frac{dh\; W(h)}{\Phi(h)} \sum_{n=1}^{\infty} \frac{(2n-1)^2 b_{2n-1}^2}{(2n-1)^2\left[\frac{\pi}{\Phi(h)}\right]^2 - z^2} \tag{206}$$

Here $\Phi(h)$ is the period[39] of $\mu_{\parallel}(\varphi)$, b_{2n-1} are the Fourier amplitudes in the trigonometric expansion [GT, Eq. (3.33b)] of $\mu_{\perp}(\varphi)/\mu \equiv \sin\theta(\varphi) \approx \theta(\varphi)$:

$$\theta(\varphi, h) = \sum_{n=0}^{\infty} b_{2n-1} \sin\frac{(2n-1)\pi\varphi}{\Phi(h)}, \quad b_{2n-1} \approx \frac{4}{\Phi(h)}\int_0^{\frac{1}{2}\Phi(h)} \theta(\varphi, h) \sin\left[\frac{(2n-1)\pi\varphi}{\Phi(h)}\right] d\varphi \tag{207}$$

[39]In GT, pp. 155–157, $\Phi(h)$ is denoted as $\breve{\Phi}$ and the librational dimensionless frequency $\breve{p} = \pi/\breve{\Phi}$ is introduced.

Comparing (206) with (205a) we should set

$$\breve{Q}(h) = \frac{\pi^2}{2\Phi(h)} \sum_{n=1}^{\infty} \frac{(2n-1)^2 b_{2n-1}^2}{(2n-1)^2[\pi/\Phi(h)]^2 - z^2} \tag{208}$$

The function (208), found for motion in a plane with $l \equiv 0$, is employed in this approximation for an ensemble of dipoles, characterized by both phase variables, h and l, distributed in the phase region (198).

To find the period $\Phi(h) \equiv \breve{\Phi}(h)$, we set $\theta = 0$ at $\varphi = 0$ and employ Eq. (169). We have

$$\frac{\breve{\Phi}(h)}{2} = \frac{\beta f}{\sqrt{h}} + \frac{\pi}{2S} = \frac{\beta f}{\sqrt{h}}[1 + \sigma(h)], \qquad \sigma(h) \equiv \frac{\pi}{2}\frac{1-f}{f}\sqrt{\frac{h}{u}} \tag{209}$$

since a *half* of this period comprises

(i) The time $\beta f/\sqrt{h}$, during which a dipole, rotating freely with angular frequency $\sqrt{h}$, turns over the well's flat part βf

(ii) The time $\pi(2S)$ equal to 1/4 of the period of harmonic oscillations which a dipole would perform in the parabolic potential with the angular frequency S

Within the potential well the equations of motion in the PL approximation are

$$\begin{aligned} &\ddot{\theta} = 0, \quad \text{so that} \quad \theta(\varphi) = \sqrt{h}\varphi \qquad \text{at } 0 \le \theta \le \beta f \\ &\ddot{\theta} = -S^2(\theta - \beta f) \qquad \text{at } \beta f \le \theta \le f \end{aligned} \tag{210}$$

The derivation given in Appendix 1 yields

$$b_{2n-1} = \frac{4(\Phi S)^2 \Phi\sqrt{h}}{\pi^2(2n-1)^2} \frac{\sin\left[(2n-1)\pi\frac{\beta f}{\sqrt{h}\Phi}\right]}{[\pi(2n-1)]^2 - (\Phi S)^2} \tag{211}$$

Taking into account that $\Phi S = \pi\frac{1+\sigma}{\sigma}$ and $\Phi\sqrt{h} = 2\beta f(1+\sigma)$, we finally express the amplitudes b_{2n-1} through the parameters of the HC model as follows:

$$b_{2n-1} = -\frac{8\beta f(1+\sigma)^3 \sin\left(\frac{2n-1}{2}\frac{\pi}{1+\sigma}\right)}{\pi^2[(2n-1)^2\sigma^2 - (1+\sigma)^2]} \tag{212}$$

Combining this with Eqs. (208) and (209) we have

$$\breve{Q}(h) = \frac{64}{\pi^4}(\beta f)^3(1+\sigma)^7\sqrt{h}\sum_{n=1}^{\infty} \times \frac{\sin^2\left(\frac{2n-1}{2}\frac{\pi}{1+\sigma}\right)}{(2n-1)^2[(2n-1)^2\sigma^2-(1+\sigma)^2]^2\{(2n-1)^2h-[2z(\beta/\pi)f(1+\sigma)]^2\}} \tag{213}$$

In view of the boundaries (198) and (199) and Fig. 27, Eq. (205a) can be written as follows:

$$\begin{aligned} \breve{L}(z) &= 2C\left[\int_0^u \breve{Q}(h)e^{-h}\,dh\int_0^{\beta f\sqrt{h}} dl + \int_0^{\frac{u}{1-f^2}} \breve{Q}(h)\,e^{-h}\,dh\int_{\beta\sqrt{h-u}}^{\beta f\sqrt{h}} dl\right] \\ &= 2C\beta f\int_0^u \sqrt{h}\,e^{-h}\breve{Q}(h)\,dh + \delta\breve{L}(z) \end{aligned} \tag{214}$$

where

$$\delta\breve{L}(z) \equiv 2C\beta\int_u^{\frac{u}{1-f^2}} \breve{Q}(h)\,e^{-h}(f\sqrt{h}-\sqrt{h-u})\,dh \tag{215}$$

with the normalizing coefficient C given by Eq. (197c). If we retain in Eq. (214) only the first term (it is reasonable, since $u(1-f^2)^{-1} \gg 1$), then we get Eqs. (171)–(173) for the librators' SF.

5. *Spectral Function of Precessors*

Unlike Section V.E, now we consider directly the spectral function of an isotropic *spatial* ensemble, for which an average over the phase variable h and l should be found. In this case $L(z) = K_{\|}(z) + 2K_{\perp}(z)$. Omitting again the first term, we have in view[40] of Eqs. (3.62) and (3.71) in GT:

$$L(z) = 2K_{\perp}(z) = \int_{[h,l]} dh\,dl\frac{4W(h)}{\Phi(h,l)}\sum_{n=-\infty}^{\infty}\frac{[\pi n-\lambda(h,l)]^2 b_n^2(h,l)}{[\pi n-\lambda(h,l)]^2\left[\frac{2}{\Phi(h,l)}\right]^2 - z^2} \tag{216}$$

[40]In GT, p. 173, the notation $\breve{p} \equiv 2\pi/\Phi$ is used for the dimensionless fundamental frequency describing the periodic $q_{\|}(\varphi)$ function.

where Φ is the period of $\mu_{\|}(\varphi), \lambda \equiv \phi(\Phi/2)$ and in accord with GT, p. 167 the Fourier amplitudes b_n, determined as $\frac{1}{2}(c_n - s_n)$, are given by

$$b_n = \frac{1}{\Phi}\int_0^{\Phi} \sin(\theta)\cos\left(\phi_{\text{per}} + \frac{n\pi\varphi}{\Phi}\right) d\varphi \approx \frac{1}{\Phi}\int_0^{\Phi} \bar{\theta}\cos\left(\frac{n\pi\varphi}{\Phi}\right) d\varphi = \bar{\theta}$$
$$\text{at } n = 0; \qquad b_n = 0 \quad \text{at } n \neq 0 \tag{217}$$

Here $\phi_{\text{per}}$is a part of the azimuthal coordinate ϕ, periodically changing in time and which on the conditions of the RP approximation vanishes, since in this approximation $\phi(\varphi)$ is proportional to time φ and $\theta(\varphi) \equiv \bar{\theta}$. Comparing Eqs. (216) and (205b), we find

$$\tilde{Q}(h) = \frac{4}{\Phi(h)}\frac{\lambda^2(h)\bar{\theta}^2(h)}{[2\lambda(h)/\Phi(h)]^2 - z^2} = \frac{\Phi(h)[l^2/\bar{\theta}^2(h)]}{l^2/\bar{\theta}^4(h) - z^2} \tag{218a}$$

The last equality follows from Eqs. (163) and (164) and from the definition of λ, since for regular precession (RP) we should set $\lambda = \frac{1}{2}\dot{\phi}\,\Phi = \frac{1}{2}\Phi(h)[l/\bar{\theta}^2(h)]$. We denote the l value in integrand of Eq. (205b) by $l \equiv l_\phi$. This l_ϕ corresponding to the energy $h = V_{\min}(\bar{\theta})$, in our RP approximation is determined by Eq. (166). Thus,

$$l_\phi^2 = \frac{u\bar{\theta}^3(\bar{\theta} - \beta f)^2}{\beta^2(1-f)^2} \tag{218b}$$

Restoring again the symbol $\sim$ for the precessors, so that $\Phi \equiv \tilde{\Phi}$, we rewrite Eq. (218a) as

$$\tilde{Q}(h) = \frac{4}{\tilde{\Phi}(h)}\frac{\lambda^2(h)\bar{\theta}^2(h)}{[2\lambda(h)/\tilde{\Phi}(h)]^2 - z^2} = \frac{\bar{\theta}^2(h)\tilde{\Phi}(h)[l^2/\bar{\theta}^4(h)]}{l^2/\bar{\theta}^4(h) - z^2} \tag{219}$$

Taking into account Eqs. (199)–(201) and neglecting the contribution to $\tilde{L}(z)$ for the part of the $\mathscr{P}$ region, where $h > h_m$, we represent (205b) as

$$\tilde{L}(z) \cong 2\int_0^{\frac{u}{1-f^2}} \tilde{Q}(h)W(h)\,dh \int_{\sqrt{h}\beta f}^{l_\phi(h)} dl = 2\int_0^{\frac{u}{1-f^2}} \tilde{Q}(h)[l_\phi(h) - \sqrt{h}\beta f]W(h)\,dh \tag{220}$$

We shall find now the period $\tilde{\Phi}$, involved in Eq. (218a), for the motion of dipole in the effective potential $V(\theta) \approx V(\bar{\theta})$ with an energy h *slightly different* from $h_{\min} = V(\bar{\theta})$. (Otherwise the period has no physical sense.) In view of Eqs. (165) and (166) we have

$$V(\theta) \approx l^2\bar{\theta}^{-2}\left[1 + 2\frac{\theta - \bar{\theta}}{\bar{\theta}} + \left(\frac{\theta - \bar{\theta}}{\bar{\theta}}\right)^2\right]^{-1} + S^2(\theta - \bar{\theta} + \bar{\theta} - \beta f)^2$$

$$\times V(\bar{\theta}) + \frac{u}{(1-f)^2\beta^2}\frac{4\bar{\theta} - \beta f}{\bar{\theta}}(\theta - \bar{\theta})^2$$

$$\approx V(\bar{\theta}) + 2[S^2(\bar{\theta} - \beta f) - l^2\bar{\theta}^{-3}](\theta - \bar{\theta}) + (3l^2\bar{\theta}^{-4} + S^2)(\theta - \bar{\theta})^2 \tag{221}$$

Thus, in such a potential the harmonic oscillations occur with the period $\tilde{\Phi}$ given by

$$\tilde{\Phi} = 2\pi(1-f)\beta\sqrt{\frac{\bar{\theta}}{(4\bar{\theta} - 3\beta f)u}} \tag{222}$$

Since we have no explicit expression for the function $\bar{\theta}(h)$, it is convenient to replace the integration over h in Eq. (220) for integration over other variable ψ related to $\bar{\theta}$. We take

$$\psi(\bar{\theta}) = [\bar{\theta} - f\beta][\beta(1-f)]^{-1} \tag{223}$$

so that

$$\bar{\theta}(\psi) = \beta[f + (1-f)\psi] \tag{224}$$

Equating the left-hand side of Eq. (201) to h, we have in accord with Eqs. (223) and (224)

$$h(\psi) = \frac{u\psi}{1-f}[2(1-f)\psi + f] \tag{225}$$

Then, as shown in Appendix 2, integral (220) yields the spectral function (174).

6. *Spectral Function of Rotators*

We use again the general theory concerning a spatial ensemble of linear molecules. In an isotropic ensemble $K_{\parallel} = K_{\perp}$, then $L(z) = 3K_{\parallel}(z)$. Now $\Phi \equiv \mathring{\Phi}$. In accord with Eq. (5.90) in VIG and with Eq. (205c) we have

$$\mathring{Q} = 3\frac{2}{\Phi(h,l)}\sum_{n=1}^{\infty}\frac{(\pi n a_n)^2}{[2\pi n/\Phi(h,l)]^2 - z^2} \tag{226}$$

where, taking into account Eqs. (7.1), (7.2), and (7.10) in VIG we should set

$$p = 2\pi/\mathbf{\Phi} \quad \text{and} \quad a_1 = \sqrt{1 - l^2/h}, \qquad a_n \equiv 0 \quad \text{for } n > 1 \tag{227a}$$

$$\overset{\circ}{\Phi} = \Phi = 2\pi/\sqrt{h} \tag{227b}$$

where p is the normalized rotation frequency and Φ is the corresponding period. These formulae were derived in VIG for free rotation in a *zero* potential. In such an ensemble the squares of the reduced frequency p and the maximal l value are equal to the reduced kinetic energy h, which in this case is equal to full energy. We observe now that in our hat-curved model the reduced potential energy, being constant, is equal to u, so Eqs. (227a) and (227b) can be applied to our model, if we replace h by $h - u$ and take $l_{\max} = \sqrt{h-u}$ as the upper limit of the integral over l (its lower limit $l_{\min} = 0$). In view[41] of Fig. 27, the lower limit of the integral over h is u. Inserting these limits into Eq. (205c), using Eq. (197) for the norm C, and the corrected expression for $\overset{\circ}{\Phi}$, we have

$$\overset{\circ}{\Phi} = \frac{2\pi}{\sqrt{h-u}} \tag{228a}$$

$$\overset{\circ}{Q} = 3\frac{2}{2\pi}\sqrt{h-u}\,\frac{\pi^2\left(1 - \frac{l^2}{h-u}\right)}{h-u-z^2} \tag{228b}$$

$$\overset{\circ}{L}(z) = 3K_{\parallel}(z) = \frac{3u}{\pi D}\int_u^{\infty} \exp(-h)\,dh \int_0^{\sqrt{h-u}} \left(1 - \frac{l^2}{h-u}\right)\frac{\sqrt{h-u}}{\pi(h-z^2)}\,dl \tag{228c}$$

Integration over l gives the formula (179) for the spectral function $\overset{\circ}{L}$ of the rotators.

7. *Statistical Averages*

Now we shall calculate the integrals (197b). Taking into account that integration over φ_0 reduces to multiplication by the period Φ, we rewrite (197b) by analogy with (205):

$$\breve{\Sigma} \equiv 2\int_{[\breve{h},\breve{l}]} \breve{\Phi}\exp(-h)\,dh\,dl; \quad \tilde{\breve{\Sigma}} \equiv 2\int_{[\tilde{h},\tilde{l}]} \breve{\Phi}\exp(-h)\,dh\,dl\,d\varphi_0,$$

and
$$\overset{\circ}{\Sigma} \equiv \int_{[\overset{\circ}{h},\overset{\circ}{l}]} \overset{\circ}{\Phi}\exp(-h)\,dh\,dl\,d\varphi_0 \tag{229a}$$

[41] In this figure, only the phase region of the hindered rotators is shown. For the energy of free rotors we should take the same lower limit ($h_{\min} = u$).

Comparing this with (205a) we find that calculation of the integrals (229a) is analogous to calculation of the corresponding spectral functions, if to replace $C\breve{Q}, C\tilde{Q}, C\mathring{Q}$ respectively by $\breve{\Phi}, \tilde{\Phi}, \mathring{\Phi}$.

For the librators by analogy with (214), omitting the second (small) integral in $[\cdot]$ and using Eq. (209) for the period $\breve{\Phi}$, we have

$$\breve{\Sigma} \equiv 2 \int_{[\breve{h},\breve{L}]} \breve{\Phi} \exp(-h)\, dh \int_0^{\beta f\sqrt{h}} dl = 4\beta^2 f \int_0^u \left[f + \frac{\pi}{2}(1-f)\sqrt{\frac{h}{u}} \right] \exp(-h)\, dh \tag{229b}$$

Correspondingly for the *precessors* by analogy with (220) we write

$$\tilde{\Sigma} \equiv 2 \int_0^{\frac{u}{1-f^2}} \exp(-h)\, dh \int_{\beta f\sqrt{h}}^{l_{\max}(h)} \tilde{\Phi}(h)\, dl = 2 \int_0^{\frac{u}{1-f^2}} \exp(-h)\, \tilde{\Phi}(h) \left\{ l_{\max}(h) - \beta f\sqrt{h} \right\} dh$$

Changing integration over h for integration over ψ we have

$$\tilde{\Sigma} = 2 \int_0^{\psi_{\max}} \exp[-h(\psi)] \tilde{\Phi}(\psi) \left\{ l_{\max}[h(\psi)] - \beta f \sqrt{h(\psi)} \right\} \frac{\partial h(\psi)}{\partial \psi}\, d\psi \tag{230}$$

where for functions in the integrand we should take Eqs. (222)–(225).

For the rotators we analogously have the following after taking Eq. (228a) into account:

$$\mathring{\Sigma} = \int_u^{\infty} \exp(-h)\, dh \int_0^{\sqrt{h-u}} \frac{2\pi}{\sqrt{h-u}}\, dl = 2\pi \exp(-u) \tag{231}$$

Thus, we have Eq. (180) for D and Eq. (184) for proportions of three sub-ensembles.

APPENDIX 1. CALCULATION OF FOURIER AMPLITUDES b_{2n-1} FOR LIBRATORS

To find the Fourier amplitudes b_{2n-1} we integrate Eq. (207) by parts:

$$\begin{aligned} b_{2n-1} &= \frac{4}{\pi(2n-1)} \left\{ -\theta(\varphi) \cos\left[(2n-1)\pi \frac{\varphi}{\Phi}\right] \Big|_0^{\Phi/2} + \int_0^{\Phi/2} \dot{\theta}(\varphi) \cos\left[(2n-1)\pi\frac{\varphi}{\Phi}\right] d\varphi \right\} \\ &= \frac{4\Phi}{\pi^2(2n-1)^2} \left\{ \dot{\theta}(\varphi) \sin\left[(2n-1)\pi\frac{\varphi}{\Phi}\right] \Big|_0^{\Phi/2} - \int_0^{\Phi/2} \ddot{\theta}(\varphi) \sin\left[(2n-1)\pi\frac{\varphi}{\Phi}\right] d\varphi \right\} \end{aligned}$$

We substitute Eqs. (210) and continue calculations by using the same scheme:

$$b_{2n-1} = \frac{4\Phi S^2 R_{2n-1}}{\pi^2 (2n-1)^2} \tag{A1a}$$

$$R_{2n-1} \equiv \int_{\beta f/\sqrt{h}}^{\Phi/2} [\theta(\varphi) - \beta f] \sin\left[(2n-1)\pi\frac{\varphi}{\Phi}\right] d\varphi \tag{A1b}$$

$$= \frac{\Phi}{\pi(2n-1)}\left\{ -[\theta(\varphi) - \beta f]\cos\left[(2n-1)\pi\frac{\varphi}{\Phi}\right]\Big|_{\beta f/\sqrt{h}}^{\Phi/2} + \int_{\beta f/\sqrt{h}}^{\Phi/2} \dot{\theta}(\varphi) \times \cos\left[(2n-1)\pi\frac{\varphi}{\Phi}\right] d\varphi\right\}$$

$$= \frac{\Phi^2}{\pi^2(2n-1)^2}\left\{ \dot{\theta}(\varphi)\sin\left[(2n-1)\pi\frac{\varphi}{\Phi}\right]\Big|_{\beta f/\sqrt{h}}^{\Phi/2} - \int_{\beta f/\sqrt{h}}^{\Phi/2} \ddot{\theta}(\varphi) \times \sin\left[(2n-1)\pi\frac{\varphi}{\Phi}\right] d\varphi\right\}$$

The first nonintegral term is zero: At lower limit $\theta = \beta f$ and at upper limit $\cos\left[(2n-1)\pi\frac{\varphi}{\Phi}\right] = 0$. The second nonintegral term vanishes only at the upper limit, since its velocity $\dot{\theta}$ vanishes at $\varphi = \Phi/2$ (namely, at the instant when a dipole is reflected from the wall of the well). Thus,

$$R_{2n-1} = \frac{\Phi^2}{\pi^2(2n-1)^2}\left\{ -\sqrt{h}\sin\left[(2n-1)\pi\frac{\beta f}{\sqrt{h}\Phi}\right] + S^2\int_{\beta f/\sqrt{h}}^{\Phi/2} [\theta(\varphi) - \beta f]\sin\left[(2n-1)\pi\frac{\varphi}{\Phi}\right] d\varphi\right\}$$

Combining this with expressions (A1a) and (A1b), we obtain Eq. (211).

APPENDIX 2. TRANSFORMATION OF INTEGRAL FOR SPECTRAL FUNCTION OF PRECESSORS

Expressions (222) and (224) give

$$\tilde{\Phi}(\psi) = \frac{2\pi\beta}{\sqrt{u}}(1-f)\sqrt{\frac{(1-f)\psi + f}{4(1-f)\psi + f}}$$

Taking into account Eqs. (218) and (224) we have

$$\frac{\bar{\theta} - \beta f}{\bar{\theta}} = \frac{(1-f)\psi}{f + (1-f)\psi} \quad \text{and} \quad \frac{l_\phi^2}{\bar{\theta}^4} = \frac{u\psi}{\beta^2(1-f)[f + (1-f)\psi]} \equiv p^2(\psi)$$

Then we find from Eq. (219)

$$\tilde{Q}(\psi) = \frac{\tilde{\Phi} p^2 \bar{\theta}^2}{p^2 - z^2} = 2\pi\beta\sqrt{u}\,\psi\sqrt{\frac{[f + \psi(1-f)]^3}{f + 4\psi(1-f)}}[p^2(\psi) - z^2]^{-1}$$

so integral (220) reduces to

$$\tilde{L}(z) \cong 2C\int_0^{\psi_{\max}} \tilde{Q}(\psi)[l_\phi(\psi) - \beta f\sqrt{h(\psi)}]\exp[-h(\psi)]\frac{\partial h(\psi)}{\partial \psi}\,d\psi \qquad \text{(A2)}$$

where $h(\psi)$ is given by Eq. (225) and consequently

$$\frac{\partial h}{\partial \psi} = \frac{u}{1-f}[4(1-f)\psi + f]$$

while Eq. (218b) yields

$$l_\phi(\psi) = \frac{\bar{\theta}(\psi) - \beta f}{\beta(1-f)}\sqrt{u\,\bar{\theta}^3(\psi)} = \sqrt{\frac{u\psi}{1-f}}\beta[(1-f)\psi + f]^{3/2}$$

The upper limit $\psi_{\max}$ in Eq. (A2) is found from Eq. (225) as

$$\psi_{\max} = \psi\left(h = \frac{u}{1-f^2}\right) = \frac{1}{4(1-f)}\left(\sqrt{f^2 + 8\frac{1-f}{1+f}} - f\right)$$

Substituting all variables into integral (A2) and taking into account Eq. (197d), we finally derive Eqs. (174)–(178) for the spectral function of the precessors.

APPENDIX 3. OPTICAL CONSTANTS OF LIQUID WATER

1. Water H_2O at 1°C, 27°C, and 50°C

In Table IX we present the list of optical constant of ordinary (H_2O) water [42] at 27°C covering very wide range of frequencies (from 10 cm^{-1} until 1000 cm^{-1}). For two other temperatures (1°C and 50°C) we present in Table X such constants recorded in Ref. 53 for a narrower region from 400 cm^{-1} to 820 cm^{-1}. Both tables comprise the absorption maximum of the librational band, and the first one includes also the maximum in the R-band. For lower frequencies we can use the empirical formulas of Liebe et al. [17]. They are represented in Section G.2.a. Note that the absorption coefficient α is determined by the imaginary component

TABLE IX
Real (n) and Imaginary (κ) Parts of the Refraction Index n^* and Absorption α (cm^{-1})[a]

ν	n	κ	α	ν	n	κ	α
10	2.600	1.090	137	510	1.462	0.421	2709
20	2.225	0.718	192	520	1.451	0.423	2779
30	2.150	0.527	210	530	1.441	0.425	2842
40	2.110	0.460	240	540	1.431	0.426	2903
50	2.070	0.438	290	550	1.419	0.427	2964
60	2.040	0.444	360	560	1.407	0.427	3022
70	2.020	0.450	429	570	1.396	0.428	3077
80	2.010	0.466	509	580	1.385	0.427	3126
90	2.000	0.487	594	590	1.372	0.425	3167
100	1.997	0.507	678	600	1.361	0.423	3203
110	1.982	0.532	773	610	1.348	0.420	3234
120	1.960	0.557	872	620	1.335	0.417	3259
130	1.929	0.577	967	630	1.324	0.412	3276
140	1.890	0.593	1065	640	1.313	0.408	3291
150	1.848	0.608	1165	650	1.303	0.403	3301
160	1.801	0.622	1266	660	1.289	0.397	3307
170	1.746	0.629	1358	670	1.277	0.392	3308
180	1.689	0.618	1412	680	1.264	0.386	3307
190	1.640	0.597	1437	690	1.249	0.379	3298
200	1.600	0.571	1445	700	1.236	0.373	3287
210	1.657	0.539	1434	710	1.223	0.365	3263
220	1.542	0.505	1407	720	1.213	0.356	3231
230	1.528	0.469	1364	730	1.201	0.347	3192
240	1.525	0.436	1323	740	1.189	0.338	3150
250	1.529	0.414	1310	750	1.182	0.328	3100
260	1.532	0.398	1311	760	1.171	0.317	3040
270	1.534	0.385	1317	770	1.157	0.305	2969
280	1.537	0.375	1331	780	1.142	0.292	2883
290	1.539	0.368	1351	790	1.138	0.277	2760
300	1.541	0.361	1374	800	1.134	0.260	2618
310	1.543	0.357	1401	810	1.130	0.243	2467
320	1.546	0.353	1432	820	1.130	0.226	2309
330	1.550	0.352	1472	830	1.132	0.208	2143
340	1.552	0.356	1532	840	1.131	0.192	1987
350	1.552	0.359	1593	850	1.132	0.176	1833
360	1.552	0.363	1658	860	1.132	0.159	1692
370	1.549	0.368	1724	870	1.135	0.144	1533
380	1.545	0.372	1793	880	1.139	0.130	1396
390	1.541	0.377	1862	890	1.143	0.118	1270
400	1.537	0.382	1933	900	1.149	0.107	1165
410	1.532	0.386	2004	910	1.156	0.0973	1064
420	1.527	0.390	2072	920	1.162	0.0898	993
430	1.521	0.394	2143	930	1.168	0.0828	927
440	1.515	0.397	2210	940	1.174	0.0764	866
450	1.510	0.401	2280	950	1.181	0.0707	817
460	1.504	0.404	2347	960	1.189	0.0661	770
470	1.496	0.408	2423	970	1.194	0.0622	733
480	1.488	0.411	2494	980	1.202	0.0589	702
490	1.480	0.415	2565	990	1.208	0.0557	673
500	1.470	0.418	2638	1000	1.214	0.0534	651

[a]Liquid water H_2O at 27°C.

Source: Ref. 42.

TABLE X

Real (n) and Imaginary (κ) Parts of the Refraction Index[a] n^*

	1°C		50°C			1°C		50°C	
ν (cm^{-1})	n	κ	n	κ	ν (cm)$^{-1}$	n	κ	n	κ
400	1.54	0.35	1.52	0.39	610	1.39	0.42	1.31	0.42
410	1.54	0.36	1.51	0.40	620	1.37	0.42	1.29	0.41
420	1.54	0.36	1.51	0.40	630	1.36	0.42	1.28	0.41
430	1.54	0.37	1.50	0.41	640	1.35	0.42	1.27	0.40
440	1.53	0.37	1.49	0.41	650	1.34	0.41	1.26	0.39
450	1.53	0.38	1.48	0.41	660	1.33	0.41	1.25	0.38
460	1.53	0.38	1.48	0.42	670	1.31	0.41	1.24	0.38
470	1.52	0.39	1.47	0.42	680	1.30	0.40	1.23	0.37
480	1.51	0.39	1.46	0.42	690	1.28	0.40	1.21	0.36
490	1.51	0.39	1.45	0.42	700	1.27	0.39	1.20	0.35
500	1.50	0.40	1.44	0.43	710	1.26	0.39	1.19	0.34
510	1.49	0.40	1.43	0.43	720	1.24	0.38	1.18	0.33
520	1.48	0.41	1.42	0.43	730	1.23	0.38	1.17	0.32
530	1.47	0.41	1.41	0.43	740	1.21	0.37	1.16	0.31
540	1.47	0.41	1.40	0.43	750	1.20	0.36	1.15	0.30
550	1.46	0.41	1.39	0.43	760	1.19	0.35	1.14	0.29
560	1.45	0.42	1.37	0.43	770	1.17	0.34	1.13	0.27
570	1.43	0.42	1.36	0.43	780	1.16	0.32	1.12	0.26
580	1.42	0.42	1.35	0.43	790	1.15	0.31	1.12	0.24
590	1.41	0.42	1.34	0.43	800	1.14	0.29	1.12	0.22
600	1.40	0.42	1.32	0.42	810	1.14	0.27	1.12	0.21
					820	1.14	0.26	1.12	0.19

[a]Liquid water H_2O at $T = 1°C$ and 50°C.

Source: Ref. 53.

κ of the complex refraction index and the complex permittivity ε^*, as well as by the real component n, as follows:

$$\alpha(\nu) = 4\pi\nu\kappa(\nu), \qquad \varepsilon^*(\nu) = \sqrt{n(\nu) + i\kappa(\nu)} \tag{A3}$$

In Fig. 28a we show for $T = 27°C$ the frequency dependence of ε' and ε'', which comprises the Debye and FIR regions. We see three maxima on the loss curve 2. The frequency dependences obtained from the empirical formulas (shown by lines) agree well the measurement data (to obtain such an agreement at the lowest frequency $\nu = 20\,\text{cm}^{-1}$ we changed a little the values of the optical constants [42] at this frequency). The evolution of the recorded quasi-resonance absorption spectra with temperature is illustrated by Fig. 28b. For the highest temperature (50°C), curve 3, there is some disagreement with the empirical

2. Double Debye Approximation for Complex Permittivity of Heavy Water

In this section we present material needed for comparison of our molecular theory with experimental data. For this aim we combine the recorded data [51] with useful information [54] and with empirical formula [17].

In spite of numerous studies, the properties of liquid water are still far from been understood at a molecular level. For instance, large isotope effects are seen in some properties, such as the temperature of maximum density, which occur at 277.2 K in liquid H_2O and 284.4 K in D_2O. The isotope shift 7.4 K will be used below with the purpose to employ the Liebe et. al. formula [17] for calculation of the low-frequency dielectric permittivity of D_2O in analogous way as it used for H_2O.

In accordance with Ref. 54 both for liquid H_2O and D_2O the double Debye approximation is applicable in the frequency range up to 2 THz (i.e., up to $\approx 70\,\text{cm}^{-1}$) and in the temperature range from 273 K to 303 K.

$$\varepsilon^* = \varepsilon_\infty + \frac{\varepsilon_s - \varepsilon_1}{1 - 2\pi i c \nu \tau_D} + \frac{\varepsilon_1 - \varepsilon_\infty}{1 - 2\pi i c \nu \tau_2} \tag{A4}$$

For *ordinary water* we shall use the empirical formula (152).

In view of Ref. 54 for *heavy water* the *static permittivity* is given by

$$\varepsilon_S(D_2O) = 78.25\left[1 - 4.617\frac{t_C - 25}{1000} + 12.2\left(\frac{t_C - 25}{1000}\right)^2 - 27\left(\frac{t_C - 25}{1000}\right)^3\right],$$
$$t_C \equiv T - 273.15 \tag{A5}$$

while for the Debye relaxation time τ_D we employ the above-mentioned formula (152), in which the temperature is shifted on -7.2 K:

$$\tau_D^{D_2O}(T) = \tau_D^{H_2O}(T - 7.2), \qquad \text{so that} \quad \tau_D(D_2O) = [2\pi\gamma_1(T - 7.2)]^{-1} \tag{A6}$$

where $\gamma_1(T)$ is defined in Eq. (152).

The "end" of the second Debye region and the "fast relaxation time" τ_2 will be found just as for liquid H_2O [see Eq. (152)]:

$$\varepsilon_\infty(\mathrm{T}) = 3.52 + 7.52\phi(T), \qquad \tau_2(T) = [2\pi\gamma_2(T)]^{-1} \tag{A7}$$

To find the permittivity corresponding to the "end" of the first Debye region for $t_C \equiv T - 273.15 = 22.2°\mathrm{C}$, we account for the experimental data [51, 54] and set

$$\varepsilon_1(D_2O) = 2.1 + \varepsilon_\infty(T - 7.2\,K) \qquad \text{for liquid } D_2O \text{ at } T = 295.35\,\text{K} \tag{A8}$$

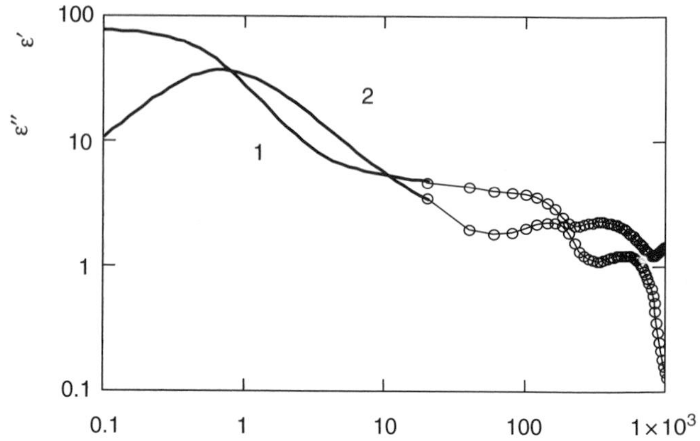

(*a*)

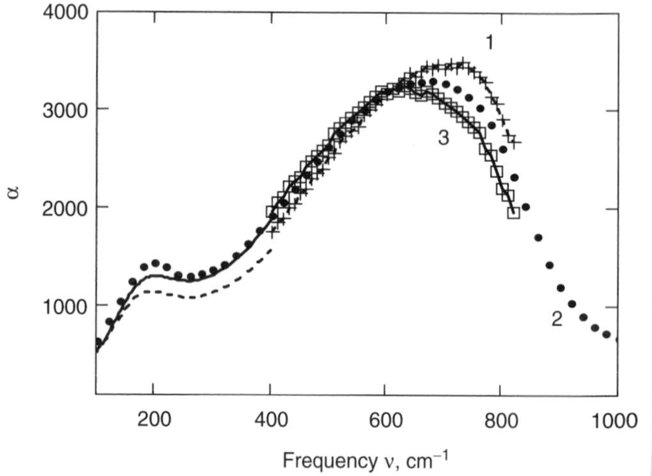

(*b*)

Figure 28. Experimental frequency dependences of dielectric parameters recorded water (a) Real (curve 1) and imaginary (curve 2) parts of the complex permittivity at 27°C are from Refs. 42 (solid lines) and 17 (circles). (b) Absorption coefficient. Solid line and refer to 1°C; filled circles 2 refer to 27°C; dashed line and squares 3 refer to 50°C. For lin from Ref. 17 were employed, for circles the data are from Ref. 42, for crosses and squar are from Ref. 53.

representation by Liebe et al., which possibly is due a crudeness of the the FIR range, where the coefficients of the double-Lorentz represent regarded as independent on temperature. The data presented here are Sections V–VII and X for the comparison of our theory with the expe

Therefore, for the comparison of our theory with the experiment at 22.2°C in the low frequency region we use the following expression:

$$\varepsilon^*(D_2O) = \varepsilon_\infty(T) + \frac{\varepsilon_s(D_2O) - \varepsilon_1(D_2O)}{1 - 2\pi i c\nu\tau_D(D_2O)} + \frac{\varepsilon_1(D_2O) - \varepsilon_\infty(T)}{1 - 2\pi i c\nu\tau_2(T)} \quad \text{for } T = 293.35\,\text{K} \tag{A9}$$

For the temperature 22.2°C the experimental values [51] of the imaginary (κ) and real (n) parts of the refraction index n are presented in Table XI; their frequency dependences are illustrated by Fig. 29. In the R-band range (near 200 cm^{-1}) the optical constants of both isotopes practically coincide but they noticeably diverge at the librational band. Figure 30 demonstrates the loss (Fig. 30a) and ε' (Fig. 30b) frequency dependences calculated from Eq. (A9) in the low-frequency region and from the Table XI in the FIR range. Both presentations agree near the boundary frequency $\nu = 20\,cm^{-1}$, which is marked in Fig. 30 by the vertical line. Thus now we have in the region from 0 to 600 cm^{-1} the experimental data for the complex permittivity $\varepsilon^*(\nu)$ and absorption $\alpha(\nu)$ of ordinary and heavy water.

3. Water H_2O and D_2O at 22.2°C

In Table XI we present the recorded data [51], which are illustrated by Figs. 29 and 30.

VI. SPECIFIC INTERACTIONS IN WATER

This section presents the continuation of Section V. In the latter a new model [10] termed the hat-curved model was described, where a rigid dipole reorients in a hat-like intermolecular potential well having a rounded bottom. This well differs considerably from the rectangular one, which is extensively applied to polar fluids. Now the theory of the hat-curved model will be generalized, taking into account the *non-rigidity* of a dipole; that is, a *simplified* polarization model of water is described here.

A. Problem of Specific Interactions

Observation of wideband dielectric/FIR spectra stretching over the frequency range from 0 to 30 THz (namely, from 0 to 1000 cm^{-1}) in such strongly absorbing liquid as water presents a serious technological problem. Its solution demands development of various spectroscopic techniques. Especially inaccessible was the submillimeter wavelength range[42] (SWR), where it is difficult to

[42]With respect to water we shall conditionally extend the SWR from 10 to 300 cm^{-1}; this frequency region falls betweem the Debye relaxation range and the librational band.

TABLE XI
Imaginary (κ) and Real (n) Parts of the Complex Refractive Index[a]

ν (cm^{-1})	κ (H_2O)	n (H_2O)	κ (D_2O)	n (D_2O)
19.531	0.7107	2.2657	0.6189	2.2193
23.437	0.6424	2.2138	0.5593	2.1639
27.344	0.5935	2.1750	0.5154	2.1249
31.250	0.5571	2.1442	0.4821	2.0955
35.156	0.5288	2.1188	0.4560	2.0724
39.062	0.5063	2.0974	0.4352	2.0537
42.969	0.4880	2.0792	0.4182	2.0383
46.875	0.4740	2.0630	0.4043	2.0258
50.781	0.4612	2.0488	0.3930	2.0156
54.687	0.4508	2.0369	0.3839	2.0075
58.594	0.4423	2.0267	0.3770	2.0012
62.500	0.4358	2.0180	0.3729	1.9963
66.406	0.4312	2.0105	0.3700	1.9920
70.312	0.4283	2.0040	0.3691	1.9887
74.219	0.4272	1.9981	0.3702	1.9859
78.125	0.4278	1.9926	0.3731	1.9833
82.031	0.4300	1.9871	0.3777	1.9805
85.937	0.4335	1.9814	0.3838	1.9772
89.994	0.4383	1.9753	0.3912	1.9734
93.750	0.4441	1.9687	0.3996	1.9687
97.656	0.4508	1.9613	0.4089	1.9630
101.563	0.4580	1.9531	0.4189	1.9563
121.094	0.4972	1.8982	0.4709	1.9062
140.625	0.5267	1.8228	0.5115	1.8317
160.156	0.5333	1.7372	0.5268	1.7447
179.687	0.5136	1.6557	0.5134	1.6617
199.219	0.4727	1.5909	0.4779	1.5967
218.750	0.4209	1.5501	0.4331	1.5564
242.187	0.3613	1.5334	0.3853	1.5386
261.719	0.3259	1.5394	0.3609	1.5395
281.250	0.3075	1.5521	0.3517	1.5432
300.781	0.3024	1.5629	0.3531	1.5434
320.312	0.3049	1.5688	0.3600	1.5383
339.844	0.3114	1.5703	0.3696	1.5287
359.375	0.3201	1.5681	0.3808	1.5151
378.906	0.3300	1.5629	0.3931	1.4976
398.437	0.3405	1.5551	0.4062	1.4755
421.875	0.3533	1.5431	0.4208	1.4415
441.406	0.3641	1.5312	0.4297	1.4067
460.937	0.3750	1.5177	0.4331	1.3664
480.469	0.3863	1.5026	0.4288	1.3224
500	0.3979	1.4854	0.4153	1.2772
519.531	0.4094	1.4657	0.3920	1.2338
539.062	0.4204	1.4432	0.3599	1.1955
558.594	0.4302	1.4176	0.3207	1.1648
578.125	0.4380	1.3891	0.2771	1.1554
601.562	0.4435	1.3513	0.2234	1.1310

[a]Liquid water H_2O and D_2O at 22.2°C.

Source: Ref. 51.

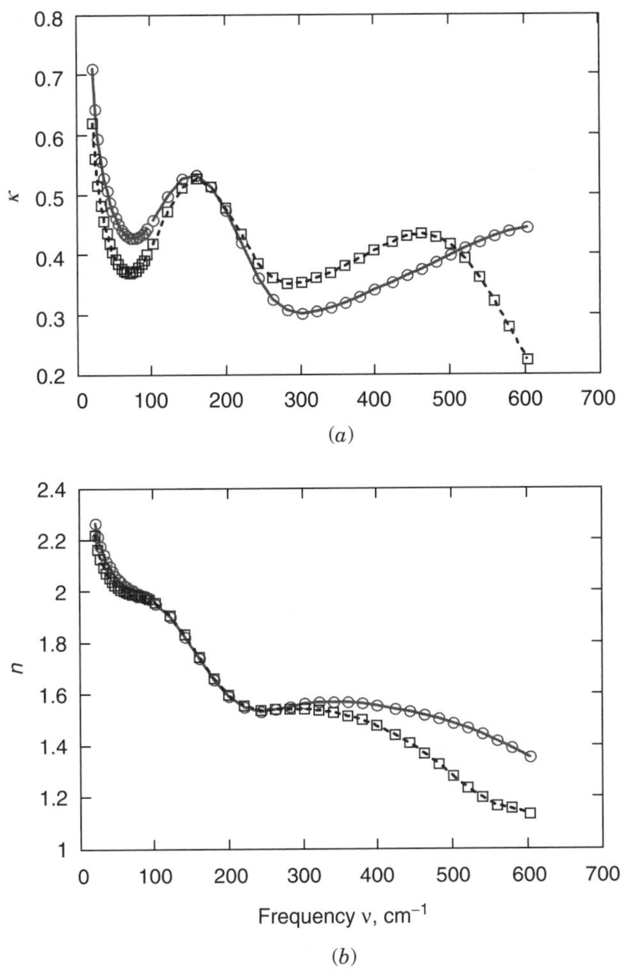

Figure 29. Imaginary (a) and real (b) parts of the complex refraction index at 22.2°C. Ordinary water is represented by solid lines and circles, heavy water is represented by dashed lines and boxes. In the low-frequency region (for $\nu < 20\,\mathrm{cm}^{-1}$), calculation is performed using approximation 17 modified as described in Appendix 3.2; in the rest region, it is performed using the recorded data [51] given in Table XI.

employ usual microwave or optical devices. Eventually, application of modern technique allowed to solve this problem (see, e.g., Refs. 51 and 54). On the other hand, although interpretation of the water spectra in SWR is useful for understanding of some key aspects of water structure and dynamics, studies of molecular mechanisms, underlying SWR, advanced rather slowly. Nevertheless,

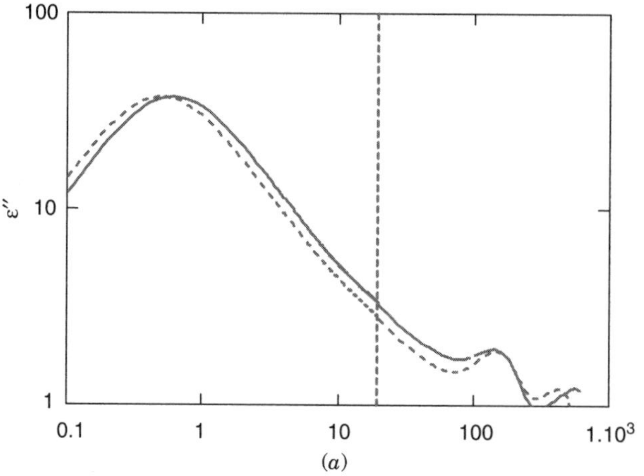

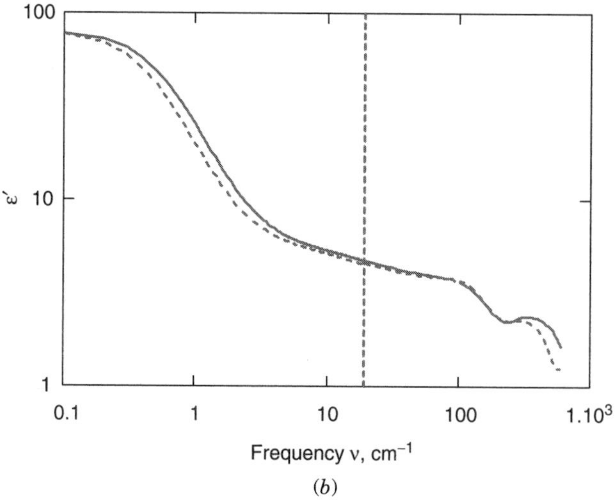

Figure 30. Imaginary (a) and real (b) parts of the complex permittivity of liquid water H_2O at 22.2°C. Ordinary water is represented by solid lines, heavy water is represented by dashed lines. To the left from vertical lines (for $\nu < 20\,\mathrm{cm}^{-1}$), calculation is performed using approximation [17] modified as described in Appendix 3.2; in the rest region, it is performed using the data 51 given in Table XII.

in the course of these studies, qualitatively new ideas have appeared. In particular, it became clear that spectral features of water, common with those relevant to "standard" simple strongly polar liquids, interfere in the SWR range with the features pertinent to the properties of a solid body.

Spectral properties of water in the R-band range (near 200 cm^{-1}), which are *directly* determined by H-bonding of water molecules, were termed [16] as stipulated by "specific" interactions. All other ones we shall term "unspecific" interactions. The latter were considered in Section V devoted to a "pure" hat-curved model capable of description of the Debye, librational, and partly R-bands (the ongoing background was considered in items 1–6 in Section V.A.1).

Here we try to study specific interactions in water in terms of *slightly modified* hat-curved model with a simplified account of collective (cooperative) effects in water in relation to SWR spectra. Below, in items A–D, we shall shortly describe how the problem of these effects was *gradually* recognized in our publications [6–9, 11]. At first, we shall draw attention on a small isotope shift of the R-band—that is, on practical coincidence of the peak absorption frequencies $\nu_R \approx 200\,\text{cm}^{-1}$ for both ordinary (H_2O) and heavy (D_2O) water.

A. Previous models of water (see 1–6 in Section V.A.1) and also the hat-curved model itself *cannot* describe properly the R-band arising in water and therefore cannot explain a small isotope shift of the center frequency ν_R. Indeed, in these models the R-band arises due to *free rotors*. Since the moment of inertia I of D_2O molecule is about twice that of H_2O, the estimated center of the R-band for D_2O would be placed at $\approx \sqrt{2}$ lower frequency than for H_2O. This result would contradict the recorded experimental data, since $\nu_R(D_2O) \approx \nu_R(H_2O) \approx 200\,\text{cm}^{-1}$. The first attempt to overcome this difficulty was made in GT, p. 549, where the *cosine-squared* (CS) potential model was formally (i.e., irrespective of a physical origin of such potential) applied for description of dielectric response of "rotators" moving "above" the CS well (in this work the "librators" were assumed to move in the rectangular well). The nonuniform CS potential yields a rather narrow absorption band; this property agrees with the experimental data [17, 42, 54]. The absorption-peak position ν_{CS} depends on the field parameter p of the model given by

$$p = \sqrt{U_{CS}/(k_B T)} \tag{232}$$

where U_{CS} is the well depth of the CS potential. If we identify ν_R with ν_{CS}, it is possible to get a correct position of ν_R by fitting the p-parameter. Note that mathematical simplifications used in this investigation were very rough.

B. In Ref. 7 an approach was presented, resembling that given in item A, but now a *preliminary* physical interpretation of the model was for the first time presented and the mathematical theory used was more rigorous. *Each* water molecule was assumed to participate, like a solid body, in the motions directed along different axes characterized by different projections μ_j of a molecule dipole moment $\boldsymbol{\mu}$. In Ref. 7, *two* potential wells were introduced, rectangular and cosine squared.

(i) The first well, deep and rather narrow, has a rectangular profile (such as was considered in Section IV), with depth $U_{\rm HAT} \gg k_{\rm B}T$. Dielectric response of dipoles moving in such a potential was described in terms of the spectral function $L_{\rm HAT}(z)$ found in VIG, Chapter 9 and in Ref. 3. The majority of the dipoles, termed *librators*, rotate via inertia between the rectangular walls being elastically reflected by them. A small number of the dipoles, termed *rotators*, perform free or hindered complete rotation.

(ii) The second gently sloping cosine squared potential well has a smaller well depth ($U_{\rm CS} < U_{\rm HAT}$). Here motion of dipoles in a *plane* was considered. The square of a dipole moment (μ^2) was presented as the sum $\mu^2_{\rm HAT} + \mu^2_{\rm CS}$, and the ratio

$$r_{CS} \equiv \mu^2_{\rm CS}(\mu^2_{\rm CS} + \mu^2_{\rm HAT})^{-1} \tag{233}$$

characterizing specific interactions of the H-bonded molecules is small ($r_{\rm CS} \ll 1$).

The cosine-squared potential model was simplified in terms of the so-called "stratified approximation," for which the spectral function $L_{\rm CS}(Z)$ is given in GT, p. 300 and in VIG, p. 462. We remark that the dielectric spectra calculated rigorously for the CS model agree with this approximation, while simpler quasi-harmonic approximation (GT, p. 285; VIG, p. 451) used in item A yields for $p > 1$ a *too narrow* theoretical absorption band.

The normalized complex frequency of vibrations is determined as

$$Z = x + iY, \qquad \text{where } Y = \eta/\tau_{\rm vib} \tag{234}$$

where the normalized collision frequency Y connected with vibrations differs from that (y) connected with reorientations (the parameterization of the model have shown that $Y \gg y$).

(iii) The complex permittivity χ^* was described as some generalization of the representation (142), in which *two* above-mentioned spectral functions $L_{\rm HAT}(z)$ and $L_{\rm CS}(Z)$ are involved and the isothermal collision model[43] is used instead of the Gross one:

$$\chi^*(x) = gG\frac{z(1 - r_{\rm CS})\,L_{\rm HAT}(z) + 2Z\,r_{\rm CS}L_{\rm CS}(Z)}{gx + iy(1 + gxz)(1 - r_{\rm CS})L_{\rm HAT}(z) + 2iY(1 + gxZ)r_{\rm CS}L_{\rm CS}(Z)} \tag{235}$$

[43]See description of this collision model in Sections VII.C.2 and VII.C.3 (on the example of a parabolic potential well).

Here multiplier 2 approximately accounts for doubling of integrated absorption due to *spatial* motion of a dipole, which is more realistic than motion in a plane to which $L_{CS}(Z)$ corresponds. For representation (235), only *one* (Debye) relaxation region with the relaxation time τ_D is characteristic. At this stage of molecular modeling it was not clear (a) why the CS potential, which affects motion of a dipole in a *separate* potential well, is the right model of specific interactions and (b) what is physical picture corresponding to a solid-body-like dipole moment μ_{CS}.

C. An analogous model was considered in Ref. 12b, but an important new step was made. Now it was assumed that the stochastic processes with two *different relaxation times* correspond to types of motion described by two wells. Two different complex susceptibilities were calculated, which have split Eq. (235) by two similar expressions for reorientation and vibration processes:

$$\chi^*_{or}(x) = \frac{g_{or}\, G(1 - r_{CS}) z\, L_{HAT}(z)}{g_{or} x + i y(1 + g_{or}\, x\, z) L_{HAT}(z)} \tag{236a}$$

$$\chi^*_{vib}(x) = \frac{g_{vib}\, Z G r_{vib}\, 2L_{CS}(Z)}{g_{vib} x + i Y(1 + g_{vib} x\, Z) 2L_{CS}(Z)} \tag{236b}$$

It was shown for the first time that (i) the relaxation time τ_{Dvib} corresponding to Eq. (236b) is much smaller than the Debye relaxation time τ_D and (ii) its value (about 0.3 ps) agrees with the "fast" relaxation time found in Refs. 17 and 54 from the empirical double Debye frequency dependences. In accord with a rather speculative reasoning, given in Ref. 12b, the *bending vibrations* of water molecules described by the CS potential well were considered as a *spectroscopic active type* of motion. This type arises as quasi-periodic (stochastic) oscillations of a polar molecule *perpendicular to the direction* of the H-bond, with a small dipole moment $\mu_{vib} \equiv \mu_{CS}$ being ascribed to this process ($\mu_{vib} \ll \mu_{or} \equiv \mu_{HAT}$). The larger dipole moment, μ_{or}, characterizes librational motion of a dipole in a rectangular well. Both modes of motion were assumed to be independent of each other and were performed in *different* potential wells. The results of Ref. 12b could explain, in terms of the parameters (232) and (233), a small isotope shift of the R-band.

It was noted in Ref. 12b that such important physical characteristic exists as *elasticity* of the spatial H-bond network, which is usually employed [15, 16, 19] for calculations of water spectra. As is intuitively clear, this elasticity should be somehow related to the R-band spectrum, since the stretching vibration, determined by the H-bond elasticity, is believed [16, 35, 51] to present the origin of this band in water. As a basic mechanism, one could regard an additional power loss due to interaction with the a.c. field of the H-bond vibrations. However in Ref. 7, as well as in Ref. 12b, a physical picture relating the CS well to bending vibrations was *not* established.

We regard the concept of elasticity to be the key aspect of the problem of specific interaction. A preliminary study [12, 12a] cast light on this problem in terms of dynamics of water molecules and of the relevant vibration frequencies. We consider this question in Section IX. On the other hand, the work [12b] will be improved in Section VII in terms of the hat-curved model.

D. Quite another approach, as compared with Refs. 7 and 12b was proposed in Refs. 6 and 8 in terms of a semiphenomenological molecular model capable of describing the wideband dielectric and far-infrared spectra of ordinary and heavy water. In the model the total dipole-moment vector was presented as a sum of two components. The absolute value $\bar{\mu}$ of the first component is set constant in time; the second component, $\tilde{\mu}(t)$, changes with time harmonically. Such rather formal presentation of a total dipole moment μ_{tot} is possibly a simplest step in taking account of the collective effects, since a time-varying dipole moment $\tilde{\mu}(t)$ arises due to *cooperative* motion of nearby polar water molecules.

The main purpose of this section is consideration of the FIR spectra due to the second dipole-moment component, $\tilde{\mu}(t)$. However, for comparison with the experimental spectra [17, 42, 51] we should also calculate the effect of a total dipole moment μ_{tot}. In Refs. 6 and 8 the modified hybrid model[44] was used, where reorientation of the dipoles in the rectangular potential well was considered. In this section the effect of the $\tilde{\mu}(t)$ electric moment will be found for the *hat-curved* potential, which is more adequate than the rectangular potential pertinent to the hybrid model. In Section VI.B we present the formula for the spectral function of the hat-curved model modified by taking into account the $\tilde{\mu}(t)$ term (derivation of the relevant formula is given in Section VI.E). The results of the calculations and discussion are presented, respectively, in Sections VI.C and VI.D.

B. Modified Spectral Function *R*(*z*)

Water has a complex structure determined by hydrogen bonding of molecules. In view of the above discussion we assume now that a dipole is *not rigid*. Introducing some changes in our recent works [6, 8], we represent a total dipole moment μ_{tot} as a superposition of the constant part $\bar{\mu}$ and of a small decaying component $\tilde{\mu}(t)$ due to fast vibration of the H-bonded molecules:

$$\mu_{\text{tot}} = \mu + \tilde{\mu}(t); \quad \tilde{\mu}(t) \equiv \mu_m \exp(-t/\tau_{\text{vib}}) \sin(\Omega t + \psi) \tag{237}$$

Note that here the scalar (not vector!) summation is employed, since both components of a dipole vector are assumed to have the same direction. In

[44] A brief description of the hybrid model based on application of the rectangular well potential was given in Section V.E.

Eq. (237) $\Omega = 2\pi c \nu_R$ is the angular frequency of the R-band peak, ψ is an arbitrary phase of vibration, $\psi \in [0, 2\pi]$; the vibration lifetime τ_{vib} and the amplitude μ_m are free parameters of the model. Later we shall introduce dimensionless quantities; see Eqs. (244)–(246). Equation (237) is underlain by the experimental fact that a dipole moment μ_0 of an isolated water molecule *differs* from an average electric moment μ of a molecule in a liquid. It is natural assuming that a polar molecule, moving in tight surroundings of other molecules, is accompanied by a time-varying component of a dipole moment. Therefore, we consider here a *rough* polarization model of water, in which we use a *given* time dependence (237). Two alternative approaches, in which nonrigidity of a polar molecule is treated in terms of classical linear-response theory, is described in Section 7.

Let $L(z)$ be the spectral function (SF) of the hat-curved model (see Section V), namely, at vanishing μ_m, and let $R(z)$ be the SF modified by the presence of the $\tilde{\mu}(t)$ term. In Section V.E we shall prove that the following scheme holds:

$$\left\{ \begin{matrix} \text{OLD SPECTRAL} \\ \text{FUNCTION } L(z) \end{matrix} \right\} \rightarrow \text{ACCOUNT OF } \tilde{\mu}(t) \rightarrow \left\{ \begin{matrix} \text{MODIFIED} \\ \text{SPECTRAL FUNCTION } R(z) \end{matrix} \right\} \tag{238}$$

where z is the reduced complex frequency

$$z = x + iy \tag{239}$$

$$x = \eta\omega, \qquad y = \eta/\tau \tag{240}$$

$$\eta \equiv \sqrt{I/(k_B T)} \tag{241}$$

I is a moment of inertia of a dipole (just as in Section IV.G.1 we consider a linear molecule);

$$R(z) = L(z) + sS(z) \tag{242}$$

$$S(z) \approx \frac{b - iw}{b - iw - z} + \frac{b + iw}{b + iw + z} \tag{243}$$

s is proportional to the amplitude μ_m squared:

$$s = \frac{1}{2} \langle \tilde{\mu}^2 \rangle / \langle \bar{\mu}^2 \rangle \tag{244}$$

$\langle \cdot \rangle$ denotes ensemble average, the dimensionless parameters b and w are introduced as

$$w \equiv \eta/\tau_{vib} \tag{245}$$

$$b \equiv \eta\,\Omega \tag{246}$$

For $L(z)$ in Eq. (242) we employ the same formulas as in Section V, [see Eqs. (170)–(173) and (179)], where f is the form factor, $0 \leq f \leq 1$, defined as the ratio of the flat-part of the hat-curved well on its bottom to the whole width of this well. Assuming that the hat-curved potential well is deep and narrow, so that the reduced depth

$$u = U_0/(k_B T) \tag{247}$$

is large $(u \gg 1)$ and that the angular half-width β of the well is small $(\beta \ll 1)$, we approximately represent $L(z)$ as a sum of two terms

$$L(z) = \breve{L}(z) + \mathring{L}(z) \tag{248}$$

We neglect, just as in Section V, the contribution to spectra due to the "precessors," since this contribution is very small.

It is shown in Section VI.E that if the time-varying part of a dipole moment is small $(s \ll 1)$, then the complex susceptibility χ^* is related to the modified SF $R(z)$ in the same fashion

$$\chi^*(x) = g\,G\,z\,R(z)\,[gx + i\,y\,R(z)]^{-1} \tag{249}$$

as χ^* is related to the spectral function L, when $s = 0$—that is, in accordance with Eq. (142):

$$\chi^*(x) = g\,G\,z\,L(z)\,[gx + iy\,L(z)]^{-1} \tag{250}$$

Here g is the Kirkwood correlation factor (146), which is determined by the static permittivity ε_s, permittivity n_∞^2 at the HF edge of the librational band and by the reduced concentration of the dipoles G. The total number of free parameters of our model is now six:

$$\begin{matrix} u, \beta, \tau, f \\ s, w \end{matrix} \quad \text{parameters of} \quad \begin{cases} \text{hat-curved potential} & (251\text{a}) \\ \text{nonrigidity of a dipole} & (251\text{b}) \end{cases}$$

It might seem at first glance that virtually *any* experimental prediction is possible with such a large set of the parameters. This opinion would be erroneous, since the frequency band under consideration is extremely wide (several orders of magnitude), in which absorption coefficient α, real (ε') and imaginary (ε'') components of the complex permittivity present many-peaked frequency dependences. It is clear that describing such complex spectra using only a few parameters is problematical. The parameters (251a) and (252b) control different parts of the wideband spectra, namely, four parameters (251a) determine the susceptibility $\chi^*(x)$ in the Debye relaxation region (with $0 \le \omega\tau_D \le 5$) and in the librational band, where $x > 0.1$, while two parameters (251b) determine the $\chi^*(x)$ dependence in a very narrow R-band, in which

$$0.5\nu_{tr} \le \nu \le 2\nu_{tr} \tag{252}$$

Therefore, the sets (251a) and (251b) somehow interfere (with respect to the spectral dependences they govern) *only* in the R-band region. This property facilitates parameterization of the model.

For a *low-frequency* region, one can neglect the dependence of the spectral function R on x. Then Eq. (249) describes the Debye-like frequency dependence, where the relaxation time can be estimated as

$$\tau_D^{est} = g\eta/(yR_0) \tag{253}$$

where $R_0 \equiv \text{Re}[R(x+iy)]_{x=0} = \text{Re}[R(iy)]$. Here we have set $\text{Re}[R(iy)]$ instead of $R(iy)$, since the "additional" spectral function $S(z)$ has a very small complex part at $x = 0$ provided by the iw terms in Eq. (243).

On the other hand, *in the high-frequency region*, where R- and librational quasi-resonance bands arise, we have $gx \gg y|L(z|$. It follows from Eq. (249) that in this region the susceptibility $\chi^*(x)$ becomes proportional to the spectral function $R(z)$.

The optimization procedure of the parameters (251a) described in Section V.D.1 remains valid. In particular, (a) if we set the right-hand side of Eq. (253) to be equal to the experimental relaxation time, then Eq. (253) can be used as an equation connecting the parameters (251a) and (251b) of our model. We can estimate the number of reorientation cycles performed during the lifetime τ from Eq. (192a). The number m_{vib} of vibration cycles performed during the lifetime τ_{vib} can be found as a ratio

$$m_{vib} \approx \tau_{vib}/T_{vib} = c\nu_{tr}\tau_{vib} \tag{254}$$

of the corresponding lifetime to the period T_{vib} of oscillations.

Finally, the proportion of the rotators is determined by Eq. (193).

To understand the effect of the parameters (251b) on the reduced absorption spectrum

$$\gamma(x) = x\,\text{Im}[R(z)] \tag{255}$$

we shall consider an example (Fig. 31), in which the chosen parameters (251a) and b are typical for water (b is commensurable with x in SWR).

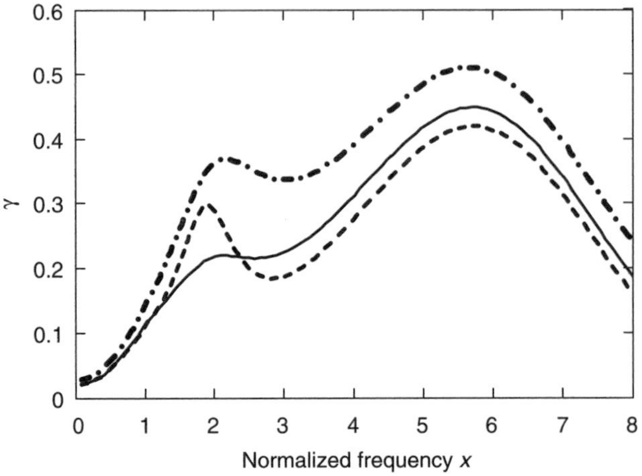

Figure 31. Normalized absorption versus normalized frequency. Solid curve: Calculation for the parameters of the hat-curved model typical for liquid D_2O (these parameters are presented in Table XIII, lower line.) Dashed curve: The lifetime τ_{vib} becomes two times longer, other parameters remain the same. Dashed-and-dotted curve: Calculation, when s becomes two times larger, other parameters remain the same.

The solid curve in Fig. 31 shows the frequency dependence of absorption [Eq. (255)]. We see the shoulder in the R-band region (at $x \approx 2$) and the main (librational) absorption peak at $x \approx 5.5$. This dependence (solid curve) agrees with the experimental data [42, 51] (a more detailed analysis is given in the next section). If the vibration lifetime τ_{vib} were twice as large, we would get an unreasonably large and narrow R-peak (dashed curve). On the other hand, if the parameter s were twice as large, the intensity of the R-band would increase unreasonably (dash-and-dotted curve). This example shows convincingly that the form of the FIR water spectra *sharply* determines the parameters (251b) of our polarization model.

C. Calculated Spectra

Let us calculate the broadband spectra of liquid water H_2O and D_2O. The adopted experimental data are presented in Table XII. In accord with the scheme (238), we use Eq. (249) for the complex susceptibility χ^* and use Eqs. (242) and (243) for the modified spectral function $R(z)$. All other expressions used in these calculations are the same as were employed in Section V.

The fitted parameters of the hat-curved model are given in Table XIIIA; the parameterization procedure was described in Section V.D. The solid lines in

TABLE XII
Experimental Data Pertaining to Liquid Water

Liquid	T°C	τ_D (ps)	ε_s	$\rho\,(g \times cm^{-3})$	$\nu_{lib}\,(cm^{-1})$	$\nu_R\,(cm^{-1})$	$\alpha_{lib}\,(cm^{-1})$	$\alpha_R\,(cm^{-1})$	k_μ
H_2O	27	7.85	77.6	0.9986	690	200	3298	1445	0.92
H_2O	22	9.3	80.0	0.9986	648	201	3403	1208	0.97
D_2O	22	11.5	80.0	1.1	500	~200	2609	1196	0.95
H_2O	$\mu_0 = 1.84\,D, n_\infty^2 = 1.7, I = 1.483 \times 10^{-43}, M = 18$								
D_2O	$\mu_0 = 1.84\,D, n_\infty^2 = 1.7, I = 2.765 \times 10^{-43}, M = 20$								

Figs. 32a–c illustrate the *absorption* spectra, calculated, respectively, for water H_2O at 27°C, water H_2O at 22.2°C, and water D_2O at 22.2°C; dotted lines show the contribution to the absorption coefficient due to vibrations of nonrigid dipoles. The latter contribution is found from the expression which follows from Eqs. (242) and (255). The experimental data [42, 51] are shown by squares. The dash-and-dotted line in Fig. 32b represents the result of calculations from the empirical formula by Liebe et al. [17] (given also in Section IV.G.2) for the complex permittivity of H_2O at 27°C comprising double Debye–double Lorentz frequency dependences.

In the *librational* band we have attained now a satisfactory agreement of the theoretical and experimental absorption frequency dependences. Comparing Figs. 32a and 32b with Figs. 26a and 26c calculated in Section V for a "pure"

TABLE XIII
Fitted and Estimated Parameters of the Hat-Curved Model

A. Fitted Parameters

		Parameters Relevant to a Rigid Dipole Moment				Parameters Relevant to a Nonrigid Dipole Moment		
Liquid	T (K)	τ (ps)	u	β (deg)	f	b	τ_{vib} (ps)	s
H_2O	300.15	0.32	5.6	23	0.75	1.3	0.06	0.028
H_2O	295.35	0.31	5.92	23	0.75	1.4	0.09	0.017
D_2O	295.35	0.46	5.6	22	0.72	1.8	0.09	0.025

B. Estimated Parameters

Liquid	T (°C)	Y	w	m_{or}	m_{vib}	$\mathring{r}$ (%)	μ_m/μ
H_2O	300.15	0.13	0.71	6.6	0.36	5.1	0.32
H_2O	295.35	0.14	0.47	5.9	0.54	3.8	0.25
D_2O	295.35	0.13	0.73	7.3	0.48	5.9	0.30

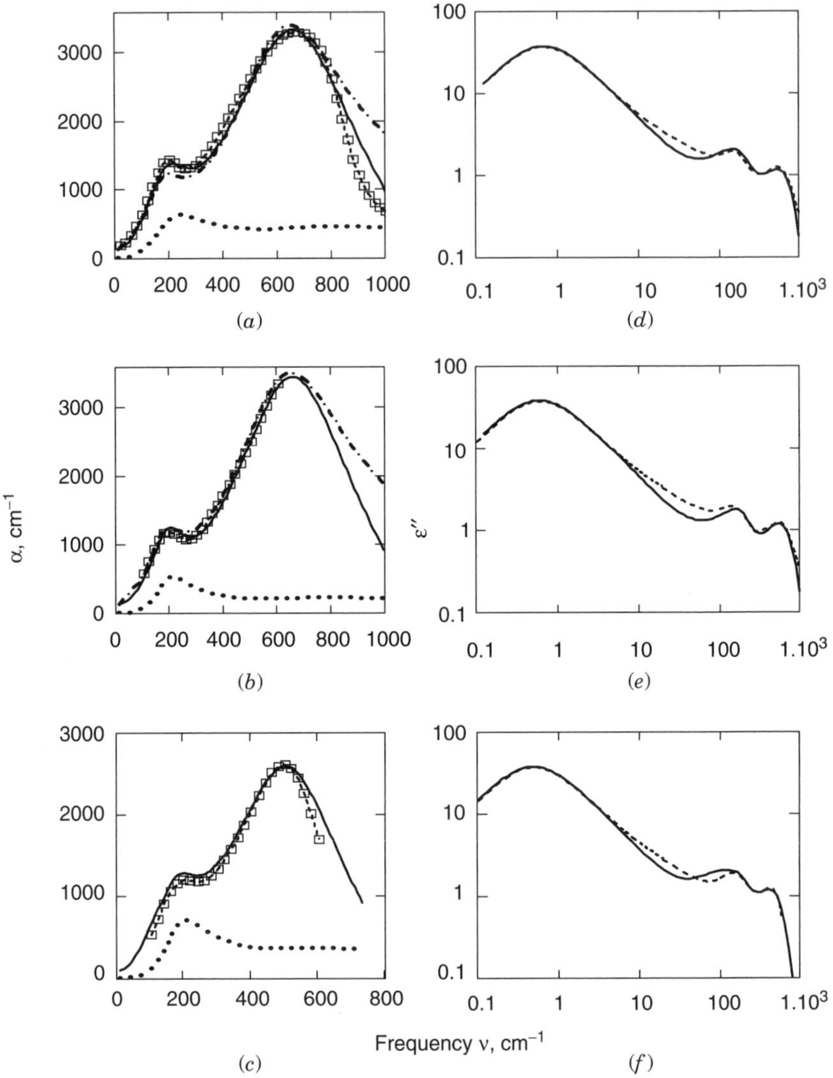

Figure 32. Absorption coefficient (a, b, c) and dielectric loss (d, e, f). Water H_2O at 27°C (a, d), water H_2O at 22.2°C (b, e), and water D_2O at 22.2°C (c, f) Solid lines: Calculation for the hat-curved model; experimental [42, 51] values of absorption (squares) and loss (dashed lines), calculation from empirical formula [17] (dashed-and-dotted lines). Contribution to absorption due to nonrigidity of dipoles is shown by dots.

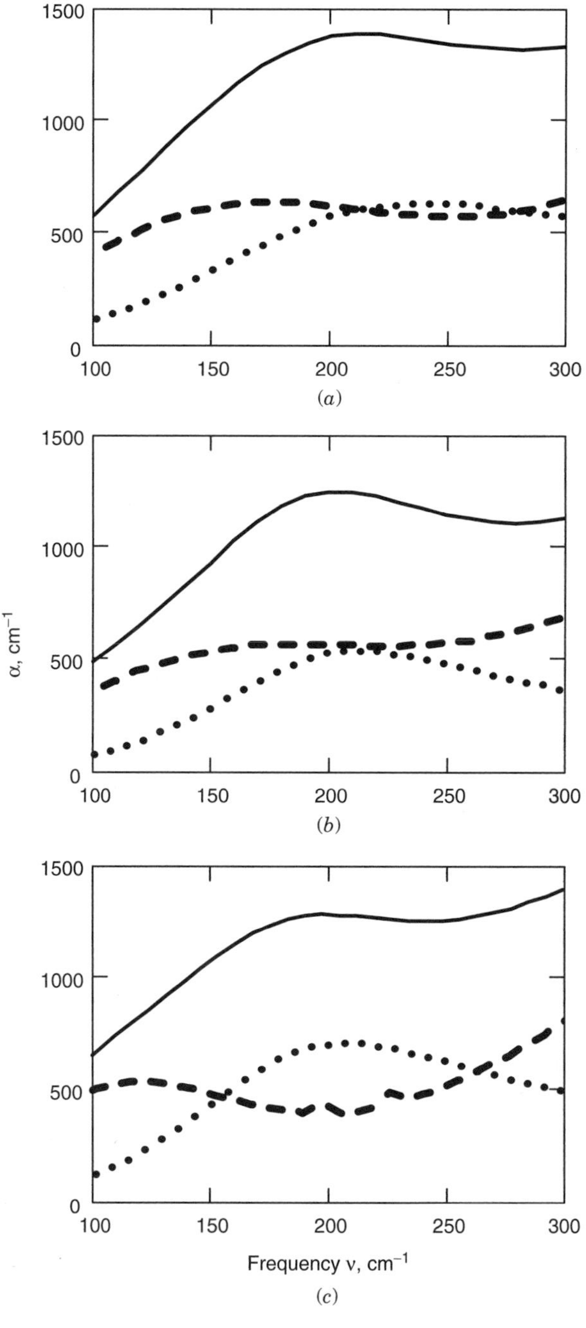

Figure 33. Absorption coefficient in the R-band calculated for the hat-curved model (solid lines), the contributions to this coefficient due to a constant (dashed lines), and the time-varying (dots) parts of a dipole moment. Water H_2O at 27°C (a), Water H_2O at 22.2°C (b), and water D_2O at 22.2°C (c).

hybrid model, we conclude that this agreement is evidently better than that obtained without account of dipoles' nonrigidness, when Eq. (250) was used instead of Eq. (249). The progress is governed by additional absorption (arising mainly in the R-band) shown by dash-and-dotted curves in Figs. 32a–c.

It is interesting to compare the results obtained for ordinary and heavy water. To interpret the difference, we show in Fig. 33 by solid curves the total absorption attained in the R-band (i.e., near the frequency 200 cm^{-1}). Dashed curves and dots show the components of this absorption determined, respectively, by a constant (in time) and by a time-varying parts of a dipole moment. In the case of D_2O, the R-absorption peak ν_R is stipulated mainly by nonrigidity of the H-bonded molecules, while in the case of H_2O both contributions (due to vibration and reorientation) are commensurable. Therefore one may ignore, in a first approximation, the vibration processes in ordinary water *as far as* it concerns the wideband absorption frequency dependences (actually this assumption was accepted in Section V, as well is in many other publications (VIG), [7, 12b, 33, 34]. However, in the case of D_2O, where the mean free-rotation-frequency is substantially less than in the case of H_2O, neglecting of the vibrating mechanism due to nonrigid dipoles appears to be nonproductive.

As for the main (librational) absorption band, new possibilities opened by the hat-curved model and described in Section V are also confirmed here, when the modified spectral function $R(z)$ is applied instead of $L(z)$. In Fig. 32c we see that our theory agrees satisfactorily with the experiment [51] recorded for D_2O, while such description given in terms of the rectangular-well potential is poor [9]. Thus, the rectangular potential well hardly could be applied for liquid D_2O, since it gives too wide an absorption band. In the case of liquid H_2O, our model also has evident advantages—if we compare it, for example, with the empirical description [17]. The right wing of the librational absorption band calculated in terms of the hat-curved model agrees better with the experiment [42] than the band determined [17] by using the Lorentz-like curve. Indeed, as we see in Figs. 32a and 32b, the latter gives, unlike our approach, a very wide absorption band (cf. solid and dash-and-dotted lines).

Figures 32d–f, placed on the right-hand side of Fig. 32, demonstrate a *wideband dielectric-loss ε'' frequency dependence.* This loss is calculated (solid lines) or measured [17, 42, 51, 54] (dashed lines) for water H_2O and D_2O at the same temperatures, as correspond to the absorption curves shown on the left-hand side of Fig. 32. Our theory gives a satisfactory agreement with the experimental data, obtained for the *Debye region, R- and librational bands*, to which three peaks (from left to right) correspond. However, in the submillimeter wavelength region (namely, from 10 to 100 cm^{-1}) the calculated loss is *less* than the recorded one. The fundamental reason for this difference will be discussed at the end of the next section.

D. Discussion

Our theory of complex permittivity, elaborated for water, drastically differs from descriptions based on the empirical formulas—given, for example, in Refs. 17 and 54—which usually are accompanied only by vague interpretation. A long way had been passed from a primitive confined rotator model VIG to the hat-curved model. The latter *alone* gives (Section V) a satisfactory description of the Debye relaxation region and of the librational band in terms of four parameters [Eq. (251a)]. As we have seen above, our approach becomes more effective, if this model is modified in accordance with the formula (238) in order to roughly take the collective effects into consideration. Then in view of Eqs. (251a) and (251b) the total number of free parameters becomes six; the theory describes the entire water spectrum from 0 to 1000 cm^{-1}. A reservation should be made that some fine details of the submillimeter water spectrum could be explained if the calculation scheme is changed once more in a way described in Section VII. Roughly speaking, dielectric response of water is determined by contributions due to rotation of individual molecules and due to their collective motion. It is possible to ascribe definite regions of the spectrum, where one or the other mechanism predominates. In this section we have employed a simple polarization model of water, presented again in terms of the hat-curved potential well.

1. *Role of Specific Interactions*

First, we shall analyze the implications arising due to nonrigidity of the dipoles. The authors strongly support a widespread viewpoint concerning principal importance of cooperative motions in water. In a number of previous works (see Section VI.A) we tried to find a description of collective motions in terms of a relevant model. In this section our method [6, 8] was applied for this aim in terms of the hat-curved model. We have shown that modification of the spectral function in terms of the scheme (238) could account for (at least formally) *specific interactions*, which are emphasized in the case of heavy water. The simplified account of the collective effects arising in water, based on consideration of a time-varying component $\tilde{\mu}(t)$ of a total dipole moment, demands only a small change of the calculation scheme described in Section V: We apply Eq. (249) instead of Eq. (250). Two additive parameters w and s [Eq. (251b)] involved in a modified spectral function $R(z)$ affect *only* the R-band (100–300 cm^{-1}). Note that the latter presents only a small part of a wideband (ranging from 0 to 30 THz) water spectrum under interest. These additional parameters can easily be fitted, if the experimental data are available in this frequency range. Indeed, s, w, and b determine, respectively, the reduced intensity, width, and center of the R-band. The lifetime τ_{vib} of the $\tilde{\mu}(t)$-state fitted for water is substantially less than the local-order lifetime τ. Our estimations show that *vibrations are extremely damped*, if we compare them with reorientations of rigid dipoles. Indeed, in view of Table XIIIB, less than one vibration cycle is occurred during the mean vibration lifetime τ_{vib}, while about a half dozen of the cycles are performed during a much longer reorientation time τ. The time τ comprises a part of a picosecond, while τ_{vib} falls into a nanosecond region ($\approx$ 60–90 ns). The amplitude μ_m of the time-varying part of a dipole moment, which is estimated here in accord with the formulas given in Ref. 8, comprises about 30% of the total dipole moment, so the assumption $s \ll 1$ holds. The fitted s value is actually determined by a relative intensity of the R-band with respect to intensity of the librational band.

2. *Role of Unspecific Interactions*

In view of the calculations considered in Section V and in other publications (VIG), these interactions, giving rise to FIR absorption and to low-frequency Debye loss, resemble interactions pertinent to strongly polar nonassociated liquids. However, if we compare water with a nonassociated liquid (e.g., CH_3F), then we shall find that in the latter (i) the R-band is absent; (ii) the number m_{vib} of the reorientation cycles is much less, so that the reduced collision frequency y is substantially greater; thus, molecular rotation is more damped and chaotic; and (iii) the fitted form factor f is greater.

We may suggest that the quantity $(1-f)$ characterizes the *spread of short-range forces governing interactions in a polar fluid.* In D_2O such forces are spread over a larger distance than in H_2O, since the fitted $f(D_2O) < f(H_2O)$. On the other hand, the angular width β (characterizing water structure) is about the same in H_2O and in D_2O. This result of our estimations confirms the known fact [16] concerning proximity of the structures of heavy and ordinary water.

The key aspect of the suggested model is that one can control the width of the librational band by changing the form factor f. Let us compare the solid and dash-and-dotted curves in Figs. 32a and 32b describing liquid H_2O. These curves are described, respectively, by the hat-curved and rectangular[45] potentials. We see that application of the hat-curved model with $f = 0.75$ substantially weakens the main drawback of the rectangular-well model, since the right wing of the librational band becomes narrower. An even more important result, Fig. 31c, is obtained for liquid D_2O with $f = 0.72$, for which also the left wing of the librational band becomes narrower. Note that the results obtained for D_2O by using the rectangular-well model are poor [9].

[45]The dash-and-dotted curves mentioned here are obtained from the empirical description [17] of the FIR spectra. However, approximately the same contour of the libration band is relevant [9] for the rectangular potential well.

3. *Next Step: Composite Model Characterized by Two Relaxation Times*

A principal drawback of the hat-curved model revealed here and also in Section V is that we cannot *exactly* describe the submillimeter $\varepsilon^*(\nu)$ spectrum of water (cf. solid and dashed lines in Figs. 32d–f). It appears that a plausible reason for such a difference is rather fundamental, since in Sections V and VI a dipole is assumed to move in one (hat-curved) potential well, to which *only one* Debye relaxation process corresponds. We remark that the decaying oscillations of a nonrigid dipole are considered in this section in such a way that the law of these oscillations is taken *a priori*—that is, without consideration of any *dynamical* process.

Another approach will be considered in Section VII, where *two independent relaxation processes* characterized by different relaxation times are taken into account. Thus, two types (modes) of molecular motion will be considered in terms of a linear response theory. These are (a) reorientation of *rigid* dipole in a hat-curved potential representing reorientation of an isolated water molecule and (b) stretching or bending vibrations of an H-bonded polar molecule. The latter is characterized by few additional free parameters similar to the parameters (251b) and referring to fast vibrations. We shall see that the second type of motion is characterized by a fast relaxation time (≈ 0.3 ps). After such improvement of the model, a Lorentz-like addition $sS(z)$ in Eq. (242) to the main spectral function $L(z)$ found above, will reveal, figuratively speaking, as *another* spectral function $L_{\text{vib}}(z)$, which, as we shall prove, noticeably contributes to the complex permittivity in the submillimeter wavelength region but *not* only near the frequency $\nu_{\text{tr}} = 200\,\text{cm}^{-1}$, as the term $sS(z)$ does. Then a very good agreement of our theory with the experimental complex permittivity could be obtained in the range 10–100 cm^{-1} for ordinary and heavy water.

E. Modified Spectral Function (Derivation)

If the time-varying moment $\tilde{\mu}$ is neglected, the complex susceptibility could be represented as a sum of two terms stipulated, respectively, by the a.c. perturbations of a dipole's trajectory and of the steady-state distribution:

$$\chi^* = \chi^*_{\text{dyn}} + \chi^*_{\text{st}} \tag{256}$$

The term χ^*_{dyn} is found for the homogeneous induced distribution ($F = 1$) [GT, Eqs. (3.4) and (3.18)]:

$$\chi^*_{\text{dyn}} = \frac{z}{x} G\, L(z) \tag{257}$$

$$L(z) = 3\left\langle q(0)\left[q(0) + iz\int_0^\infty q(\varphi)\exp(iz\varphi)d\varphi\right]\right\rangle \tag{258a}$$

where $q(\varphi) = \mu_E(\varphi)/\mu$, $\varphi = t/\eta$. The other term can be expressed through χ^*_{dyn} in the same way[46] as is described in VIG, Section 6.1 or in GT, Eq. (4.3):

$$\chi^*_{\text{st}}(x) = \chi^*(x) - \chi^*_{\text{dyn}}(x) \tag{258b}$$

Now shall suppose that $\tilde{\mu} \neq 0$.

1. *Purely Harmonically Changing* $\tilde{\mu}(t)$

First we shall omit the decaying term $\exp(-t/\tau_{\text{vib}})$ in Eq. (237). To modify the χ^*_{dyn} term, we project Eq. (237) onto the direction of the a.c. field:

$$\frac{\mu_E}{\bar{\mu}} = \frac{\bar{\mu}_E}{\bar{\mu}} + \frac{\tilde{\mu}_E}{\bar{\mu}} \tag{259}$$

where μ_E denote now the projection of the *total* electric moment. Let σ denote the reduced time-varying part of electric moment. Since the *directions* of this and of a constant part of a dipole moment are assumed to be the same, we may write

$$\sigma \equiv \frac{\tilde{\mu}_E}{\bar{\mu}_E} = \frac{\tilde{\mu}}{\bar{\mu}} \tag{260}$$

Then

$$\mu_{\text{tot}} = \bar{\mu}(1+\sigma) \quad \text{and} \quad \frac{\mu_E}{\bar{\mu}} = q + \frac{\tilde{\mu}_E}{\bar{\mu}_E}\frac{\bar{\mu}_E}{\bar{\mu}} = (1+\sigma)q \tag{261}$$

where

$$q \equiv \frac{\bar{\mu}_E}{\bar{\mu}} = \frac{\mu_E}{\mu_{\text{tot}}} \tag{262}$$

We denote by $R(z)$ the modified spectral function and make the replacements

$$L(z) \to R(z), \qquad G \to \bar{G} \equiv \bar{\mu}^2 N/(3k_B T) \tag{263}$$

in the formulas (257) and (258a), and (258b) and express the spectral function $R(z)$ analogously with (258a) as a function of the product $q(1+\sigma)$:

$$\chi_1^* = \bar{G}(z/x)R(z) \tag{264}$$

$$R(z) = 3\left\langle [1+\sigma_0]q(0)\left\{[1+\sigma_0]q(0) + iz \times \int_0^\infty [1+\sigma(\varphi,\sigma_0,\dot{\sigma}_0)]q(\varphi)\exp(iz\varphi)\,d\varphi\right\}\right\rangle \tag{265}$$

[46]See also Section I. In Section VII.C, such derivation is presented with all details for an example of the harmonic oscillator.

Here

$$\dot{\sigma}(\varphi) \equiv d\sigma(\varphi)/d\varphi, \qquad \sigma_0 \equiv \sigma(0), \qquad \dot{\sigma}_0 \equiv \mathrm{d}\sigma(\varphi)/d\varphi|_{\varphi=0} \tag{266}$$

For quantity σ, harmonically changing with time, we may write the equations

$$\sigma[\varphi, \sigma_0, \dot{\sigma}_0] = \sigma_0 \cos b\varphi + \frac{\dot{\sigma}_0}{b} \sin b\varphi = \xi \sin(\varphi + \psi) \tag{267a}$$

$$b = \eta\Omega, \quad \sigma_0 = \xi \sin\psi, \quad \dot{\sigma}_0 = b\xi\cos\psi \tag{267b}$$

which are equivalent to Eq. (237).

The time dependence of $\tilde{\mu}$ and consequently of σ in Eq. (267a) is actually stipulated by a motion of some microscopic object, to which we *conditionally* may assign the "coordinate" $\sigma(\varphi)$ and the "velocity" $\dot{\sigma}(\varphi)$. In view of Eqs. (237), (267a), and (267b), this motion can equivalently be characterized by the phase ψ and "energy"

$$\xi^2 = [\sigma(\varphi)]^2 + [\dot{\sigma}(\varphi)]^2 b^{-2} \tag{268}$$

where ξ is also dimensionless oscillation amplitude. Introducing formally σ_0 and $\dot{\sigma}_0$ as *additional* conjugated canonical variables, we designate as $\langle\cdot\rangle$ the average over all phase variables. Using the Boltzmann-like distributions w (actually it is the Gauss distribution), we consider two variants: (i) distributions over "coordinates" σ and "velocities" $\dot{\sigma}$ and (ii) distribution over "energies" ξ^2:

$$w(\sigma, \dot{\sigma}) \propto \exp\left[-\left(\sigma^2 + \dot{\sigma}^2 b^{-2}\right)/\varsigma^2\right] \tag{269a}$$

and

$$w(\xi) \propto \exp\left(-\xi^2/\varsigma^2\right) \tag{269b}$$

where the conditional "temperature" ς^2 controls the width of these distributions. Note that both distribution functions do not depend on initial "phase" ψ.

In view of Eq. (269a) the ensemble averages in Eq. (265) of the quantities, proportional to $\dot{\sigma}_0$ and σ_0, vanish. Then, taking into account that $\langle q^2 \rangle = 1/3$, we have from Eqs. (265) and (267)

$$R(z) = 3\Big\langle (1+\sigma_0)^2 q(0)^2 + iz \int_0^\infty (1+\sigma_0)\, q(0) \times (1 + \sigma_0 \cos b\varphi)\, q(\varphi) \exp(iz\varphi)\, d\varphi \Big\rangle \tag{270a}$$

$$= \left\langle 1 + 3\mathrm{i}z \int_0^\infty q(0)\; q(\varphi)\, e^{iz\varphi} d\varphi \right\rangle + \frac{\langle\sigma^2\rangle}{2} \left\langle 2 + 3\mathrm{i}z \int_0^\infty q(0)\; q(\varphi) \left[e^{\mathrm{i}(z+b)\varphi} + e^{\mathrm{i}(z-b)\varphi}\right] d\varphi \right\rangle \tag{270b}$$

In accord with definition (258a) we derive

$$R(z) = L(z) + \frac{1}{2}\langle\sigma^2\rangle\Big\langle 1 + \frac{z}{z+b} - \frac{z}{z+b} + \frac{3iz(z+b)}{z+b} \times \int_0^\infty q(0)q(\varphi)\exp[i(z+b)\varphi]d\varphi\Big\rangle + \frac{1}{2}\langle\sigma^2\rangle\Big\langle 1 + \frac{z}{z-b} - \frac{z}{z-b} + \frac{3iz(z-b)}{z-b} \times \int_0^\infty q(0)q(\varphi)\exp[i(z-b)\varphi]\,d\varphi\Big\rangle \tag{271a}$$

$$= L(z) + sS(z) \tag{271b}$$

where

$$s = \frac{1}{2}\langle\sigma^2\rangle \tag{272}$$

$$S(z) \equiv \frac{b}{b-z} + \frac{b}{b+z} + z\left\{\frac{L(z+b)}{b+z} - \frac{L(z-b)}{b-z}\right\} \tag{273}$$

For the Gross collision model, using Eq. (258b), we finally have

$$\chi^*(x) = \bar{g}\bar{G}zR(z)[\bar{g}\,x + iy\,R(z)]^{-1} \tag{274a}$$

$$\bar{g} \equiv g(\mu/\bar{\mu})^2, \qquad \bar{G} \equiv (\bar{\mu}/\mu)^2 \tag{274b}$$

It was shown in Ref. 8 that $(\bar{\mu}/\mu)^2 \approx (1+2s)^{-1}$. Therefore, since $s \ll 1$, we may replace $\bar{g}$ and $\bar{G}$ respectively by g and G. Moreover, since in the R-band region $|L(z \pm b)| \ll 1$, we may neglect in (273) the term proportional to $\{\cdot\}$. Then $S(z)$ approximately reduces to the *Lorentzian*:

$$S(z) \approx L_{\mathrm{LOR}}(z) = b(b-z)^{-1} + b(b+z)^{-1} \tag{275}$$

The same spectral dependence (the Lorentzian) is involved in empirical formula [17] for description of quasi-resonance FIR spectrum in water in the R-band region. However, now this spectral dependence, also describing the R-band, arises as a result of a linear-response theory.

2. *Account of Decaying Term* $\exp(-t/\tau_{\mathrm{vib}})$

Restoring this term in Eq. (237), we write $\exp(-t/\tau_{\mathrm{vib}}) = \exp(-w\varphi)$. Then we have instead of Eq. (267a):

$$\sigma[\varphi, \sigma_0, \dot{\sigma}_0, w] = [\sigma_0 \cos b\varphi + (\dot{\sigma}_0/b)\sin b\varphi]\exp(-w\varphi) \tag{276}$$

Inserting exponential multiplier $\exp(-w\varphi)$ into Eq. (270a) before $\cos b\varphi$, we obtain using transformations analogous to those which yield Eq. (273):

$$S(z) \equiv \frac{b - iw}{b - iw - z} + \frac{b + iw}{b + iw + z} + z\left\{\frac{L(z + b + iw)}{b + iw + z} - \frac{L(z - b + iw)}{b - iw - z}\right\} \quad (277)$$

Omitting the small last term given by $\{\cdot\}$, we finally have Eq. (242) with $S(z)$ in the form (243) instead of (275).

VII. COMPOSITE MODELS: APPLICATION TO WATER

This section presents a fundamental development of Sections V and VI. Here a linear dielectric response of liquid H_2O is investigated in terms of two processes characterized by two correlation times. One process involves reorientation of a single polar molecule, and the second one involves a cooperative process, namely, damped vibrations of H-bonded molecules. For the studies of the reorientation process the hat-curved model is employed, which was considered in detail in Section V. In this model a hat-like intermolecular potential comprises a flat bottom and parabolic walls followed by a constant potential. For the studies of vibration process two variants are employed.

(A) The harmonic oscillator (HO) model based on application of the parabolic potential, in which two charges $+\delta q$ and $-\delta q$ oscillate along the H-bond (the linear-response theory of the HO model based on the ACF method is given in detail in Section VII.C).

(B) The cosine-squared (CS) potential, in which vibration of a polar molecule is considered.

Both HC–HO and HC–CS models yield very good description of the wideband spectra in ordinary and heavy water, but interpretation of these compsite model differs.

A. Hat-Curved–Harmonic Oscillator Composite Model

1. *About Two Mechanisms of Dielectric Relaxation*

The fundamental importance of studies of water structure, of its molecular dynamics, and of wideband spectra arising due to the motion of polar molecules is governed by various scientific and technological applications of these studies. This importance is emphasized by the fact that just the structure and dynamics of water determine biological functioning of living organisms. For instance, proteins act in water surroundings as stochastic micromachines. In recent years the studies of the frequency spectrum evolved in two different directions, one towards another: (i) from the optical region to lower frequencies and (ii) in the reverse direction. These studies have now met at *submillimeter wavelengths*. It became clear that for a reasonable description of the wideband spectra of water, one should take into account not only the properties of water as a typical liquid but also those properties of a polar fluid, which *resemble a solid state of matter*.

The low-frequency Raman and far-infrared spectroscopy give, as well as other techniques, important information for a *qualitative* understanding of some physical processes in water. Following Walrafen's qualitative picture [16], the *specific* interactions arise from "the absorption from intermolecular vibrations such as hydrogen bond bending and stretching" centred at $\nu \sim 170\,\mathrm{cm}^{-1}$, while *unspecific* interactions are assumed "to arise from an extremely broad non specific absorption and, to a smaller extent, from the neighboring broad librational contour." Thus the mechanism giving rise the librational band is *excluded* in the classification [16] from unspecific interactions. On the contrary, in our treatment a simpler classification will be used: We shall *include* the librational-band mechanism into unspecific interactions, since in our theory actually *only two mechanisms* describe all the spectrum of the complex permittivity in water [see Eq. (278) below].

Thus, specific interactions *directly* determine the spectroscopic features due to *hydrogen bonding* of the water molecules, while unspecific interactions arise in all or many polar liquids and *are not* directly related to the H-bonds. Now it became clear that the basis of four different processes (terms) used in Ref. [17] and mentioned above could rationally be explained on a *molecular* basis. One may say that *specific interactions* are more or less *cooperative* in their nature. They reveal some features of a *solid state*, while unspecific interactions could be understood in terms of a liquid state of matter, if we consider chaotic gas-like motions of a *single* polar molecule, namely, rotational motions of a dipole in a dense surroundings of other molecules. The modern aspect of the spectroscopic studies leads us to a conclusion that both gas-like and solid-state-like effects are the characteristic features of water. In this section we will first distinguish between the following two mechanisms of dielectric relaxation:

(A) *Reorientation of a single polar molecule* during the mean lifetime τ_{or} in a rather narrow intermolecular potential well considered in Section V in terms of the *hat-curved* (HC) *model*.

(B) *Stretching vibration of a dipole formed by the H-bonded water molecules* considered in terms of the *harmonic oscillator* (HO) *model* [21, 37, VIG].

We shall combine the (A) and (B) mechanisms within a *composite* HC–HO model capable of describing the complex permittivity $\varepsilon^*(\nu)$ and absorption coefficient $\alpha(\nu)$ of liquid H_2O and D_2O. The theory will be given in a simple analytical form. We shall see that such a modeling could give an agreement with the experimental spectra [54–56], at least not worse and possibly better than the empirical description [17, 54–56] of the same spectra. An advantage of our modeling is (1) understanding of a nature of the described phenomena, (2) possibility to generalize *in the future* the theory capable of describing other aqueous systems (ice Ih, solutions of electrolytes, hydration of proteins, etc.), and (3) possibility to propose a calculation scheme capable of prediction the evolution of spectra due to the change in the temperature.

In Section V the *reorientation* mechanism (A) was investigated in terms of the *only* (hat curved) potential well. Correspondingly, the only stochastic process characterized by the Debye relaxation time τ_D was discussed there. This restriction has led to a poor description of the submillimeter (10–100 cm^{-1}) spectrum of water, since it is the second stochastic process which determines the frequency dependence $\varepsilon^*(\nu)$ in this frequency range. The specific *vibration* mechanism (B) is applied for investigation of the submillimetre and the far-infrared spectrum in water. Here we shall demonstrate that if the harmonic oscillator model is applied, the small isotope shift of the R-band could be interpreted as a result of a small difference of the masses of the water isotopes.

The role of specific interactions was not recognized for a long time. An important publication concerning this problem was the work by Liebe et al. [17], where a fine non-Debye behavior of the complex permittivity $\varepsilon^*(\nu)$ was discovered in the submillimeter frequency range. The new phenomenon was described as *the second Debye term* with the relaxation time τ_2, which was shown to be very short compared with the usual Debye relaxation time τ_D (note that τ_D and τ_2 comprise, respectively, about 10 and 0.3 ps). A physical nature of the processes, which determines the second Debye term, was not recognized nor in Ref. [17], nor later in a number works—for example, in Refs. 54–56, where the "double Debye" approach by Liebe et al. was successfully confirmed.

So, the goal of this section is to set forward investigation of both (A) and (B) mechanisms in the frames of one treatment. In sections VII.A.2–VII.A.4 the hat-curved and harmonic oscillator models will be described with the details sufficient for understanding the employed method of calculations. The results of the latter will be described in Section VII.A.5 and will be discussed in Section VII.A.6.

2. *Two Components of Complex Permittivity*

We employ the linear response theory based on a phenomenological molecular model of water. In the proposed *composite* HC–HO model the complex permittivity is represented as the sum

$$\varepsilon^*(\nu) = \varepsilon^*_{or}(\nu) + \Delta\varepsilon^*(\nu) \tag{278}$$

where the first and the last terms account for, respectively, the (A) and (B) mechanisms of molecular interactions/dielectric response mentioned in Section VII.A.1. We have from here the following expression for the *static* permittivity due to these mechanisms:

$$\varepsilon_s = \varepsilon_{s,\mathrm{or}} + \Delta\varepsilon_s \tag{279}$$

where $\Delta\varepsilon_s$ could be estimated from the experimental data as the difference

$$\Delta\varepsilon_s = \varepsilon_1 - \varepsilon_\infty \tag{280a}$$

of the "end points" of two relaxation regions described in an empirical representation as[47] [17, 54–56]:

$$\varepsilon^* = \varepsilon_\infty + \frac{\varepsilon_s - \varepsilon_1}{1 - 2\pi i c \nu \tau_D} + \frac{\varepsilon_1 - \varepsilon_\infty}{1 - 2\pi i c \nu \tau_2} \tag{280b}$$

We assume that a water molecule may participate like a solid body in two modes/types of motions (intermolecular interactions).

(1) The first type is a gas-like reorientation determined by a dipole moment $\boldsymbol{\mu}$ of a molecule in a liquid. For simplicity, just as in Sections IV–VI we consider that H_2O and D_2O are linear molecules, with the moment of inertia I.

(2) The second type of motion is *stretching vibration* along the H-bond direction. But *neutral* water molecules oscillating along some direction are *not* electrically active. To solve the problem of a relevant dielectric response, we propose here the *concept of an effective nonrigid dipole*; that is, we propose that the length r(O–H) of the covalent bond increases [57, 58] in liquid H_2O comparatively with this length in an isolated water molecule. For instance, in accord with [58] the bond length increases from 0.96 Å to 1.02 Å. Consequently, the distribution of the charges constituting the O–H bond *changes* in water compared with such distribution in an isolated water molecule. We suppose that in the space region, pertinent to O–H···O or O–D···O oscillating hydrogen bond, a time-varying dipole-moment component $\tilde{\mu}(t)$ arises. Such a component $\tilde{\mu}(t)$ was successfully employed for preliminary calculations in Refs. 6 and 8 and, with more details, in Section VI. We recognize that so far we have no possibility to specify *directly* a *microscopic* picture of charge/mass distribution from consideration of the water structure in terms of the harmonic oscillator model (or in any other terms). Thus, *we postulate*

[47] See also Eq. (A4) given in Appendix 3 of Section V.

existence charges $\pm\,\delta q$ *which constitute an effective nonrigid dipole* oscillating along the H-bond direction, although we do not know a mechanism of its formation. The ratio $\delta q_{\mathrm{vib}}^2/m_{\mathrm{vib}}$ of the charge squared to the vibrating mass serves as a fitting parameter of the HO model. Further estimations show that we possibly may identify m_{vib} with the mass of a water molecule. Then we may suggest that an effective nonrigid dipole introduced in our work is formed by *two water molecules*.

3. Reorientation Process

Taking into account the internal-field correction, replacing χ^* and ε^* by χ_{or}^* and $\varepsilon_{\mathrm{or}}^*$, we use the same relationships connecting the complex susceptibility and permittivity as were employed in Sections IV–VI:

$$\chi_{\mathrm{or}}^*(\nu) = \frac{\varepsilon_{\mathrm{or}}^*(\nu) - n_\infty^2}{4\pi}\,\frac{2\varepsilon_{\mathrm{or}}^*(\nu) + n_\infty^2}{3\varepsilon_{\mathrm{or}}^*(\nu)} \tag{281a}$$

$$\varepsilon_{\mathrm{or}}^* = \frac{1}{4}\left[12\pi\chi_{\mathrm{or}}^* + n_\infty^2 + \sqrt{(12\pi\chi_{\mathrm{or}}^* + n_\infty^2)^2 + 8n_\infty^4}\right] \tag{281b}$$

where n_∞ is the refraction index near the transparency region (for water $n_\infty^2 = 1.7$). Thus all optical permittivity, n_∞^2, which is assumed not to depend on frequency ν, is included into the reorientation term $\varepsilon_{\mathrm{or}}^*(\nu)$. The reorientation susceptibility $\chi_{\mathrm{or}}^*(\nu)$ is determined, taking into account the dissipation mechanism by the spectral function (SF) of the hat-curved model, as follows:

$$\chi_{\mathrm{or}}^*(x) = \frac{g_{\mathrm{or}}\, G_{\mathrm{or}}\, z\, L_{\mathrm{or}}(z)}{g_{\mathrm{or}}\, x + i\, y L_{\mathrm{or}}(z)} \tag{282}$$

$$x = \omega\eta, \qquad y = \eta/\tau, \qquad z = x + iy, \qquad \eta \equiv \sqrt{I/(2k_{\mathrm{B}}T)} \tag{283}$$

$$G_{\mathrm{or}} = \mu^2 N/(3k_{\mathrm{B}}T) \tag{284}$$

$$N = (\rho N_A)/M \tag{285}$$

$$\mu = (1/3)\mu_0 k_\mu (n_\infty^2 + 2) \tag{286}$$

The simplified formula for the spectral function $L_{\mathrm{or}}(z)$ is the same as used in Sections V and VI; see Section V, Eqs. (170)–(173), (179), (194a), and (194b). In Eq. (282) the Kirkwood correlation factor is given by

$$g_{\mathrm{or}} = \frac{(\varepsilon_s - \Delta\varepsilon_s) - n_\infty^2}{4\pi G_{\mathrm{or}}}\,\frac{2(\varepsilon_s - \Delta\varepsilon_s) + n_\infty^2}{3(\varepsilon_s - \Delta\varepsilon_s)} \tag{287}$$

If we set $\Delta\varepsilon^* = 0$, $\Delta\varepsilon_s = 0$, all formulas (281a)–(287) reduce to those given in Section IV.G.1, if we replace g_{or} by g, G_{or} by G, and χ_{or}^* by χ^*.

Equations (281b) and (282) determine the frequency dependence of the reorienting complex permittivity $\varepsilon^*_{\text{or}}(\nu)$. One can estimate the principal (Debye) relaxation time by using the relation

$$\tau_{\text{D}}^{\text{est}} = g\eta/[yL_{\text{or}}(iy)] \tag{288}$$

where $L_{\text{or}}(iy) \equiv L_{\text{or}}(z)_{x=0}$ is the "static" value of the spectral function. The right-hand side of Eq. (288) depends on the parameters of the hat-curved model. Hence, we can relate these parameters to the experimental data by using the equation

$$\tau_{\text{D}}^{\text{est}}(\text{parameters of the HC model}) = \tau_{\text{D}}(\text{from an experiment}) \tag{289}$$

4. *Vibration Process*

The expression for the vibration susceptibility is analogous to Eq. (281a) but we should cancel the optical permittivity n^2_∞, since it was included into the term χ^*_{or}. Hence,

$$\chi^*_{\text{vib}}(\nu) = \frac{1}{6\pi}\Delta\varepsilon^*(\nu) \tag{290a}$$

$$\chi^*_{\text{vib}}(0) = \frac{1}{6\pi}\Delta\varepsilon_s \tag{290b}$$

We use the harmonic oscillator model [18], (VIG, p. 27), which will be described in detail in Section VII.C. The vibration spectral function is given by the formula

$$L_{\text{vib}}(Z) \equiv L_{\text{HO}}(Z) = (1 - Z^2)^{-1} \tag{291}$$

where the reduced frequencies are expressed as

$$Z = X + iY, \qquad X = \omega/\Omega, \qquad Y = (\Omega\tau_{\text{vib}})^{-1} \tag{292}$$

and τ_{vib} is the mean lifetime of the vibrations. The relation of the complex susceptibility $\chi^*_{\text{vib}}(\nu)$ to the SF (291) is analogous to that given by Eq. (282) but with other parameters:

$$\chi^*_{\text{vib}}(x) = \frac{g_{\text{vib}}\, G_{\text{vib}} Z\, L_{\text{vib}}(Z)}{g_{\text{vib}}\, X + i\, Y\, L_{\text{vib}}(Z)} \tag{293}$$

where

$$G_{\text{vib}} = \frac{\delta q^2_{\text{vib}}(N_{\text{vib}}/2)}{3\bar{m}_{\text{vib}}\Omega^2} \approx \frac{\delta q^2_{\text{vib}} N_{\text{vib}}}{3m_{\text{vib}}\Omega^2} \approx \frac{\delta q^2_{\text{vib}} N}{3m_{\text{vib}}\Omega^2} \tag{294}$$

Here we assume that masses of two oscillating charged particles are approximately equal, so that the reduced mass $\bar{m}_{\text{vib}} = \frac{m_{1\text{vib}} m_{2\text{vib}}}{m_{1\text{vib}} + m_{2\text{vib}}}$ is about half of the mass m_{vib} of one such particle; for definiteness we suggest the concentration N_{vib} of oscillating charges to be equal to the concentration N of water molecules.[48]

The correlation factor g_{vib} of the hat-curved model could be related to the fitting parameter R_{vib} of the model [see Eqs. (296) and (297)]. We take the intrinsic oscillator frequency Ω near the center of the R-band, $\Omega = \sqrt{k/\bar{m}} = 2\pi c \nu_k$, by setting $\nu_k = 190\,\text{cm}^{-1}$. In the vicinity of this resonance frequency (at $X \approx 1$, or at $\omega \approx \Omega$), Eq. (293) is close to the Lorentz line

$$\chi^*_{\text{vib}}(Z) \approx \chi_{\text{LOR}}(Z) = G_{\text{vib}} L_{\text{HO}}(Z) = G_{\text{vib}}(1 - Z^2)^{-1} \tag{295}$$

while at *lower* frequencies Eq. (293) describes the complex susceptibility pertinent to the submillimeter wavelength region. The example will be given further.

For the parameterization of the HO model we introduce a dimensionless ratio $R_{\text{vib}} = G_{\text{vib}}/G_{\text{or}}$. Taking into account Eqs. (284), (285), and (294) we have

$$R_{\text{vib}} = \frac{G_{\text{vib}}}{G_{\text{or}}} = \frac{\delta q^2_{\text{vib}}}{m_{\text{vib}}} \frac{k_B T}{\mu^2 \Omega^2} \tag{296}$$

The vibration process could be characterized by a dimensionless charge squared to mass ratio

$$\gamma_{\text{vib}} = \frac{\delta q^2_{\text{vib}}/m_{\text{vib}}}{e^2/m} = \frac{\delta q^2_{\text{vib}}/m_{\text{vib}}}{e^2/(m_H M)} \tag{297}$$

where e is charge of electron and m_H is the mass of a proton. Fitting R_{vib} from the comparison of the theoretical and experimental spectra, we can estimate the ratio (297) from the relation

$$\gamma_{\text{vib}} = R_{\text{vib}} \left(\frac{\mu \Omega}{e} \right)^2 \frac{m_H M}{k_B T} \tag{298}$$

Taking into account Eqs. (290b), (296) and setting $x = 0$ in Eq. (293), we have

$$\chi_{\text{vib}}(0) = \frac{\Delta \varepsilon_s}{6\pi} = g_{\text{vib}} G_{\text{vib}} = g_{\text{vib}} G_{\text{or}} R_{\text{vib}} \tag{299}$$

[48] It is seen from Eq. (294) that the same value of the factor G_{vib} can be obtained for different ratios $N_{\text{vib}}/m_{\text{vib}}$.

Next, we can express the vibration correlation factor $g_{\rm vib}$ through the fitting parameter $R_{\rm vib}$ as

$$g_{\rm vib} = \frac{\Delta\varepsilon_s}{6\pi G_{\rm or} R_{\rm vib}} = \frac{3k_{\rm B}T\,\Delta\varepsilon_s}{6\pi R_{\rm vib}\mu^2 N} = \frac{k_{\rm B}TM\,\Delta\varepsilon_s}{2\pi R_{\rm vib}\mu^2 N_A \rho} \tag{300}$$

where $\Delta\varepsilon_s$ may be related to the experimental low-frequency water spectra from Eq. (280a).

The frequency dependence of the complex permittivity pertinent to the vibration process is described by Eqs. (290a) and (293):

$$\Delta\varepsilon(\nu) = \frac{6\pi g_{\rm vib}\, GR_{\rm vib} Z(\nu)\, L_{\rm vib}(Z)}{g_{\rm vib}\, X + i\,Y L_{\rm vib}(Z)} \tag{301}$$

$$Z \equiv Z(\nu) = (2\pi c\nu/\Omega) + i\,Y \tag{302}$$

Having found the loss-maximum frequency $\nu_{\rm Dvib}$, (if it arises),

$$\mathrm{Im}\{\Delta\varepsilon^*(\nu)|_{\nu=\nu_{\rm Dvib}}\} = \max \tag{303}$$

we may roughly estimate the relaxation time of the second stochastic process: $\tau_{\rm Dvib} = (2\pi\, c\nu_{\rm Dvib})^{-1}$.

5. *Results of Calculations*

a. Free and Statistical Parameters. Just as in Sections V and VI, we assume that the angular half-width β of the well be small ($\beta \ll \pi/2$) and the well depth $U_{\rm or}$ substantially exceeds $k_{\rm B}T$, so that the reduced well depth $u = U_{\rm or}/(k_{\rm B}T)$ is large ($u \gg 1$). We use spherical coordinates: polar angle θ and azimuthal coordinate ϕ, where θ is deflection of a dipole-moment vector $\boldsymbol{\mu}$ from the symmetry axis of the potential and ϕ is the precession angle, denoting angular displacement of $\boldsymbol{\mu}$ about this axis in a plane perpendicular to the latter. In the whole arc 2π of the circle we introduce two identical wells with oppositely directed symmetry axes. Below we set θ to vary in a quarter of the arc, $0 \le \theta \le \pi/2$. The potential $U(\theta)$ has a *flat* part $f\beta$ on the bottom of the well, and in its remainder part U depends on θ by the *parabolic* law. At the edge of the well, where $\theta = \beta$, U reaches its maximum value $U_{\rm or}$. Outside the well the potential is constant, $U(\theta) \equiv U_{\rm or}$. We introduce the form factor f defined as a ratio of the flat part to the whole well width 2β, where f may vary from 0 to 1. So, the potential profile is given by

$$U(\theta)/(k_{\rm B}T) = 0 \quad \text{for } 0 \le \theta \le \beta f \qquad \text{(flat part inside the well)} \tag{304a}$$

$$U(\theta)/(k_{\rm B}T) = S^2(\theta - \beta f)^2 \quad \text{for } \beta f \le \theta \le \beta \qquad \text{(parabolic part of the well)} \tag{304b}$$

$$U(\theta)/(k_{\rm B}T) = u \quad \text{for } \beta \le \theta \le \pi/2 \qquad \text{(flat part outside the well)} \tag{304c}$$

where $S \equiv \sqrt{u}(\beta - \beta f)^{-1}$ is termed the steepness of the parabolic part of the well. The profile of the well calculated for the case of water is illustrated further by Figs. 40a and 40b.

The main advantage of the hat-curved potential is that it is possible to narrow the width $\Delta\nu_{or}$ of the librational absorption band by decreasing the form factor f. Indeed, $\Delta\nu_{or}$ attains its maximum value when $f = 1$. Note that $f = 1$ is just the case of the hat flat or its simplified variant, the hybrid model, both of which were described in Section IV. The latter was often applied before (VIG) and is characterized by a rather wide absorption band, especially in the case of heavy water. In another extreme case, $f \to 0$, the linewidth $\Delta\nu_{or}$ becomes very low. When $f = 0$, we have the case of the parabolic potential well, whose dielectric response was described, for example, in GT and VIG. Thus, when the form factor f of the hat-curved well decreases from 1 to 0, the width $\Delta\nu_{or}$ decreases from its maximum to some minimum value.

All the dipoles could be distributed over three groups comprising "librators," "precessors," and "rotators." Our calculations are made in the *planar libration–regular precession* approximation. For the set of parameters relevant to water (with $\beta \ll 1$ and $u \gg 1$) the contribution of the precessors to the spectral function is negligible, then we represent L_{HC} as the sum

$$L_{or}(z) = L_{HC}(z) \approx \breve{L}(z) + \mathring{L}(z) \tag{305}$$

of two *simple* integrals over energy from elementary functions; see formulas (170)–(173), (179), (194a), and (194b) given in Section V. In view of Eqs. (281b) and (282), this expression determines the complex permittivity $\varepsilon^*(\nu)$ of reorienting particles.

Using Eqs. (278), (281b), (282), and (301), we can calculate the total complex permittivity $\varepsilon^*(\nu)$ from the above theory. After that we can find the absorption coefficient $\alpha(\nu)$, its components $\alpha_{or}(\nu)$ and $\alpha_{vib}(\nu)$ due to the reorientation and vibration processes, and the refraction index n:

$$\alpha(\nu) = \alpha_{or}(\nu) + \alpha_{vib}(\nu) = 2\pi\nu\frac{\varepsilon''(\nu)}{n(\nu)} \tag{306a}$$

$$\alpha_{or} = 2\pi\nu\frac{\varepsilon''_{or}(\nu)}{n(\nu)}, \qquad \alpha_{vib} = 2\pi\nu\frac{\Delta\varepsilon''(\nu)}{n(\nu)} \tag{306b}$$

$$n(\nu) = \mathrm{Re}\sqrt{\varepsilon^*(\nu)} = \frac{1}{\sqrt{2}}\sqrt{\sqrt{\varepsilon'^2 + \varepsilon''^2} + \varepsilon'} \tag{306c}$$

Our composite model is characterized by six free parameters having definite physical meanings:

$$\underbrace{u, \beta, f, \tau_{or};}_{\text{for the hat-curved model}} \qquad \underbrace{\tau_{vib}, R_{vib}}_{\text{for the harmonic oscillator model}} \tag{307}$$

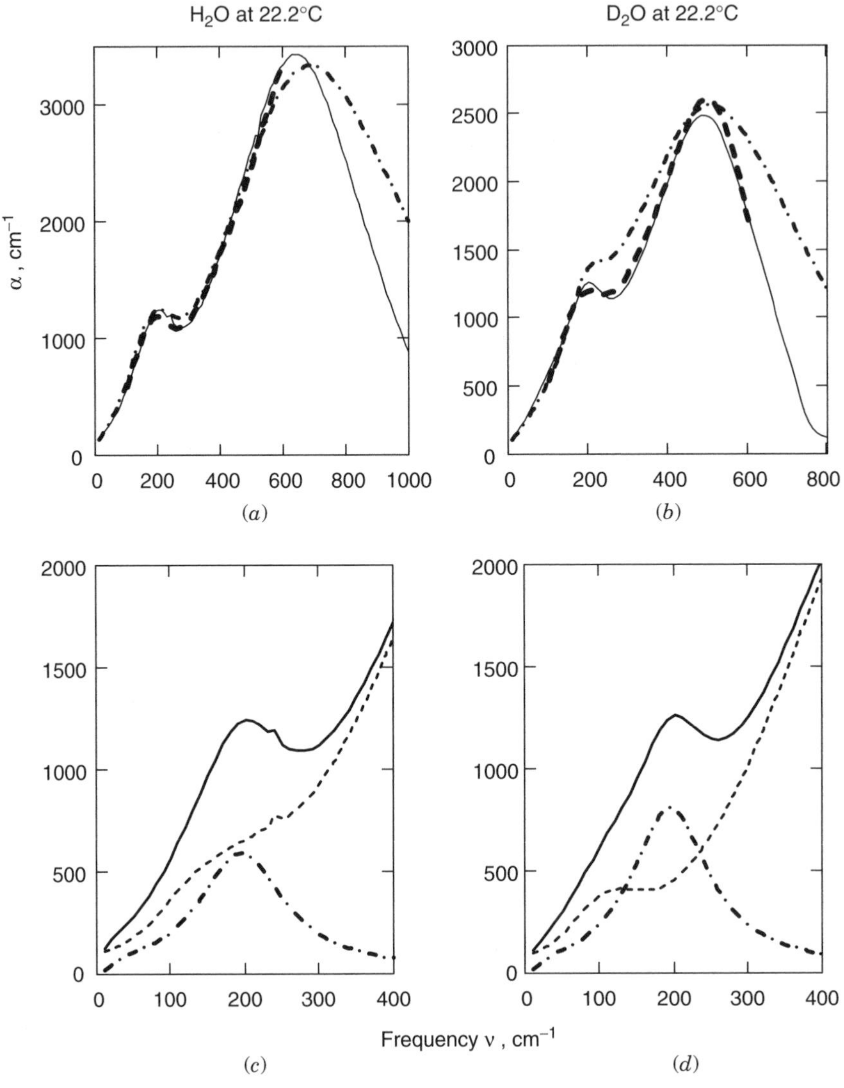

Figure 34. (a, b) Wideband frequency dependence of the absorption coefficient of ordinary (a) and heavy (b) water. Calculation for the hat-curved–harmonic oscillator model (solid lines) and for the hybrid–cosine-squared potential model (dashed-and-dotted lines). The experimental data [51] are shown by dashed lines. (c, d) The absorption coefficient versus frequency dependence in the R-band region. Calculation for ordinary (c) and heavy (d) water in terms of the hat-curved–harmonic oscillator model (solid lines); contributions to this absorption due to reorientations (dashed lines) and dut to stretching vibrations (dashed-and-dotted lines). Temperature 22.2°C.

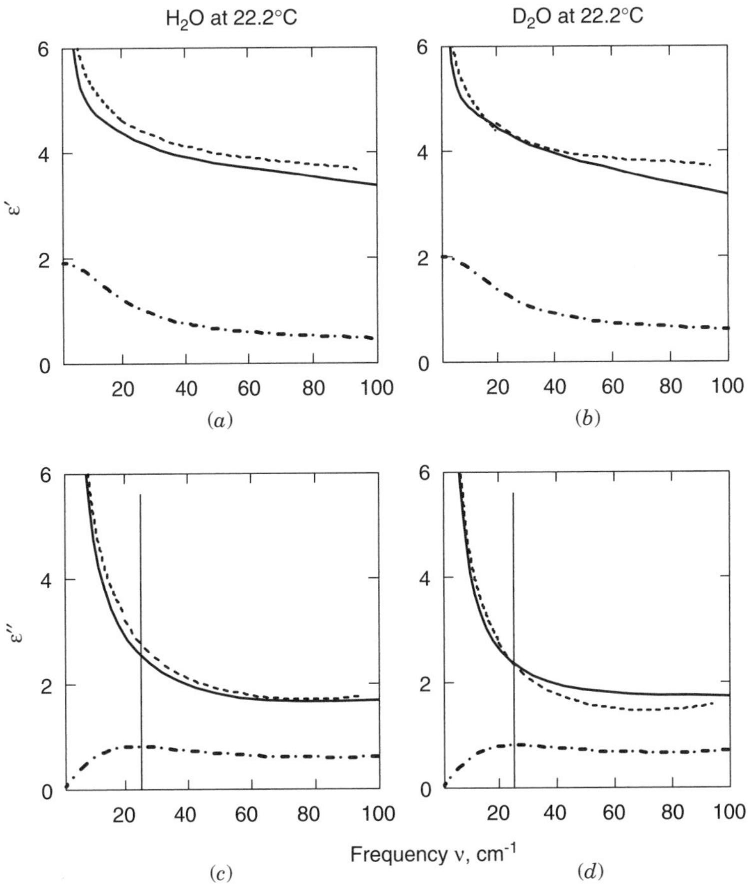

Figure 35. Frequency dependence in the submillimeter wavelength region of the real (a, b) and imaginary (c, d) parts of the complex permittivity. Solid lines: Calculation for the composite HC–HO model. Dashed lines: Experimental data [51]. Dashed-and-dotted lines show the contributions to the calculated quantities due to stretching vibrations of an effective non-rigid dipole. The vertical lines are pertinent to the estimated frequency ν_{Dvib} of the second stochastic process. Parts (a) and (c) refer to ordinary water, and parts (b) and (d) refer to heavy water. Temperature 22.2°C.

The first group of the parameters determine the frequency dependences in the Debye and librational bands and the second group—in the submillimeter wavelength range and in the R-band. A few statistical parameters of the composite model are determined by the same formulas as were given in Sections V and VI.

The parameters (307) affect the theoretical spectra in a complicated manner. It is convenient to start from the parameterization of the hat-curved model in order to obtain first a rough description of the librational and Debye bands. Then one could roughly describe the R-band and the submillimetre spectrum by fitting the parameters of the harmonic oscillator model. After that, one may return to the librational and Debye bands, and so on.

Now we apply the composite model to liquid H_2O and D_2O water. The results of the calculations are demonstrated in Figs. 34–39 by solid lines; Figs. 34–36 refer to ordinary and heavy water at 22.2°C, and Figs. 37–39 refer to ordinary water at 27°C. The used experimental data were given in Table XII. The fitted/estimated parameters are presented in Table XIV. In the latter, ν_D and ν_{Dvib} are the frequencies $(2\pi c\tau_D)^{-1}$ and $(2\pi c\tau_{Dvib})^{-1}$ corresponding to the Debye and "fast" relaxation times.

b. Ordinary and Heavy Water: Comparison of Calculated Spectra at 22.2°C. In the R-band the *experimental* absorption curve $\alpha(\nu)$ [51] shown by dashed curves exhibits in the case of H_2O a distinct maximum (Fig. 34a), and the *shoulder* in the case of D_2O (Fig. 34b). Here we see that the calculated (solid lines) absorption $\alpha(\nu)$ agrees reasonably in the far infra-red region with the recorded spectra. The libration band calculated by using the hat-curved model is noticeably narrower than that calculated for the rectangular potential (namely, for the hybrid model; cf. solid and dashed-and-dotted lines). Thus the hat-curved modification of the rectangular-well model gives a satisfactory description of the wideband spectra also in the case of liquid D_2O.

In Fig. 34c (for H_2O) and Fig. 34d (for D_2O), solid lines show the absorption coefficients α calculated in the *R-band*; the estimated contributions to this $\alpha(\nu)$ due to reorientations and vibrations are shown by dashed and dashed-and-dotted curves, respectively. The resonance *peak* at $\nu_R \approx 200\,\text{cm}^{-1}$ found for both water isotopes is actually determined by a vibrating nonrigid dipole. The distinction between the curves calculated for ordinary and heavy water is substantial. In the case of H_2O, Fig. 34c, the contributions of reorientations and vibrations to the resulting $\alpha(\nu)$ curve are commensurable near the center of the R-absorption peak, while in the case of D_2O (Fig. 34d) the main contribution to $\alpha(\nu)$ near the R-band peak comes from *vibration* of the H-bonded molecules.

An account of vibrations—namely, of the term $\Delta\varepsilon^*(\nu)$ in Eq. (278)—noticeably improves the agreement between the spectra calculated and the recorded in the *submillimeter* wavelength range (cf. the solid and dashed lines in Fig. 35). Our theory would give [9] a poor description of the experiment [51], especially in the case of liquid D_2O [140], if we ignore the vibration (second) term in Eq. (278). The vibrations' contribution to the dielectric loss shown in Figs. 35c and 35d by dashed-and-dotted lines is substantial *just* in submillimeter wavelength region. Since the vibration loss curve is almost flat in this region, we conclude that the *second stochastic (vibration) process is overdamped.* This

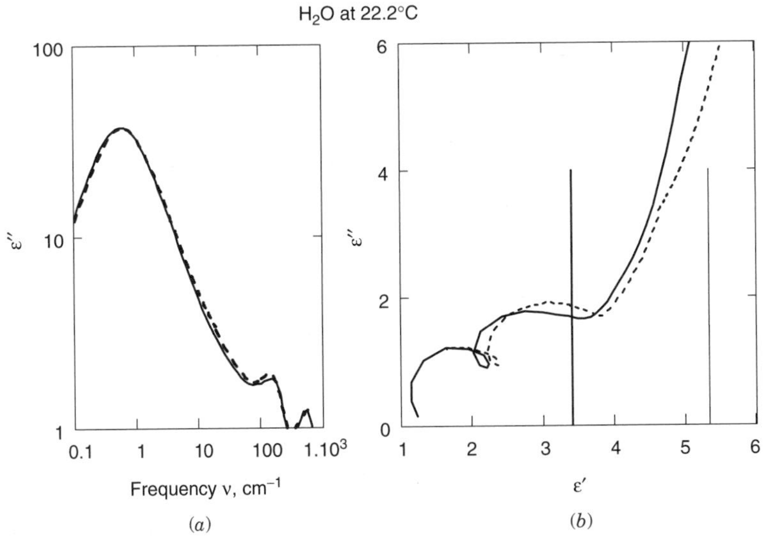

Figure 36. The wideband loss frequency dependence (a) and the far-infrared part of Cole–Cole diagram (b) calculated (solid lines) and measured [51] (dashed lines) for liquid H_2O at 22.2°C. The right and left vertical lines refer to the "ends" of the Debye and of the second-relaxation regions, respectively.

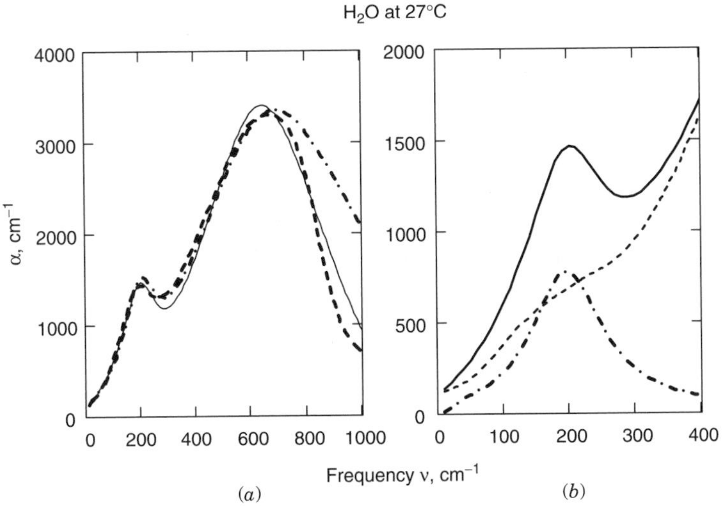

Figure 37. Absorption-frequency dependence, water H_2O at temperature 27°C. Calculation for the HC–HO model (solid line) and for the hybrid–cosine-squared potential model (dashed-and-dotted line). Dahsed curve: Experimental data [42]. (b) Same as in Fig. 34c but refers to $T = 300\,K$.

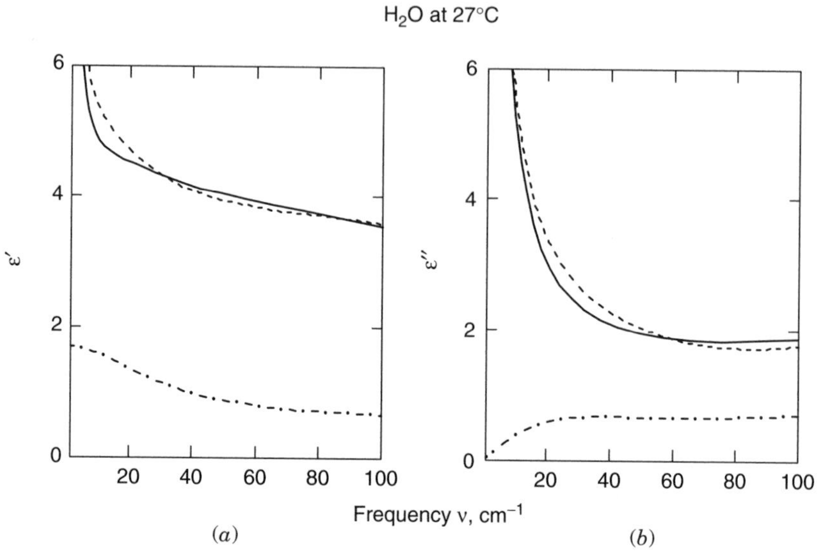

Figure 38. The (a) and (b) dependences are the same as Figs. 35a and 35c, respectively, but refer to water H_2O at $T = 300\,K$ (the vertical line is omitted).

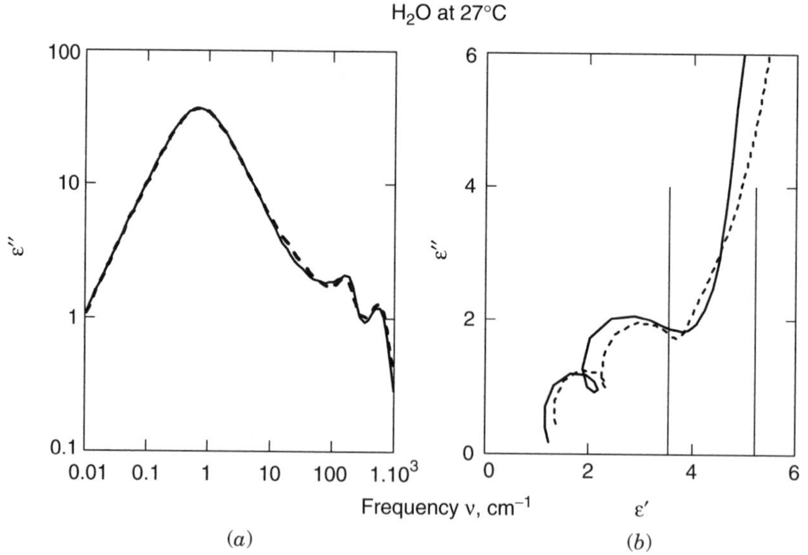

Figure 39. The (a) and (b) dependences are the same as in Figs. 36a and 36b, respectively, but refer to water H_2O at $T = 300\,K$.

TABLE XIV
Fitted and Estimated Parameters of the Composite Hat-Curved–Harmonic Oscillator Model

A. Fitted Parameters

		Parameters of the Hat-Curved Model				Parameters of the HO Model	
Water	°C	u	β (deg)	f	τ_{or} (ps)	τ_{vib} (ps)	Rvib
H_2O	27	5.8	23	0.8	0.22	0.065	0.025
H_2O	22.2	5.8	23	0.8	0.25	0.065	0.020
D_2O	22.2	5.75	25	0.7	0.28	0.07	0.027

B. Estimated Parameters

Water	°C	τ_{Dvib}(ps)	γ_{vib}	y	Y	ν_D (cm)$^{-1}$	ν_{vibD} (cm)$^{-1}$	$\mathring{r}$ (%)	m_{or}	m_{vib}
H_2O	27	—	0.55	0.19	0.43	0.68	—	3.9	4.4	0.37
H_2O	22.2	0.22	0.44	0.17	0.43	0.60	23.7	3.5	5.0	0.37
D_2O	22.2	0.18	0.59	0.21	0.40	0.49	30	4.1	4.0	0.40

result additionally confirms analogous conclusion made in Section VI in terms of a primitive composite model.

We also demonstrate for H_2O a good coincidence of the calculated/measured *wideband* loss spectra and of the high-frequency part of the Cole–Cole diagram (see, respectively, Figs. 36a and 36b. In the latter the right and left vertical lines mark, respectively, the end points of the Debye and the second relaxation regions, which correspond to the double Debye empirical representation, Eq. (280b). Note that the Debye loss frequency $\nu_D = (2\pi c\tau_D)^{-1}$ is about 40 times lower than that, $\nu_{Dvib} = (2\pi c\tau_D \text{vib})^{-1}$, which refers to the above-mentioned concave loss maximum pertinent to submillimeter wavelengths. It is evident that the form of the Cole–Cole diagram referring to the 10- to 100-cm^{-1} frequency region has nothing common with the "usual" Cole–Cole diagram, which has the form of a semicircle and is recorded in the Debye relaxation region—that is, at microwaves.

c. Wideband Spectra of Water H_2O at 27°C. Figures 34–36 refer to 22°C while Figs. 37–39 refer to 27°C. Taking Table XIV into account and comparing Figs. 34a,c with Figs. 37a,b, Figs. 35a,c with Figs. 38a,b, and Figs. 36a,b with Figs. 39a,b, one can get preliminary information on how the temperature influences the spectra of water and how the parameters of our composite model reflect this influence.

For 27°C we compare in Fig. 37a the theoretical absorption coefficient (solid lines) with the experimental one (dashed lines) measured [42] in a very wide frequency range, namely, from 10 cm^{-1} to 800 cm^{-1}. (For lower frequencies the empirical formula [17] given by Eq. (152) is used for such a comparison.) We see that a satisfactory agreement between the experiment and the theory elaborated for our composite model is obtained also for this temperature (27°C). As Fig. 37a demonstrates, some discrepancy between the calculated and the measured absorption curves still remains in the frequency region from 800 to 1000 cm^{-1}. Indeed, the theoretical absorption $\alpha(\nu)$ curve falls off to the transparency region more slowly than the recorded one. However, this discrepancy is less than that which was found in terms of the hybrid model (see Section IV.G.b). The progress is seen, if one compares the solid and dashed-and-dotted lines shown in Fig. 37a. Similar to the latter line is the result of the empirical approximation by Liebe et al. [17] based (in this range) on the last Lorentz term. This empirical approach gives a substantially worse description of the high-frequency wing of the librational band than does the hat-curved model. The "mathematical" reason for such a discrepancy is that the Lorentz line tends to zero too slowly at $\nu \to \infty$.

When the temperature rises, the Debye relaxation time τ_D and the fitted mean reorientation time τ_{or} decrease, since intermolecular interactions weaken and become more chaotic.

6. *HC–HO Model: Discussion*

A general picture of the dielectric spectra described in Section VI is confirmed here in main features, when a more advanced composite hat-curved–harmonic oscillator model was employed. Better agreement of theory with experimental data is obtained. Not less important is change of the conncepts underlying *interpretation* of specific interactions in water.

a. Narrowing of Librational Band. The key aspect of the model described here is that we can control the *width* of the librational band by changing the form factor f. Application to *ordinary water* of the hat-curved potential instead of the rectangular one allows to narrower the librational band; see solid lines in Fig. 34a for 22.2°C and in Figs. 37a for 27°C. The calculated absorption band reasonably agrees with the experimental data, if the fitted form factor f is taken $\approx$0.8. This result is more emphasized in the case of liquid D_2O (Fig. 34b), where the fitted $f \approx 0.7$ and both wings of the calculated librational band become narrower than for $f = 1$. Thus, when the hat-curved model is applied to *heavy* water, the calculated dielectric–far-infra red spectra generally agree with the experimental spectra [51], whereas for the rectangular profile $(f = 1)$ a rather wide absorption band is characteristic (VIG) [9], especially in the case of heavy water. Thus, an *a priori* idea underlying the hat-curved model is confirmed in this work for liquid H_2O at two temperatures and for liquid D_2O at 22.2°C. Note that estimation (190a) gives a correct order of magnitude of the experimental librational-band center frequency ν_{lib}.

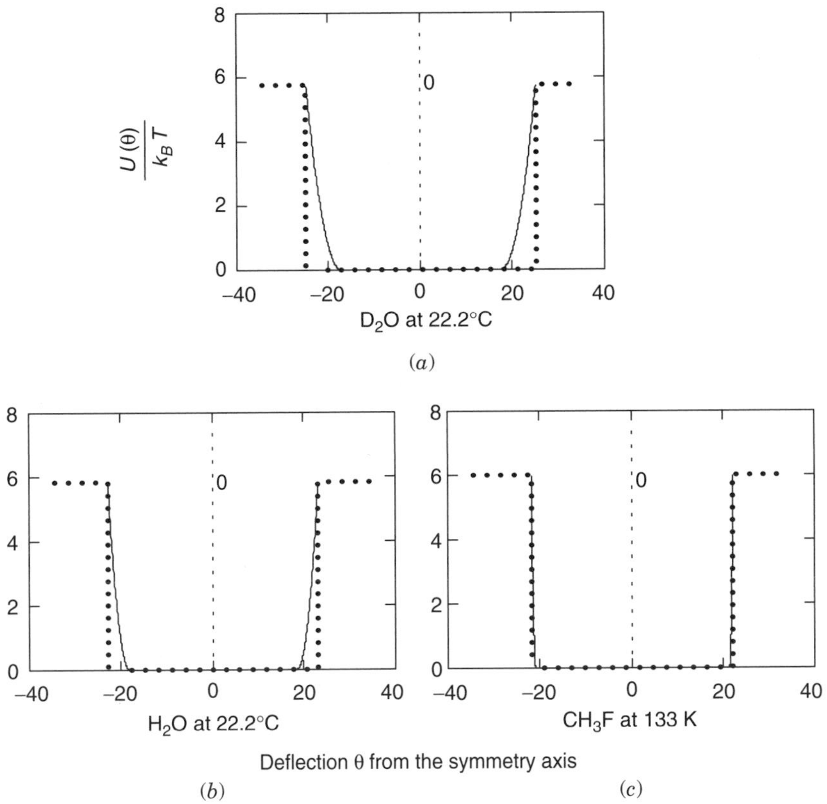

Figure 40. The hat-curved potential profile pertinent to D_2O at 22.2°C (a), H_2O at 22.2°C (b), and CH_3F at 133K (c). Calculation for the hat-curved model with the following parameters: $f = 0.7, u = 5.75, \beta = 25°$ in (a); $f = 0.8, u = 5.8, \beta = 23°$ in (b); and $f = 0.96, u = 6, \beta = 22°$ in (c).

b. Potential Profile. In Fig. 40 we compare the forms of the potential well fitted in this work for D_2O (Fig. 40a) and H_2O (Fig. 40b) and also (see Section IV.G.3) fitted for a strongly polar nonassociated liquid CH_3F at 133 K—that is, near the triple point (Fig. 40c). The stronger the interactions in a liquid, the smaller the fitted form factor f—that is, the more the potential profile declines from the rectangular one. Unspecific interactions characteristic for a nonassociated liquid, which give rise to the far-infrared absorption and to the low-frequency Debye loss, resemble similar frequency dependences characteristic for liquid water. If we apply the same (hat-curved) model for calculation of the absorption spectra, then the main distinctions between a nonassociated (e.g., CH_3F) and associated (water) fluids become as follows. In the case of the first medium (e.g., CH_3F):

(i) The number of the reorientation cycles m_{or} is much lower than in the case of water, thus indicating that molecular rotation is more damped and chaotic than in associated liquids.

(ii) There is an absence of specific interactions (and correspondingly of the R-band) arising due to H-bonding of the molecules.

(iii) It has a greater form factor: $f(CH_3F) \approx 0.96$, while $f(H_2O) \approx 0.8$ and $f(D_2O) \approx 0.7$.

We suggest to employ the parameter f as some measure of localization of intermolecular interactions—that is, of a *spread of short-range forces* in a liquid: The lower the value of f, the stronger these forces at larger distances. We estimate roughly this spread as a linear length of the curved part $(1-f)\beta$ of the well:

$$\langle \text{spread} \rangle = (1-f)\,\beta\, r_{\text{mol}} \tag{308}$$

where r_{mol} is a mean radius of a reorienting molecule. Taking for water $r_{\text{mol}} = 1.5\,\text{Å}$ and the parameters f and β from Table XIV and taking for CH_3F $f = 0.96$, $\beta = 22.9°$, $r_{\text{mol}} = N^{-1/3} = 3.9\,\text{Å}$ (GT, p. 352), we find

$$\langle \text{spread} \rangle \approx \begin{cases} 0.12\,\text{Å} & \text{for } H_2O \text{ at 22 and } 27°\text{C} \\ 0.2\,\text{Å} & \text{for } D_2O \text{ at } 22°\text{C} \\ 0.03\,\text{Å} & \text{for } CH_3F \text{ at 133 K} \end{cases}$$

Thus, the parameter (308) is for heavy water about twice that estimated for ordinary water. Our result for water corresponds to a known [59, 60] statement that *in heavy water the H-bond is stronger than in ordinary water*. Moreover, one may suggest that in ice Ih the form factor f should be lower than in D_2O, since the librational band in ice is narrower than in water (the comparison of these bands are given in VIG, in Ref. 61, and in Section X. On the other hand, in a nonassociated liquid the so found spread is about 4 times lower than in the case of H_2O; this result also looks reasonable.

Now we shall point out on an *experimental* confirmation concerning the role of a cooperative motions in formation of the R-band in water. In a recent paper [62] it was inferred from an isotope shift of the R-band that the oscillations to this band are stipulated by *molecules deformations*. Next, in view of work [63], where the Raman spectra of water substituted with respect to hydrogen and oxygen were studied, one can conclude that the isotope shift is greater for the oxygen substitution. This result means that the oscillations pertinent to the

R-band are mostly connected with motions of oxygen atoms, which actually determine the harmonic-oscillator response. Hence, these facts would be difficult to interpret if we didn't accept the viewpoint that cooperative effects should be superimposed on the motions of a water molecule as a whole.

c. Two Types of Molecular Motion in Water: Physical Picture. We gave here a crude mathematical treatment, in terms of the harmonic oscillator model, of the Walrafen's idea [16] that important spectral features arise in water at submillimeter wavelengths due to vibrations along the H-bond of nonrigid O–H···O or D–H···D dipoles. In view of Table XIV, the charge squared to mass ratio $\delta q_{\text{vib}}^2/m_{\text{vib}}$ estimated from Eq. (297) is commensurable with the e^2/m ratio, where e is the charge of an electron and m is the mass of a water molecule. Since the parameter γ_{vib} [Eq. (297)] determined from *independently* fitted parameters (307) of the model is approximately the same ($\gamma_{\text{vib}} \approx 1/2$) for both water isotopes and for both temperatures studied above, it appears reasonable to identify m_{vib} with the *mass of a water molecule* and to suggest that an introduced vibrating effective nonrigid dipole is formed by *two water molecules.* Then we find that our model is capable of explaining a small difference[49] between the experimentally observed center frequencies ν_R of both water isotopes. In terms of our approach, these frequencies are determined by the masses of the vibrating water molecules.

We may suggest the following physical picture of two processes considered in this section. *Each molecule* of a liquid water, obviously, may suffer *each state* of molecular motion. It looks reasonable that rather long reorientation periods take turns with shorter vibration periods. A reorienting molecule moves in a rather hollow space (like that determined by the hat-curved potential) but in dense surrounding of other water molecules; its law of motion is determined by the effective moment of inertia I. In the second state, interaction of at least two molecules occurs in a solid-body-like fashion. The law of motion is now determined by the charge squared to the reduced mass ratio characterizing a couple of the particles. Rather long ($\sim$0.25 ps) reorientations determine the intense infrared absorption band and the complex permittivity in the Debye range. Short ($\sim$0.07 ps) damped stretching vibrations determine the weak R-band and the plateau-like loss $\varepsilon''(\nu)$ band in the submillimeter wavelength region. Specific vibration mechanism combined with unspecific reorientation mechanism yields good description of the wideband $\varepsilon'(\nu)$ and $\varepsilon''(\nu)$ spectra of liquid H_2O and D_2O (see Figs. 35 and 38). Note that the theory of the R-band given in Section VI in terms of a harmonically varying component $\mu_{\text{vib}}(t)$ of a total dipole moment does not yield, unlike the HC–HO model, a satisfactory

[49]This difference was set zero in our calculations.

description of the *submillimeter* spectrum in water, since existence of two potential wells and thus of the second stochastic process were not considered in Section VI.

The hat-curved–harmonic oscillator model, unlike other descriptions of the complex permittivity available now for us [17, 55, 56, 64], gives some insight into the mechanisms governing the experimental spectra. Namely, the estimated relaxation time of a nonrigid dipole ($\tau_{\text{Dvib}} \approx 0.2\,\text{ps}$) is close to that determined in the course of very accurate experimental investigations and of their statistical treatment [17, 54–56]. The reduced parameters presented in Tables XIVA and XIVB and the form of the hat-curved potential well (determined by the parameters u, β, f) do not show marked dependence on the temperature, while the spectra themselves vary with T in greater extent. We shall continue discussion of these results in Section X.A.

d. Other Properties

(a) The lifetime τ_{or} of reorientation is several (3.4–4) times longer than the vibration lifetime τ_{vib} (see Table XIV).

(b) The integrated absorption [18, p. 33], (VIG, p. 169)

$$\prod = \int_0^{\infty} \omega\chi''(\omega)\,d\omega = \frac{\pi}{3}\begin{Bmatrix} (N_{\text{vib}}/2)\delta q_{\text{vib}}^2/\bar{m}_{\text{vib}} \\ 2\mu^2 N_{\text{or}}/I \end{Bmatrix} \begin{cases} \text{for vibrations} & (309) \\ \text{for reorientations} & (310) \end{cases}$$

calculated for reorientations is also several times greater than that calculated for vibrations:

$$\prod_{\text{or}} \Big/ \prod_{\text{vib}} \approx \begin{cases} 0.047 & \text{for } H_2O \text{ at } 22.2°\text{C} \\ 0.112 & \text{for } D_2O \text{ at } 22.2°\text{C} \end{cases}$$

Note that in this estimation we again set $N_{\text{vib}} = N_{\text{or}} = N = N_A\rho/M$ and $\bar{m}_{\text{vib}} = m_{\text{vib}}/2$; the last ratio is about two times larger for heavy than for ordinary water.

(c) The mean-squared amplitude a of harmonic oscillations, found in Section VII.A.6, is

$$a \approx \frac{1}{2\pi c\,\nu_{\text{R}}}\sqrt{\frac{2k_{\text{B}}T}{m_H M}} \tag{311}$$

For the fitted parameters $a \approx 0.2$ Å. This estimation shows that a is *small* as compared with the mean radius of a water molecule.

Figure 41 summarizes information about the described composite model.

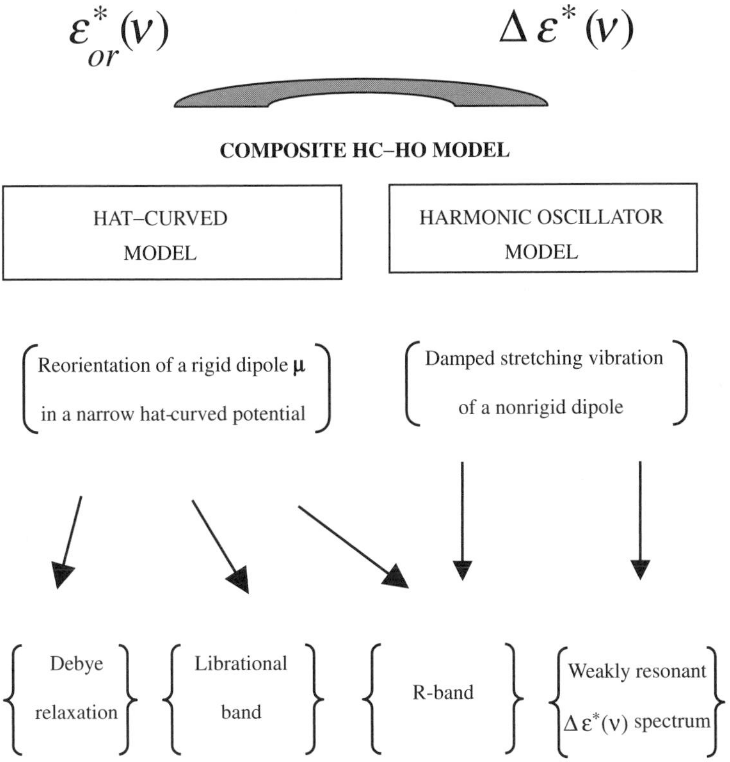

Figure 41. The scheme pertaining to the composite hat-curved–harmonic oscillator model: the contributions of various mechanisms of dielectric relaxation to broadband spectra arising in liquid water. Frequency ν is given in cm^{-1}.

B. Hat-Curved–Cosine-Squared Potential Composite Model

1. Theory and Calculated Spectra of H_2O and D_2O

We shall remove an important drawback of the polarization model described in Section VI by considering another variant of a composite model than that described in previous Section VII.A. We use again a *linear-response* theory to find the contribution of a vibrating dipole to the total permittivity ε^*. We split the total concentration N of polar molecules into the sum N_{or} and N_{vib}, where each term refers to rotation of a like *rigid* dipole (viz. with the same electric moment μ) but characterized by *different law of motion*:

$$N = N_{\text{or}} + N_{\text{vib}},\, viz.\, N_{\text{or}} = N(1 - R_{\text{vib}});\;\; R_{\text{vib}} \equiv N_{\text{vib}}/N, \tag{312}$$

where R_{vib} is actually the *proportion of the H-bonded molecules.*

As in Section VII.A, we consider *two* potential wells and represent the complex permittivity as a sum

$$\varepsilon^*(\nu) = \varepsilon^*_{\text{or}}(\nu) + \Delta\varepsilon^*(\nu) \tag{313}$$

of two terms, whose meanings are different from those in relationship (278). The first term arises due to reorientation of a larger fraction of molecules rotating in the hat-curved potential, and the second term due to vibration in a plane of a smaller fraction referring to the cosine-squared (CS) potential well. A cardinal distinction of this approach from that described in Section VII.A is that we (i) replace the parabolic well by the potential $U_{\text{or}}(\vartheta)$ *having a finite well depth* U_{vib} and (ii) consider a vibrating *dipole* instead of vibrating *charges*. The second term in Eq. (313) is characterized by the fast relaxation time τ_{Dvib}. Further results *radically* differ, as we shall see, also from those obtained in Ref. 12b, where a representation similar to Eq. (313) was employed but the first term was found for the rectangular (not for the hat-curved) well.

The intermolecular CS potential $U_{\text{CS}}(\vartheta)$ and the static field $E_{\text{CS}}(\vartheta)$ are respectively given by

$$\frac{U_{\text{CS}}(\vartheta)}{k_{\text{B}}T} = p^2(1 - \cos^2\vartheta) \quad \text{and} \quad E_{\text{CS}}(\vartheta) = \frac{U_{\text{vib}}}{\mu_{\text{vib}}}\cos\vartheta \tag{314}$$

where p is the *field parameter* of the CS well given by $p^2 \equiv U_{\text{vib}}/(k_B T)$. Two bottoms of this potential are placed at $\vartheta = 0$ and $\vartheta = \pi$; near these angles the static field changes direction. For a typical shallow well (we take $p = 0.8$) the dependences (314) are illustrated in Fig. 42.

Since the lifetime τ_{vib} of *vibrating* particles is assumed to differ from τ_{or}, we introduce reduced collision and complex frequencies $Y = \eta\tau_{\text{vib}}^{-1}$ and $Z = x + iY$ as differing, respectively, from y and z given by Eq. (283). The relationships (290a), (290b), and (293) for the vibrating complex susceptibility hold. We employ expression[50] for the spectral function (VIG, p. 462); GT, p. 300; see also Ref. 12b, for which the "stratified" approximation is used:

[50]Here we have corrected two misprints in Eqs. (12.88) and (12.91) of VIG.

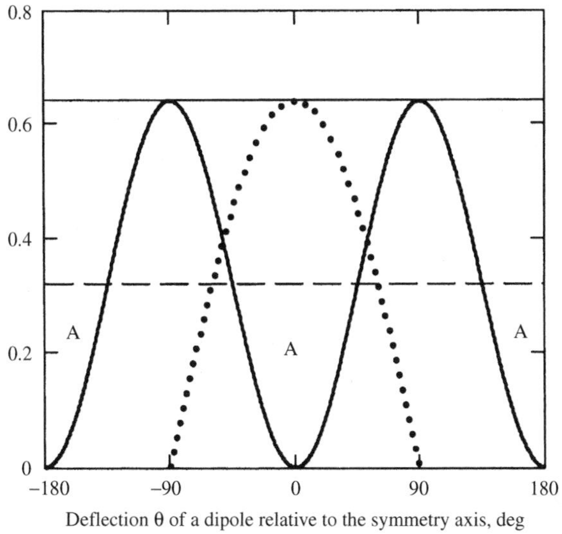

Figure 42. Form of the cosine-squared potential, in which a dipole vibrates (solid line). The horizontal dashed line marks the mean angular amplitude ($\approx 45°$) on the potential curve $U(\theta)$. Dotted curve denotes dimensionless intermolecular static electric field. For the chosen p value $(p = 0.8)$ about 46% and 56% of the dipoles perform complete rotation, respectively, in the case of ordinary and heavy water. Upper horizontal line marks the value of the potential near edge of well. In the center of regions A the potential undergoes minimum (zero value) and the absolute value of static field-maximum.

$$\begin{aligned}
L_{\text{vib}}(Z) &= 2[\check{L}(Z) + \tilde{L}(Z) + \mathring{L}(z)] \\
\check{L}(Z) &= \frac{\exp(p^2/2)}{p\sqrt{\pi}I_0(p^2/2)} \int_0^{\sqrt{3}p/2} \frac{t^3 \exp(-t^2)}{\check{\gamma}} \left\{ \frac{1}{p^2 - Z^2\check{\gamma}^2} + \frac{t^2/(4p^2)}{4p^2 - Z^2\check{\gamma}^2} \right\} dt \\
\tilde{L}(Z) &= \frac{2i\exp\left(-\frac{p^2}{4}\right)\left[1 - \exp\left(-\frac{7p^2}{12}\right)\right]}{p\pi\sqrt{\pi}I_0(p^2/2)} \left\{ \frac{p}{Z} - 2Zp \sum_{n=1}^{\infty} \frac{(-1)^n}{(pn - iZ)^2} \right\} \qquad (315) \\
\mathring{L}(Z) &= \frac{2\exp(p^2/2)}{\sqrt{\pi}I_0(p^2/2)} \int_{\frac{2p}{\sqrt{3}}}^{\infty} \frac{1 + \frac{p^2}{2t^2}}{1 - \left(\frac{p}{4t}\right)^4} \frac{t^2 \exp(-t^2)dt}{\mathring{\gamma}\left(t^2 - z^2\mathring{\gamma}^2\right)} \\
\check{\gamma} &= 1 + \frac{t^2}{4p^2}, \; \mathring{\gamma}(t) = 1 + \frac{p^2}{4t^2}
\end{aligned}$$

where $I_0(\cdot)$ is the zero-order modified Bessel function of the first kind. The factor 2 in Eq. (315) is used in order to correct the result of calculation with respect to the integrated absorption, which for a more realistic case of vibration *in space* is twice of that for motion in a plane. In Eq. (315) the superscripts ˇ and ° refer, respectively, to the low-energy librators and the high-energy rotators. The former alter direction of their angular velocity and the latter execute complete rotations during the lifetime τ_{vib}; the superscript ~ refers to the intermediate-energy dipoles that exhibit slow reorientation. The following three ranges of the normalized energy

$$h \equiv \frac{H}{k_B T} \in \left\{ \left[0, \frac{\sqrt{3}p}{2}\right], \quad \left[\frac{\sqrt{3}p}{2}, \frac{2p}{\sqrt{3}}\right], \quad \left[\frac{2p}{\sqrt{3}}, \infty\right] \right\}$$

correspond to this artificial division of the whole interval of integration over h used to simplify calculations. In the HC–CS model the functions $\varepsilon^*(\nu)$ and $\alpha(\nu)$ depend on seven parameters:

$$u, \ \beta, \ \tau_{\text{or}}, \ f \qquad \text{(for the HC model)} \qquad (316a)$$

$$p, \ \tau_{\text{vib}}, \ R_{\text{vib}} \qquad \text{(for the CS model)} \qquad (316b)$$

Let us write down the formulae (VIG, GT) that allow us to estimate the parameters of the CS model.

The proportions of vibrating librators and hindered rotators:

$$\breve{r} = \frac{2p\exp(p^2/2)}{\pi\sqrt{\pi}I_0(p^2/2)} \int_0^1 \exp(-p^2\breve{m})\mathbf{K}(\breve{k}(m))d\breve{m}; \quad \breve{k}(m) = \sqrt{m} \quad \text{and} \quad \mathring{r} = 1 - \breve{r} \qquad (317)$$

where $\mathbf{K}(\breve{k})$ is the full elliptic integral of the first kind and $\breve{k}$ is its modulus.

The mean number of cycles performed by librators and rotators during the lifetime τ_{vib}:

$$\breve{m}_{\text{vib}} = \frac{\langle\breve{p}\rangle}{2\pi Y} \quad \text{and} \quad \mathring{m}_{\text{vib}} = \frac{\langle\mathring{p}\rangle}{2\pi Y}, \qquad \text{where } \langle\breve{p}\rangle = \frac{2\,\text{sh}(p^2/2)}{\sqrt{\pi}I_0(p^2/2)\breve{r}}$$

$$\text{and} \quad \langle\mathring{p}\rangle \approx \frac{\exp(-p^2/2)}{r^\circ\sqrt{\pi}I_0(p^2/2)} \qquad (318)$$

The mean libration amplitude:

$$\langle\beta\rangle = \frac{2p\exp(p^2/2)}{\breve{r}\pi\sqrt{\pi}I_0(p^2/2)} \int_0^1 \text{arc}\sin(\sqrt{m})\mathbf{K}(\sqrt{m})\exp(-p^2m)\,dm \qquad (319)$$

In Fig. 43 the results of calculations are shown for two water isotopes (for temperature 22.2°C); the fitted/estimated parameters are presented in Table XV.

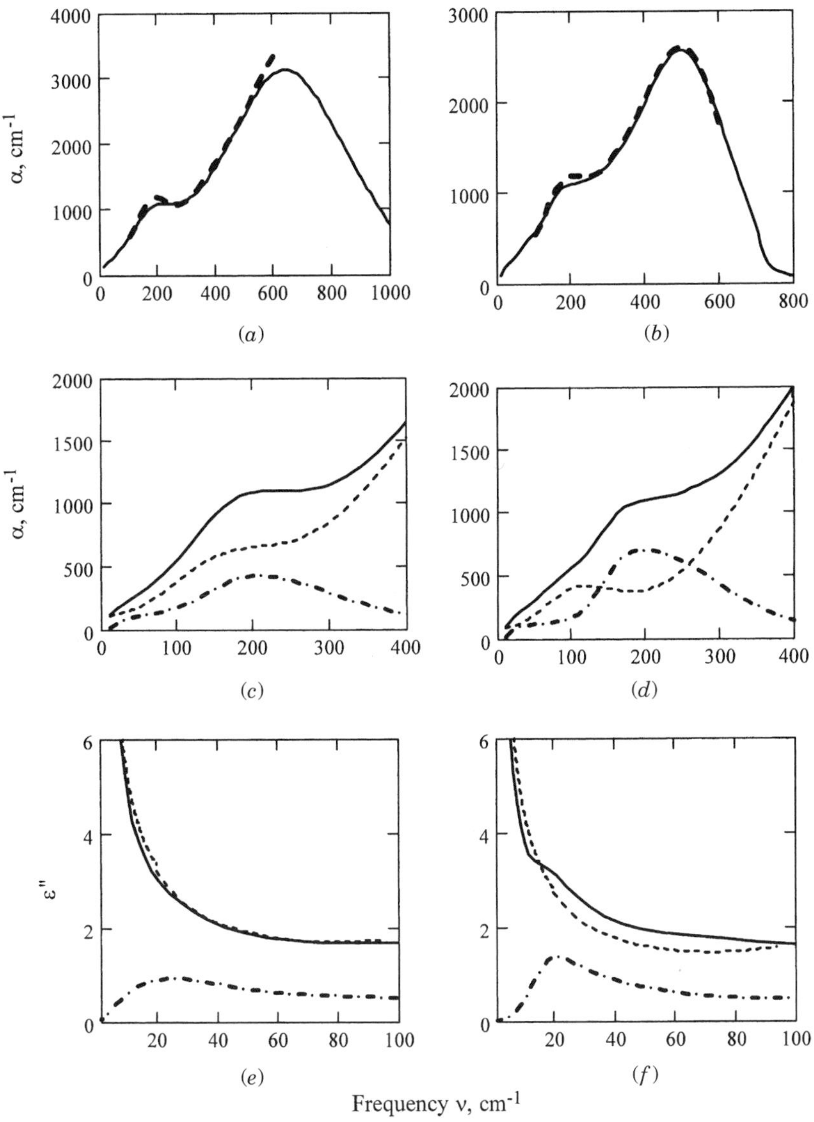

Figure 43. Wideband FIR spectra calculated for the composite hat-curved–cosine-squared potential model (solid lines); dashed-and-dotted lines mark the contribution due to dipoles vibrating in the shallow CS well. Water H_2O (a, c, e) and water D_2O (b, d, f) at 22.2°C. Absorption coefficient (a–d) and dielectric loss (e, f); in Figs. a, b, e, f, dashed lines refer to the experiment [17, 51, 54]. In Figs. c, d dahsed lines mark the contribution to absorption due to dipoles reorienting in a deep hat-curved well.

TABLE XV
Fitted (A) and Estimated (B) Parameters of the Hat-Curved–Cosine-Squared Potential Model[a]

		Reorientation				*Vibration*		
		τ_{or}	u	β	f	τ_{vib}	p	R_{vib}
A	H_2O	0.25	5.7	23	0.75	0.11	0.8	0.040
	D_2O	0.31	5.7	23	0.65	0.30	0.8	0.055
		y	m_{or}	$\overset{\circ}{r}_{or}$	ν_D	m_{vib}		τ_{Dvib}
B	H_2O	0.17	4.7	4.6	0.60	0.48		0.224
	D_2O	0.19	4.4	5.6	0.49	1.30		0.258
						$\overset{\circ}{r}_{vib}$	ν_{Dvib}	$\langle\beta_{vib}\rangle$
C	H_2O					46	23.7	45.1
	D_2O					56	20.5	45.1

[a]Frequency is given in cm^{-1}, time in ps, proportion in percent, angle in degrees.
Note: $k_\mu(H_2O) = 1, k_\mu(D_2O) = 0.973$.

The theoretical FIR absorption curves (Figs. 43a,b), as well as the calculated loss (Figs. 43e,f), coincide with the experimental $\alpha(\nu)$, $\varepsilon''(\nu)$ dependences not worse than it was demonstrated in Section VII.A for the HC–HO model. Dashed-and-dotted curves in Figs. 43c,d demonstrate that vibrations substantially contribute to total absorption in a rather narrow band around the frequency $200\,cm^{-1}$. The maximum "vibration" absorption is larger in D_2O and is less in H_2O than the contribution to total absorption due to reorienting particles (shown by dashed curves). In view of Figs. 43e,f, the *loss* due to vibrating dipoles exhibits a maximum at $\nu \approx 20\,cm^{-1}$, to which the estimated fast relaxation times 0.22 ps for H_2O and 0.26 ps for D_2O correspond. Close values were found experimetnally [17, 51] from the double Debye dependence (280b).

2. *HC–CS Model: Discussion*

The theory of wideband complex permittivity of water described in the review drastically differs from the empirical double Debye representation [17, 54] of the complex permittivity given for water by formula (280b). Evolution of the employed *potential profiles*, in which *a dipole* moves, explored by a dynamic linear-response method can be illustrated as follows:

0. Rectangular, with one rigid dipole → (Section IV)	**1**. Hat-curved, with one rigid dipole → (Section V)	**2**. Hat-curved, with one nonrigid dipole $\bar{\mu} + \tilde{\mu}(t)$ → (Section VI)	**3**. Hat-curved + cosine-squared, with two rigid dipoles → (Section VII.B)

The position "0," marked above, represents our starting point (the rectangular potential well). A considerable time elapsed between the primitive confined rotator model described in 1981 (VIG, GT, [18]) and the hat-curved model.

1. In terms of the four parameters in Eq. (316a) the latter *alone* gives a satisfactory description of the Debye-relaxation and librational bands of various liquids. *One can control the width of the librational band* of liquid H_2O by changing the form factor *f*.

2. The hat-curved model becomes more effective (see Section VI) if it is modified in accordance with the formulas (237) and (238), which roughly account for collective effects in terms of a nonrigid dipole. The additive parameters in Eq. (251b) affect *only* the R-band (100–300 cm^{-1}). The modified hat-curved model described in Section VI gives a good description of the FIR adsorption band. Progress is especially evident in the case of heavy water, for which the rectangular-well model yields poor results [9]. The amplitude μ_m of the time-varying part of a dipole moment, estimated in accordance with Ref. 8, comprises about 30% of the total dipole moment.

3. However, neither the hat-curved model alone nor the theory of the polarization model given in Section VI cannot describe *exactly* the submillimeter $\varepsilon^*(\nu)$ spectrum of water. This drawback arises due to the assumptions that (i) dipoles rotate under the effect of *one* (hat curved) potential, to which corresponds *only one* (Debye) relaxation band, and (ii) the quasi-harmonic time dependence (237) of a nonrigid dipole moment is chosen *without consideration of any dynamic process*. These drawbacks could be eliminated if we elaborate a linear-response theory in terms of two potential wells. Previously [12b], the dielectric response in the range from 10 to 100 cm^{-1} was rather successfully studied in terms of the composite hybrid–cosine-squared potential model. However, application of the rectangular potential, involved in the hybrid model, does not allow us to give a proper description of the FIR spectrum of heavy water. Much more success is obtained in this Section VII.B for the combination {hat-curved + cosine-squared} potentials. Namely, we have explained fine detail of the submillimeter/FIR water spectrum in terms of *two independent relaxation processes* stipulated by hindered rotation of *rigid* dipoles. The first one, characterized by the slow relaxation time τ_D (about 10 ps), is pertinent to reorientation of a dipole in the hat-curved potential; the second process refers to its vibration and is characterized by the fast relaxation time τ_{Dvib} equal to ≈ 0.2 ps, which closely agree with the experimental τ_2 value [17, 54]. Two employed spectral functions, $L_{or}(z)$ and $L_{vib}(Z)$, are pertinent to different relaxation times, τ_D and τ_{Dvib}. In the HC–CS model:

(i) We have a very good agreement between the calculated and experimental complex permittivity $\varepsilon^*(\nu)$ in a wide band ranging from 0 to 1000 cm^{-1}, which includes *also the submillimeter wavelength region*.

(ii) The lifetimes τ_{vib} and τ_{or} fitted for the vibrating and reorienting particles unlike the HC–HO well are commensurable (see Table XV). Thus the behavior of the H-bonded particles vibrating in a finite (CS in our case) well substantially differs from that concerning the parabolic well. Being interesting from a physical viewpoint, this result possibly could allow us to reduce the number of the fitting parameters, since basically one may set $\tau_{vib} = \tau_{or}$. Moreover, since the factor k_{μ} is close to 1 (see the note to Table XV), then setting $k_{\mu} = 1$ in Eq. (286) we may employ the Kirkwood–Frohlich expression for a dipole moment μ.

(iii) The parameterization of the HC–CS model shows (Table XV) that the cosine-squared potential well is shallow (see Fig. 42), since its well depth is commensurable with $k_B T$. Hence, the vibrating H-bonded molecules are *not* actually localized in the well, since the proportion $\overset{\circ}{r}_{vib}$ of the particles performing hindered rotation is commensurable with the proportion $\breve{r}_{vib}$ of the hindered librators moving inside the CS well. This picture resembles a sort of "molecular chaos." The same conclusion anticipates large (about $\pi/4$) mean angle $\langle\beta\rangle$ of the vibrating dipoles. Conversely, the dipoles reorienting in a deep ($u_{or} \approx 6\,k_B T$) hat-curved well *are localized* in the hat-curved potential, since the proportion of the rotators moving in the HC well comprises only about 5%.

(iv) A distinction between two water isotopes is emphasized in the difference of (a) the fitted form-factor f (interpretation of this effect is given in Section VII.A in terms of the spread of short-range forces) and (b) the estimated proporation R_{vib} of the vibrating dipoles. Both factors indicate once more that the H-bonds are stronger in heavy water than in ordinary water. It appears that introduction in Section VI of a nonrigid time-varying dipole moment $\tilde{\mu}(t)$ is equivalent to consideration of dielectric relaxation of a rigid dipole μ vibrating in an addtitional potential well having the cosine-squared profile. But the latter gives better description of the submillimeter spectrum of water.

3. *Conclusions*

1. Both {hat-curved–harmonic oscillator} and {hat-curved–cosine-squared potential} composite models considered in this section give excellent description of wideband spectra of water H_2O and D_2O in the range from 0 to 1000 cm^{-1}. However, it appears that the physical picture of fast vibrations of the H-bonded molecules differ for these two approaches. In the first one, where

the parabolic potential well is employed, we conclude in Section VII.A that vibrations are very chaotic, since the fitted lifetimes obey the inequality $\tau_{vib} \ll \tau_{or}$. On the contrary, in the second approach described in Section VII.B, where the *finite-depth* CS potential model is involved, the vibration lifetimes are rather close to the reorientation lifetimes. The conclusion about chaotic vibration is based on (i) a proximity of proportions of the dipoles performing librations and complete rotations in the CS well and (ii) a large mean vibration amplitude $\langle \beta_{vib} \rangle$. The calculations made in Section VII.B in terms of a *finite-depth* CS potential model possibly give a more adequate understanding of molecular dynamics in water, since a good agreement of theory with the experimental data is achieved for a more reasonable form of the potential affecting vibrating particles; at first blush this approach is more realistic than that used in the HO model since it is based on the existence of *polar molecules* in liquid water, not on the existence of effective charges constituting a nonrigid dipole.

2. However, an *independent consideration* is desirable in order to find out, which of two approaches (given in Section VII.A or VII.B) is more preferable. Our work is based on a phenomenological approach, since we employ *heuristic* (i.e., *ad hoc*) intermolecular potentials. The form of the adopted hat-curved potential, being evidently less idealized than the widely used rectangular-well form, gives much better agreement between the calculated and the recorded spectra than the latter form. However, due to inherent drawback of heuristic modeling, it is difficult to recognize *definitely* a physical nature of the employed potential well. In the context of this section, it is hardly possible to (a) present a clear description of the *physical factors*, which causes the hat-curved potential-well bottom to round, or (b) prove foundation for the parabolic potential, in which charges $\pm\delta q$ oscillate, or for the cosine-squared potential, in which a dipole μ_{vib} turns.

It appears that fortunately in the future *nonheuristic* models should appear, in which the dynamics of *water structure* (even strongly simplified) could be studied in terms of analytic theory. One may consider these models as a good alternative or at least as a useful addition to phenomenological modeling. A preliminary investigation [10, 12, 12a] shows that the FIR absorption bands arising in water around 200 cm^{-1} could be described in terms of *elastic* interactions. Correspondingly, a simplified solution of a one-dimensional dynamic equation describes the law of motion of a *rotating* water molecule, taking into account the torques arising (i) due to *expansion* of the H-bond and (ii) due to *change of its orientation*, with these torques being proportional, respectively, to proper (different) force constants. This quite new approach, which is briefly described in Section IX, basically allows us to generate "self-consistent" forms of intermolecular potentials and to estimate the forms of the absorption spectrum *only from the law of motion* of a reorienting water molecule, since the dipole moment $\boldsymbol{\mu}$ is assumed to rigidly reorient with a given water molecule. Moreover, combination of the linear-response method (VIG, GT) with the above-mentioned equations of motions principally enables us to generate also the spectra of the complex permittivity $\varepsilon^*(\nu)$, which are *not based on a priori introduced potential profiles* (as was made throughout Sections III–VII.

C. Nonrigid Oscillator: Linear-Response Theory for the Parabolic Potential

1. Equation of Motion of Harmonic Oscillator

Let us consider a pair of oppositely charged particles with masses m_1 and m_2 and charges $(+q)$ and $(-q)$, respectively, where $q > 0$. We assume that the energy of their interaction depends as the square of the distance r between the centers of mass of these charges and introduce the elastic stiffness κ characterizing restoring force applied to the charged particle (see Fig. 44). The full energy of a system comprising two particles is given by

$$H = \frac{1}{2}m_+\left(\frac{dr_+}{dt}\right)^2 + \frac{1}{2}m_-\left(\frac{dr_-}{dt}\right)^2 + \frac{\kappa r^2}{2} \tag{320}$$

The coordinate r is determined as the difference

$$r = r_+ - r_- \tag{321}$$

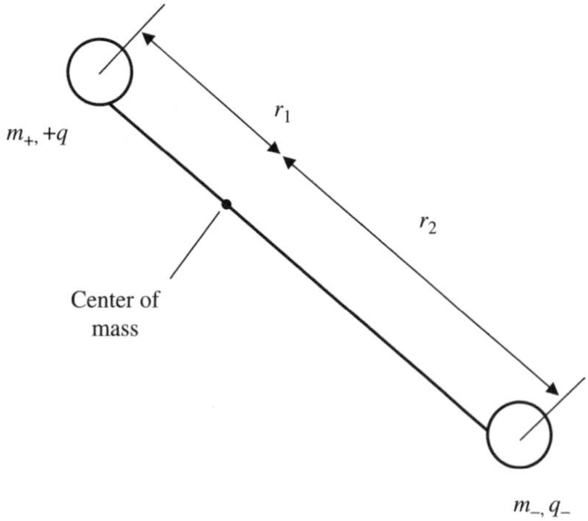

Figure 44. Configuration of a nonrigid dipole formed by two charges oscillating along a straight line.

of the particles' coordinates r_+ and r_- counted off from their common center of mass, which obeys the equation

$$m_+r_+ + m_-r_- = 0 \tag{322}$$

Excluding r_+ and r_- from Eqs. (321) and (322), we have an expression for the energy in the form

$$H = \frac{m}{2}\left[\left(\frac{dr}{dt}\right)^2 + (\Omega r)^2\right] \tag{323a}$$

where the intrinsic angular frequency Ω and the reduced mass m are, respectively, given by

$$\Omega = \sqrt{\kappa/m} \tag{323b}$$

$$m = \frac{m_+m_-}{m_+ + m_-} \tag{323c}$$

Taking into account that the full energy H is constant, we derive from (323a) the equation of motion

$$d^2r/dt^2 + \Omega^2 r = 0. \tag{324}$$

2. *Spectral Function for Back-and-Forth Motion of Two Charged Particles*

a. Complex Susceptibility χ^* in Terms of Perturbed Trajectories. In Section II we have outlined our ACF method in application to an ensemble of *rigid* reorienting dipoles. Now we shall derive, starting from Maxwell equations, a formula for the spectral function for a *nonrigid dipole.* Namely, we consider now a pair of bounded *charges* oscillating along a straight line. In a further derivation we use the same basic theorems and assumptions as those employed in Section II and take *explicitly* into account that the law of motion, which is now a harmonic function. The result for the spectral function [see Eqs. (353) and (355a)] will be also be expressed by a simple analytic formula.

Let N be concentration of the *pairs* of bounded charges. We suggest that a homogeneous plane wave represented in the form

$$\boldsymbol{E}(t) = \mathrm{Re}[\hat{\boldsymbol{E}}\exp(i\Phi)], \qquad \boldsymbol{J}(t) = \mathrm{Re}[\hat{\boldsymbol{J}}\exp(i\Phi)] \tag{325}$$

propagates through a medium. In Eq. (325), $\Phi = \omega t + \gamma - \mathbf{kr}$ is a phase; $k = k' - ik''$ is a propagation constant; γ is an "initial" phase; ω is angular frequency of a.c. field; and $\hat{\boldsymbol{E}}, \hat{\boldsymbol{J}}$ are the corresponding complex amplitudes. Then a nonhomogeneous wave equation,

$$\left(k^2 - \frac{\omega^2}{c^2}\right)\hat{\boldsymbol{E}} = -4\pi i \frac{\omega}{c^2}\hat{\boldsymbol{J}} \tag{326a}$$

equivalent to Eq. (3), follows from Maxwell equations. Denoting the complex-conjugation symbol by an asterisk (so that $\chi^* = \chi' + i\chi''$), we express the square of the propagation constant through the complex susceptibility χ^* stipulated by *perturbed* motions of charges in radiation field:

$$k^{*2} = \frac{\omega^2}{c^2}(4\pi\chi^* + n_\infty^2) \tag{326b}$$

where n_∞^2 is the dielectric permittivity at the high-frequency edge of the relevant relaxation region.[51] Transformation of the wave equation (326a) to an integral form with use of the t_0 theorem yields the expression

$$\chi^* = \frac{i}{\pi}\langle \hat{E}\hat{E}^* \rangle^{-1} \int_{\gamma,\Gamma} \delta N(\gamma, \Gamma) \int_0^{t_v} Q(t, \gamma)\, dt \tag{327a}$$

equivalent to Eq. (12). Here it is assumed that the phase γ changes in the interval $[0, 2\pi]$. The complex power Q delivered by a *moving* charge q in the course of interaction with *radiation field* is given by

$$Q(t, \gamma) = q\boldsymbol{v}(t)\hat{\boldsymbol{E}} \exp(i\,\Phi^*) = qv_E(t)\hat{\boldsymbol{E}} \exp(i\,\Phi^*) \tag{327b}$$

where the phase is given by $\Phi^* = \omega t + \gamma + i\mathbf{kr}$. Neglecting the spatial dispersion, we set further $\exp(i\Phi^*) \equiv \exp(i\omega t + i\gamma)$. The number of charges $\delta N(\gamma, \Gamma)$ born in a unit volume of a medium after the last strong collision with the phase coordinates Γ distributed in the interval $[\Gamma, \Gamma + d\Gamma]$ with phases in the interval $[\gamma, \gamma + d\gamma]$ is given by

$$\delta N(\gamma, \Gamma) = \frac{N}{\omega\tau} W(\Gamma)\, d\Gamma\, F(\gamma, \Gamma)\, d\gamma \exp\left(-\frac{t_v}{\tau}\right)\frac{dt_v}{\tau} \tag{327c}$$

t_v is the lifetime equal to the period between strong collisions, which are assumed to be instantaneous; the distribution over t_v is taken to be exponential; the instant after the last collision conditionally is set equal to 0; the mean lifetime τ, which

[51]When this theory is applied to the composite model (see Section VII.A), we set $n_\infty^2 \to 0$.

plays the role of a friction coefficient, obeys the identity $\tau \equiv \int_0^\infty t_v \times \exp(-t_v/\tau)dt_v/\tau$.

The steady-state Maxwell–Boltzmann distribution is determined as usual:

$$W(\Gamma) = C\exp[-H/(k_B T)] \tag{327d}$$

where the constant C obeys the normalizing condition

$$\langle W(\Gamma)\rangle \equiv \int_\Gamma W(\Gamma)d\Gamma = 1 \tag{327e}$$

$F(\gamma, \Gamma)$ is a nonequilibrium (*induced* by a.c. field) correction to the steady-state distribution.

We *conditionally* may refer to the susceptibility χ^* (327a) as "*translational*," since it is determined by back-and-forth motion of the bound charges. Note that Eqs. (327a)–(327f) are also applicable for calculation of the *ionic* susceptibility $\chi^*(\omega)$ stipulated by a moving *single* charge, to which the complex conductivity

$$\sigma(\omega) = -\omega\chi^*(\omega) \tag{327f}$$

corresponds (see Section VIII).

b. Complex Susceptibility χ^*_{dyn} in Terms of Trajectories Found at Equilibrium. Assuming that a.c. field is weak, we shall transform Eq. (327a) in order to express its integrand through an *unperturbed trajectory* of a particle. For the latter we mark dynamical variables by the superscript ^{0}, and the a.c. perturbation is marked by the symbol δ; thus a perturbation of a particle's velocity in the direction of a.c. field is denoted δv_E. The induced distribution is represented as the sum of a homogeneous one ($F = 1$) and of the perturbation δF, so that

$$F \equiv 1 + \delta F \tag{328}$$

Correspondingly, the susceptibility is linearly connected with two perturbed quantities: (1) with the effect of the disturbed particle's velocity calculated at $F = 1$; (2) and with the effect of the perturbation δF calculated at the steady-state velocity v_E^0 of charges, namely,

$$\chi^*(\omega) = \chi^*_{\rm dyn}(\omega) + \chi^*_{\rm st}(\omega) \tag{329}$$

Both terms in Eq. (329) are found by taking Eqs. (327a)–(327e) into account:

$$\chi^*_{\rm dyn} = \frac{2iNq}{\hat{E}^*\omega\tau}\left\langle \int_0^{2\pi} \exp(i\gamma)\frac{d\gamma}{2\pi}\int_0^{\infty}\exp\left(-\frac{t_v}{\tau}\right)\frac{dt_v}{\tau} \times \int_0^{t_v}[v_E^0(t)+\delta v_E(t,\gamma)]\exp(i\omega t)\,dt\right\rangle \tag{330}$$

$$\chi^*_{\rm st} = \frac{2iNq}{\hat{E}^*\omega\tau}\left\langle \int_0^{2\pi}(F-1)\exp(i\gamma)\frac{d\gamma}{2\pi} \times \int_0^{\infty}\exp\left(-\frac{t_v}{\tau}\right)\frac{dt_v}{\tau}\int_0^{t_v} v_E^0(t)\exp(i\omega t)\,dt\right\rangle \tag{331}$$

Introducing the complex frequency

$$\hat{\omega} = \omega + i\tau^{-1} \tag{332}$$

we obtain from (330) and (331) simple integrals after integration over t_v by parts (the nonintegral term vanishes; further we change notation assuming $t_v \to t$):

$$\chi^*_{\rm dyn} = \frac{2iNq}{\hat{E}^*\omega\tau}\left\langle \int_0^{2\pi}\exp(i\gamma)\frac{d\gamma}{2\pi}\int_0^{\infty}[v_E^0(t)+\delta v_E(t,\gamma)]\exp(i\hat{\omega}t)\,dt\right\rangle \tag{333a}$$

$$\chi^*_{\rm st} = \frac{2iNq}{\hat{E}^*\omega\tau}\left\langle \int_0^{2\pi}\delta F(\gamma,\Gamma)\exp(i\gamma)\frac{d\gamma}{2\pi}\int_0^{\infty} v_E^0(t)\exp(i\hat{\omega}t)dt\right\rangle \tag{333b}$$

We denote integration over initial phase γ as $\{\cdot\}$

$$\{f\} \equiv \int_0^{2\pi} f(\gamma)\exp(i\gamma)\frac{d\gamma}{2\pi} \tag{334}$$

Noting that an unperturbed velocity $v^0(t)$ does not yield a contribution to the integral (333a), we rewrite Eq. (333a) in the form

$$\chi^*_{\rm dyn} = \frac{2iNq}{\hat{E}^*\omega\tau}\left\langle \int_0^{\infty}\{\delta v_E(t,\gamma)\}\exp(i\hat{\omega}t)\,dt\right\rangle \tag{335}$$

In view of the average-perturbation theorem (GT, pp. 146, 374; VIG, p. 82), see also Eq. (18), the averaged perturbation $<\delta v_E(t)\rangle$ of the particles velocity is determined by the work $\mathscr{A}(t)$ of an a.c. field performed over a charged particle *moving along unperturbed trajectory*:

$$\langle \delta v_E(t,\gamma)\rangle = \frac{\langle v_E^0(t)\mathscr{A}(t,\gamma)\rangle}{k_B T} \tag{336a}$$

$$\mathscr{A}(t,\gamma) = q\int_0^t E(t',\gamma)v_E^0(t')dt' \tag{336b}$$

In accord with Eq. (325), we have

$$E(t,\gamma) = \frac{1}{2}\mathrm{Re}\left(\hat{E}e^{i\omega t+i\gamma} + \hat{E}^* e^{-i\omega t - i\gamma}\right) \tag{336c}$$

Furthermore, we (a) change notation of the velocity at equilibrium, $v^0(t) \to v(t)$; (b) omit **kr** in the exponent, since the space scale having order of a path, which a particle walks between strong collisions, is much less than the free-space wavelength λ at the frequencies under investigation; and (c) omit the exponential term proportional to $\exp[i(\omega t + \gamma)]$, since it vanishes after integration over γ. Then we have

$$\langle \delta v_E(t,\gamma)\rangle = \left\langle \frac{q\hat{E}^*}{2k_B T}\int_0^t v_E \exp(-i\omega t' - i\gamma)\, dt' \right\rangle \tag{337}$$

Substitution into Eq. (335) and averaging over γ with account of the definition (334) yields

$$\chi^*_{\mathrm{dyn}} = \frac{iq^2 N\langle s\rangle}{k_B T\omega\tau} \tag{338a}$$

$$s \equiv \int_0^\infty v_E(t)\exp(i\hat{\omega}t)\, dt \int_0^t v_E(t')\exp(-i\omega t')\, dt' \tag{338b}$$

Let $h = H/(k_B T)$ be the normalized energy of the oscillator and $\Lambda(h)$ be the period of the velocity $v_E(t)$. Taking h and t_0 as a pair of the complex-conjugated phase variables (t_0 is an arbitrary constant additive to time t) and using the property

$$\int_0^{\Lambda(h)} \frac{d}{dt_0} v_E(t')v_E(t)\, dt_0 = 0$$

we derive the expression

$$\left\langle \frac{dv_E(t)}{dt} v_E(t') + v_E(t)\frac{dv_E(t')}{dt'} \right\rangle = 0 \tag{339}$$

Integration of Eq. (338b) by parts over t' and further over t, taking into account of Eq. (339), gives

$$s = \frac{i}{\omega}\int_0^\infty v_E(t)e^{i\hat{\omega}t}\left[v_E(t)e^{-i\omega t} - v_E(0) - \int_0^t \frac{dv_E(t')}{dt'}e^{-i\omega t'}dt'\right]dt$$

$$= \frac{i}{\omega}\int_0^\infty \left\{v_E(t)e^{i\hat{\omega}t}\left[v_E(t)e^{-i\omega t} - v_E(0)\right] + \frac{dv_E(t)}{dt}e^{i\hat{\omega}t}\int_0^t v_E(t')e^{-i\omega t}dt'\right\}$$

$$= \frac{i}{\omega}\int_0^\infty v_E(t)e^{i\hat{\omega}t}\left[v_E(t)e^{-i\omega t} - v_E(0) - i\hat{\omega}\int_0^t v_E(t')e^{-i\omega t'}dt' - v_E(t)e^{-i\omega t}\right]dt$$

Comparison with the definition of s in Eq. (338b) yields

$$\left(1 - \frac{\hat{\omega}}{\omega}\right)s = -\frac{is}{\omega\tau} = -\frac{iv_E(0)}{\omega}\int_0^\infty v_E(t)\exp(i\hat{\omega}t)\,dt \tag{340}$$

Solving this equation over s and combining the solution with Eq. (338b) we have

$$\chi^*_{\text{dyn}} = \frac{iq^2N}{k_B T\omega}\left\langle v_E(0)\int_0^\infty v_E(t)\exp(i\hat{\omega}t)\,dt\right\rangle \tag{341a}$$

We have obtained a *general* integral-form expression for the autocorrelator specified for the *bounded charges*.[52] Substituting $v_E(t) = v(t)\cos\Theta$ and integrating over Θ in the interval $[0, \pi/2]$, we finally have

$$\chi^*_{\text{dyn}} = \frac{iq^2N}{3k_B T\omega}\left\langle v(0)\int_0^\infty v(t)\exp(i\hat{\omega}t)\,dt\right\rangle \tag{341b}$$

This expression *diverges* at $\omega \to 0$. Therefore, a homogeneous induced distribution, characterized by $F = 1$, does not give a physically acceptable result.

c. *Analytic Formula for the Spectral Function (Boltzmann Collision Model).* Now we shall account for the specific (harmonic) law of motion of a dipole. Double integration over parts in Eq. (341b) by using the steady-state equation of

[52]The same expression is valid [10, 11] also for an ionic model.

motion (324) gives

$$\int_0^\infty v(t)\exp(i\hat{\omega}t)\,dt = \frac{1}{i\hat{\omega}}\left(-v_0 + \int_0^\infty \frac{dv(t)}{dt}e^{i\hat{\omega}t}dt\right)$$
$$= \frac{1}{i\hat{\omega}}\left(-v_0 + \Omega^2\int_0^\infty r(t)e^{i\hat{\omega}t}dt\right)$$
$$= \frac{1}{i\hat{\omega}}\left(-v_0 - \frac{\Omega^2 r_0}{i\hat{\omega}} - \frac{\Omega^2}{i\hat{\omega}}\int_0^\infty v(t)e^{i\hat{\omega}t}dt\right)$$

Combining this expression with Eq. (341b), we have

$$\int_0^\infty v(t)\exp(i\hat{\omega}t)\,dt = (iv_0\hat{\omega} + \Omega^2 r_0)(\hat{\omega}^2 - \Omega^2)^{-1} \tag{342}$$

Hence we have derived an analytic expression for χ^*_{dyn}:

$$\chi^*_{\text{dyn}} = \frac{q^2 N\hat{\omega}}{3k_B T\omega(\Omega^2 - \hat{\omega}^2)}\langle v_0^2\rangle = \frac{q^2 N\hat{\omega}}{3m\omega(\Omega^2 - \hat{\omega}^2)} \tag{343}$$

where in view of Eq. (323a)–(323c) we have taken into account that at equilibrium the mean velocity squared is

$$\langle v_0^2\rangle = k_B T/m \tag{344a}$$

Equation (343) tends to ∞ at $\omega \to 0$, just as Eq. (341b) does.

Now we take an account of induced distribution F; that is, we shall study the effect of δF on the susceptibility. The oscillator's coordinate r presents a superposition of a steady state and of time-varying parts, r_0 and $r_1(t)$. At an "initial" instant t_0 (at the instant immediate after the last strong collision) characterized by an initial phase $\omega t_0 = \gamma$, the coordinate r and the parabolic potential U are given by

$$r(\gamma) = r_0 + r_1(\gamma), \qquad r_0 \gg \max[r_1(\gamma)] \tag{344b}$$

$$U_B = U(\gamma) = -q[r_0 + r_1(\gamma)]E(\gamma)\cos\Theta \tag{344c}$$

$$F_B(\gamma) = \exp\left[-\frac{U(\gamma)}{k_B T}\right] \approx 1 - \frac{U(\gamma)}{k_B T} \tag{344d}$$

where the subscript B means "Boltzmann" (distribution) with respect to a.c. electric field,[53] so that

$$\delta F(\gamma) = \frac{qr(\gamma)E(\gamma)\cos\Theta}{k_B T} \tag{344e}$$

Thus we introduce an approximate potential energy $U_B = U(\gamma)$, which at any instant t_0 is equal to the product of a nonrigid dipole moment $q\boldsymbol{r}(\gamma)$ and of the a.c. electric field $\boldsymbol{E}(\gamma)$, with Θ being the angle between the dipole moment $q\boldsymbol{r}(\gamma)$ and field $\boldsymbol{E}(\gamma)$. Taking into account that in Eq. (333b) the product $v_E \delta F$ is proportional to $\cos^2\Theta$, we find that averaging over Θ again gives 1/3. In our case, when the induced distribution F is taken in the form of the Boltzmann distribution $(F \equiv F_B)$, we supply the susceptibility χ^* by the subscript B:

$$\chi^*(\omega) \equiv \chi_B^*(\omega) \qquad \text{at} \quad F \equiv F_B \tag{345}$$

Since a.c. field is weak, in a first approximation we may neglect a small product of $\boldsymbol{r}_1(\gamma)\boldsymbol{E}(\gamma)$. Inserting Eq. (344d) into (333b) and averaging over γ with account of (336b), we have

$$\chi_{\text{st}}^* = \chi_B - \chi_{\text{dyn}} = \frac{iNq^2}{3\omega\tau k_B T}\left\langle r_0 \int_0^\infty v(t)\exp(i\hat{\omega}t)dt \right\rangle \tag{346}$$

Using Eq. (342), we get from (346):

$$\chi_B^* - \chi_{\text{dyn}}^* = \frac{iNq^2}{3\omega\tau k_B T}\langle \Omega^2 r_0^2 + i\hat{\omega} v_0 r_0 \rangle (\hat{\omega}^2 - \Omega^2)^{-1} \tag{347}$$

We take coordinate r_0 and velocity v_0 as the steady state phase variables; both are determined at an infinite interval $[-\infty, \infty]$. The mean $\langle v_0 \rangle = 0$; the mean $\langle r_0^2 \rangle$ is found by using Eq. (323a):

$$\langle r_0^2 \rangle = \int_{-\infty}^{\infty} dr_0 r_0^2 \exp\left[-\frac{m\Omega^2 r_0^2}{2k_B T}\right] \Big/ \int_{-\infty}^{\infty} dr_0 \exp\left[-\frac{m\Omega^2 r_0^2}{2k_B T}\right] = \frac{k_B T}{m\Omega^2} \tag{348}$$

Combining it with Eq. (347), we obtain

$$\chi_B^* - \chi_{\text{dyn}}^* = -\frac{iNq^2}{3\omega\tau m(\omega^2 - \Omega^2)} \tag{349}$$

[53]Generally, *a.c.* elcetric field could *not* unambiguously be characterized by any potential function. However, in our theory of "collision models," which is applied for description of a *low-frequency* spectrum, we may employ the quantity U_B, which has a sense of a quasi-static potential energy.

Taking account of Eq. (343), we express the total complex permittivity in the form of a *Lorentz line*

$$\chi_B^* = \frac{Nq^2}{3m\omega}\left(\hat{\omega} - \frac{i}{\tau}\right)\frac{1}{\Omega^2 - \hat{\omega}^2} = \frac{Nq^2}{3m\left(\Omega^2 - \hat{\omega}^2\right)} \tag{350}$$

The latter does *not* diverge at $\omega \to \infty$.

Denoting $a \equiv \sqrt{\langle r_0^2 \rangle}$, we estimate from Eq. (348) a mean-squared amplitude a of harmonic oscillations:

$$a = \frac{1}{\Omega}\sqrt{\frac{k_B T}{m}} \tag{351a}$$

If we (1) take a reduced mass of two oscillating molecules equal to (1/2) $m_H M$—that is, equal to half of the mass of one water molecule (m_H is mass of a proton and M is the molecular mass)—and (2) set $\Omega = 2\pi c \nu_R$, where ν_R is a peak-absorption frequency of the R-band ($\nu_{tr} \approx 190\,\mathrm{cm}^{-1}$), then we find for water the following estimation:

$$a = \approx \frac{1}{2\pi c \nu_R}\sqrt{\frac{2k_B T}{m_H M}} \approx 0.15\text{Å} \tag{351b}$$

Now we introduce a few nondimensional parameters: concentration (G), real (X), imaginary (Y), and complex (Z) frequencies:

$$G = \frac{q^2 N}{3m\Omega^2}, \qquad X = \frac{\omega}{\Omega}, \qquad Y = \frac{1}{\Omega\,\tau_{\text{vib}}}, \qquad Z = X + iY \tag{352}$$

where τ_{vib} is the lifetime of oscillations (above we have termed it "vibrations"), which generally differs from the reorientation lifetime τ_{or}. In terms of Eq. (352) the Lorentz line (350) corresponds to a Boltzmann induced distribution F_B, and the line (343) corresponds to a homogeneous induced distribution ($F = 1$). These are given by

$$\chi_B^*(Z) = GL(Z) \tag{353}$$

$$\chi_{\text{dyn}}^*(Z) = \chi_B^*(Z)\frac{Z}{X} \tag{354}$$

where $L(Z) \equiv L_{\text{LOR}}(Z)$ could be termed the *spectral function* (SF)

$$L(Z) = \frac{1}{1 - Z^2} \tag{355a}$$

In view of Eqs. (340) and (348), the latter is expressed as the following velocity autocorrelator:

$$L(z) = \frac{i}{\hat{\omega}}\left\langle \frac{v_E(0)}{r_0}\int_0^\infty \frac{v_E(t)}{r_0}\exp(i\hat{\omega}t)\,dt\right\rangle \tag{355b}$$

We should note that:

(i) Equation (355a) holds *only* for a harmonic oscillator (i.e., for two bounded charges).

(ii) Equation (355b) is valid also in a general case of motion of a single charged particle.

(iii) Equations (353) and (354) hold for *any* molecular model considered in this chapter, if we take a relevant constant multiplier G.

3. Frequency Dependences Pertinent to Different Collision Models

a. General Formulas. The induced Boltzmann distribution F_B (344d) has, evidently, a restricted importance, since the a.c. potential function $U_B(t)$, Eq. (344c), changes keeping time with a.c. field $E(t)$. For a more realistic self-consistent approach to this problem, described in GT, p. 188, or in VIG, p. 183, for a dipolar ensemble, some phase shift between harmonically changing functions $U_B(t)$ and $E(t)$ arise, so that a Debye frequency dependence becomes characteristic in a *low-frequency* range. It appears to be instructive to obtain similar relations also for an ensemble of bounded oscillating charges by using simple analytic expressions. We *directly* shall prove here a validity of assertions given in Section II that some relations between the complex susceptibility χ^* and the spectral function actually L do not depend on a specific model under consideration. We shall also give a few illustrations of a characteristic spectra $\chi^*(x)$. In view of commonality of the above-mentioned relations, we change notations (352) of dimensionless parameters by using x, y, z, τ instead of X, Y, Z, τ_{vib}. Thus in this subsection we set:

$$X \to x, \quad Y \to y, \quad Z \to z, \quad \tau_{vib} \to \tau \tag{356}$$

(a) *The isothermal induced distribution F_T (isothermal line).* Let δF be a linear function of the steady-state coordinate r_0 and velocity v_0 of the harmonic oscillator. We involve unknown proportionality coefficients $\hat{A}^*$ and $\hat{B}^*$. By analogy with (344d), we write

$$\delta F(\gamma, r_0, v_0) = \frac{q\cos\Theta}{2k_BT}\Big\{\left[\hat{E}\hat{A}^*\exp(i\gamma) + \hat{E}^*\hat{A}\exp(-i\gamma)\right]r_0 + \left[\hat{E}\hat{B}^*\exp(i\gamma) + \hat{E}^*\hat{B}\exp(-i\gamma)\right]v_0\Big\} \tag{357}$$

Then

$$\frac{1}{2\pi}\int_0^{2\pi} \delta F(\gamma, r_0, v_0)\exp(i\gamma)\frac{d\gamma}{2\pi} = \frac{q\hat{E}^*\cos\Theta}{2k_B T}\left(\hat{A}r_0 + \hat{B}v_0\right) \tag{358}$$

Averaging of Eq. (333b) over Θ gives

$$\chi^*_{\text{ind}} = \frac{iq^2N}{3\omega\tau k_B T}\left\langle \left(\hat{A}r_0 + \hat{B}v_0\right)\int_0^\infty \frac{dr}{dt}\exp(i\hat{\omega}t)\,dt\right\rangle \tag{359}$$

In view of (341), we have

$$\frac{iq^2N\hat{B}}{3\omega\tau k_B T}\left\langle v_0\int_0^\infty \frac{dr}{dt}\exp(i\hat{\omega}t)\,dt\right\rangle = \frac{\hat{B}\chi^*_{\text{dyn}}}{\tau} \tag{360}$$

while in view of (342), (343) and (348), we obtain

$$\left\langle r_0\int_0^\infty \frac{dr}{dt}\exp(i\hat{\omega}t)\,dt\right\rangle = \frac{1}{\hat{\omega}^2 - \Omega^2}\left\langle iv_0r_0\hat{\omega} + \Omega^2r_0^2\right\rangle$$
$$= -\frac{\langle\Omega^2r_0^2\rangle\chi^*_{\text{dyn}}3m\omega}{q^2N\hat{\omega}} \tag{361}$$
$$\times\frac{iq^2N}{3\omega\tau k_B T}\left\langle r_0\int_0^\infty \frac{dr}{dt}\exp(i\hat{\omega}t)\,dt\right\rangle = -\frac{i\chi^*_{\text{dyn}}}{\hat{\omega}\tau} \tag{362}$$

Returning to Eq. (359), we have

$$\chi^*(\omega) = \chi_1(\omega) + \chi_2(\omega) = \chi_1(\omega)\left[1 - \frac{i\hat{A}(\omega)}{\hat{\omega}\tau} + \frac{\hat{B}(\omega)}{\tau}\right] \tag{363}$$

We shall rewrite this formula in terms of the spectral function $L(z)$ and dimensionless parameters (352) by taking into account the change of notations, Eq. (356):

$$\chi^*(\omega) = \frac{GL(z)}{x}\left\{z - iy\left[\hat{A}(\omega) + i\Omega z\hat{B}(\omega)\right]\right\} \tag{364}$$

This formula yields the previous result (353), corresponding to a given (Boltzmann) induced distribution F_B, if we set $\hat{A} = 1$ and $\hat{B} = 0$. To find $\hat{A}$

and $\hat{B}$ for an arbitrary induced distribution $F(\gamma, \Gamma)$, we relate[54] the polarization $P(t)$ and its derivation $dP(t)/dt$ to a.c. field:

$$P(t) = \mathrm{Re}\{\hat{P}\exp(\omega t)\} = \mathrm{Re}\{\chi\hat{E}\exp(i\omega t)\} \tag{365}$$

$$\frac{d}{dt}P(t) = \frac{d}{dt}\mathrm{Re}\{\hat{P}\exp(\omega t)\} = \mathrm{Re}\{i\omega\chi\hat{E}\exp(i\omega t)\} \tag{366}$$

Using Eq. (359), we (1) introduce the factor g, which formally accounts for correlation of orientations of nonrigid dipoles (g will be found independently) and (2) express the polarization at any instant t_0 of the last collision as a total dipole moment of a unit volume containing N bound charges:

$$P(0) = gNq\langle r_E(0)F[\gamma, r_E(0), v_E(0)]\rangle = \mathrm{Re}\{\chi\hat{E}\exp(i\gamma)\} \tag{367}$$

Analogously, we rewrite Eq. (366). Now correlation of initial (arising immediately after collision) particles' velocities is neglected:

$$Nq\langle v_E^0 F[\gamma, r_E(0), v_E(0)]\rangle = \mathrm{Re}\left[i\omega\chi\hat{E}\exp(\gamma)\right] \tag{368}$$

This relationship is actually Ohm's law written in a differential form. Substituting Eq. (357) into the left-hand parts of Eq. (367) and (368) and averaging them with account of Eqs. (344a) and (348), we derive

$$\hat{A}(\omega) = \frac{\chi^*(\omega)}{Gg}, \qquad \hat{B}(\omega) = -i\frac{\chi^*(\omega)}{G\Omega^2} \tag{369}$$

Combining these formulas with (364), we finally have

$$\chi^* \equiv \chi_T^* = \frac{gGzL(z)}{gx + iyL(z)(1 + gxz)} \tag{370}$$

This is exactly expression for χ^* given in GT, p. 191 or in VIG, p. 195. If we insert here the Lorentz line (355a), we shall have

$$\chi_T^* = \frac{gGz}{gx(1 - z^2) + iy(1 + gxz)} \tag{371}$$

[54]If $\chi = \chi' - i\chi''$, where $\chi'' > 0$, then as it is easy to see, the polarization $P(t)$ induced by a.c. field actually remains behind $E(t)$, as it should be.

The subscript T in Eqs. (370) and (371) refers to the "isothermal" collision model (induced distribution F_{T}), to which the complex coefficients (369) and general relations (367) and (368) correspond.[55]

(b) *The Gross line.* If we neglect dependence of the induced distribution F on the "initial" oscillator's velocity $v(\gamma)$, then above we should replace by 0 the coefficient $\hat{B}$. Then Eq. (364) should be replaced by

$$\chi^*(\omega) = \frac{GL(z)}{x}\left[z - iy\hat{A}(\omega)\right]$$

where we should insert from (369) $\hat{A}(\omega) = \frac{\chi^*(\omega)}{Gg}$. Denoting $F = F_{\mathrm{G}}$ this (Gross) induced distribution and the corresponding susceptibility χ_G^*, we find that

$$\chi_G^* = \frac{gGzL(z)}{gx + iyL(z)} \tag{372}$$

We have got the same general expression as was found before (GT, p. 191; VIG, p. 195). For the spectral function (355a) Eq. (372) reduces to

$$\chi_G^* = \frac{gGz}{gx(1 - z^2) + iy} \tag{373}$$

The spectral dependencies (371) and (372) are characterized by the following *general properties.*

(a) *Debye-like low-frequency dependence.* Thus, when an induced distribution F (or, to be more precise, when the complex coefficients $\hat{A}$ and $\hat{B}$ are introduced), the susceptibility χ^* is expressed through $L(z)$ as a *bilinear* function. At low frequency x, the formulas (371) and (372) exhibit a maximum on the loss curve $\chi''(x)$. Such a maximum does *not* appear, when the Boltzmann induced distribution F_{B}, Eq. (353), is employed.

(b) *The static susceptibility does not depend on free parameters of a model.* At zero frequency ($x = 0$) the static susceptibility

$$\chi_s = gG \tag{374}$$

occurs to be the same in both cases, Eqs. (371) and (372), as that given by Eq. (353) at $F = F_{\mathrm{B}}$. This static susceptibility does *not* depend on a lifetime τ (in

[55]The term "isothermal" was previously used in terms of the model of kinetic equations applied to free motion of the particles between strong collisions [18, p. 126; 65]. In this particular case the collision integral $\mathrm{St}(f)$ of the Bhatnagar–Gross–Krook (BGK) model is found for $T'(q,t) = T = \text{const}$.

other words, does not depend on a dimensionless collision frequency y), as it should be. Note that this important property does not hold for the Lorentz line, Eq. (353), for which the Boltzmann induced distribution F_B is employed, which is *not* self-consistent; so, the Lorentz-line susceptibility (353) depends on y, if $x = 0$.

b. Integrated Absorption. In view of Eq. (340), the integral over ω of the product ω and of imaginary part of the susceptibility (we term such a quadrature *integral absorption*) is given by

$$\Pi = \int_{-\infty}^{+\infty} \omega\chi''_{\mathrm{dyn}}(\omega)\, d\omega = \frac{q^2 N}{3k_B T}\Im \tag{375a}$$

$$\Im \equiv \left\langle \int_{-\infty}^{\infty} d\omega\, v(0) \int_0^{\infty} v(t)\cos(\omega t)\exp\left(-\frac{t}{\tau}\right) dt \right\rangle \tag{375b}$$

Changing the order of integration and using $x = \omega/\Omega$ and $y = (\Omega\tau)^{-1}$ as a real part and an imaginary part of the dimensionless frequency z, we have

$$\Im = \Omega\left\langle v(0) \int_0^{\infty} v(t)\exp(-\Omega y t)\, dt \int_{-\infty}^{\infty} \cos(\Omega x t)\, dx \right\rangle \tag{376}$$

Since the inner integral reduces to a δ-function, we obtain

$$2 \lim_{x\to\infty} \left[\frac{\sin(\Omega x t)}{t}\right] = 2\pi\delta(t) \tag{377}$$

$$\Im = 2\pi\left\langle v(0) \int_0^{\infty} v(t)\exp(-\Omega y t)\delta(t)\, dt \right\rangle \tag{378}$$

For the chosen model potential the mean value $\langle v(0)v(t)\rangle$ is an even function of time. Indeed, we may present a general solution of Eq. (324) as

$$r(r_0, v_0, t) = r_0\cos\Omega t + (v_0/\Omega)\sin\Omega t, \qquad v(r_0, v_0, t) = -r_0\Omega\sin\Omega t + v_0\cos\Omega t \tag{379}$$

Then

$$\langle v v_0\rangle = \langle -r_0\Omega v_0\sin\Omega t + v_0^2\cos\Omega t\rangle \tag{380}$$

After integration over ensemble, the first term vanishes and the second term yields an even function of time t. Thus, we may represent $\Im$ as integral over t

taken in the interval $[-\infty, \infty]$. Accounting for Eq. (344a) we have

$$\Im = \pi \left\langle v(0) \int_0^{\infty} v(t) \exp(-\Omega y|t|) \delta(t)\, dt \right\rangle = \pi \langle v^2 \rangle = \pi k_B T/m \qquad (381)$$

$$\Pi = \int_{-\infty}^{+\infty} \omega \chi''(\omega) d\omega = q^2 N/(3m) \qquad (382)$$

The dielectric loss χ^* is presented here without a subscript, since the result (382) remains the same for *any* induced distribution $F(\gamma, \Gamma)$ found in terms of the ACF method. A fundamental property of Eq. (382) consists in *deficiency* of the dependence of Π on the parameters of the model and the temperature. For an ionic ensemble, Π is three times larger [141].

c. *Typical Frequency Dependences.* Let us consider the form of lines $\chi_T^*(x)$ and $\chi_G^*(x)$ given respectively by Eqs. (371) and (373).

The *first line*, $\chi_T^*(x)$, has a nonphysical behavior, if correlation of orientations is lacking $(g = 1)$, since the susceptibility $\chi_T^*(x)$ does not have an imaginary component in this case; moreover, $\chi_T^*(x)$ diverges at the rigorous resonance condition $(x = 1)$, when $\chi_T^* = G(1 - x^2)^{-1}$. These results mean that the correlation factor g should actually be nonzero and/or that the parabolic potential has a restricted applicability.

For the case of the *Gross line* (373), $\chi_G^*(x)$, the dielectric loss arises at *all* frequencies x, if the collision frequency $y \neq 0$. At the correlation factor $g = 1$ the susceptibility becomes equal to $\chi_G^* = G(1 - xz)^{-1}$; it diverges at $x = 1$ only in an idealized case $y = 0$.

The line forms, described by Eqs. (371) and (373), are illustrated in Figs. 45 and 46. In the first one (Fig. 45) we compare the loss (a) and absorption (b) for the isothermal, Gross, and Lorentz lines; see solid, dashed, and dashed-and-dotted curves, respectively. These curves are calculated in a vicinity of the resonance point $x = 1$. In Figure 45c we show the frequency dependences of a real part of the susceptibility; the three curves are extended also to a low-frequency region. The collisions frequency y and the correlation factor g are fixed in Fig. 45 $(y = 0.4, g = 2.5)$.

In Figs. 45 and 46 we see the frequency the dependences typical for *resonance-absorption*. The isothermal line (at F_T) is characterized by the loss peak, whose *intensity* is evidently larger and *bandwidth* is narrower than those pertinent to two other lines. The Lorentz line (at F_B) is characterized by (i) a smallest maximum loss and (ii) an absence of the loss shoulder at lower frequencies. This shoulder appears (at F_T and F_G) due to the denominator in Eqs. (370) and (372), pertinent to the self-consistent collision models. Since the plateau-like loss curve is indeed a feature characteristic for water in the

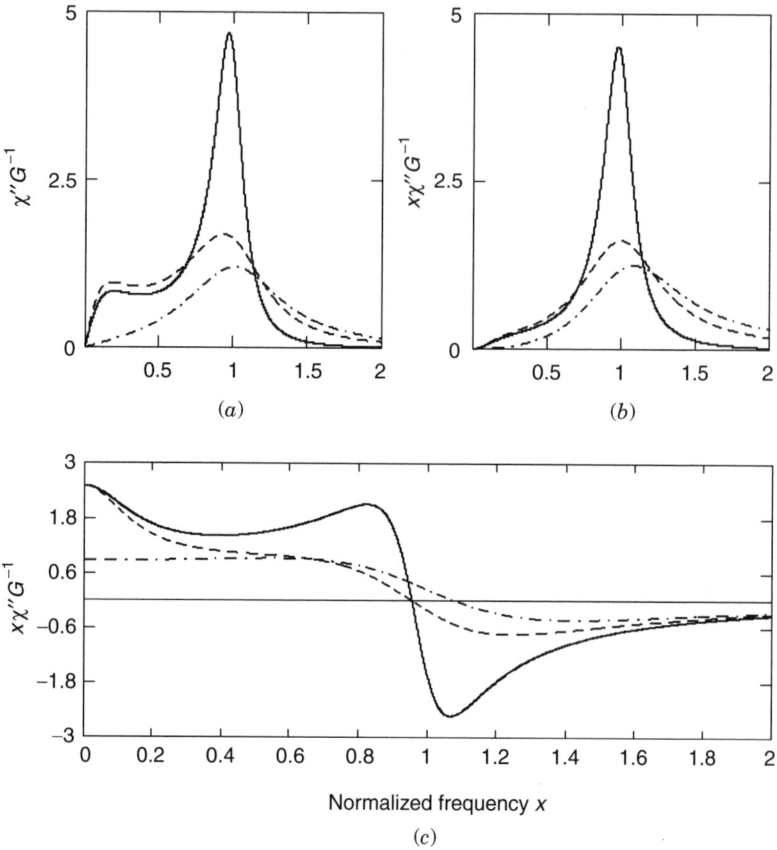

Figure 45. Frequency dependence of an imaginary (a) and of real (c) parts of the susceptibility; (b) absorption coefficient versus x. All quantities are nondimensional. Calculation for the isothermal (solid curves), Gross (dashed curves), and Lorentz (dashed-and-dotted curves) lines. The normalized collision frequency, y, is 0.4, and the correlation factor, g, is 2.5.

submillimeter wavelength region, we regard the non-self-consistent Lorentz line (arising at F_B) to be important only in the aspect of a general theory.

For calculations of the water spectra, we have applied (in Section VII.A.5) the Gross collision model, since it gives a good agreement with the experimental spectra and is more simple than the isothermal collision model.[56]

[56]We do not know now such fine experimental spectral features, which would impel us to apply the isothermal collision model.

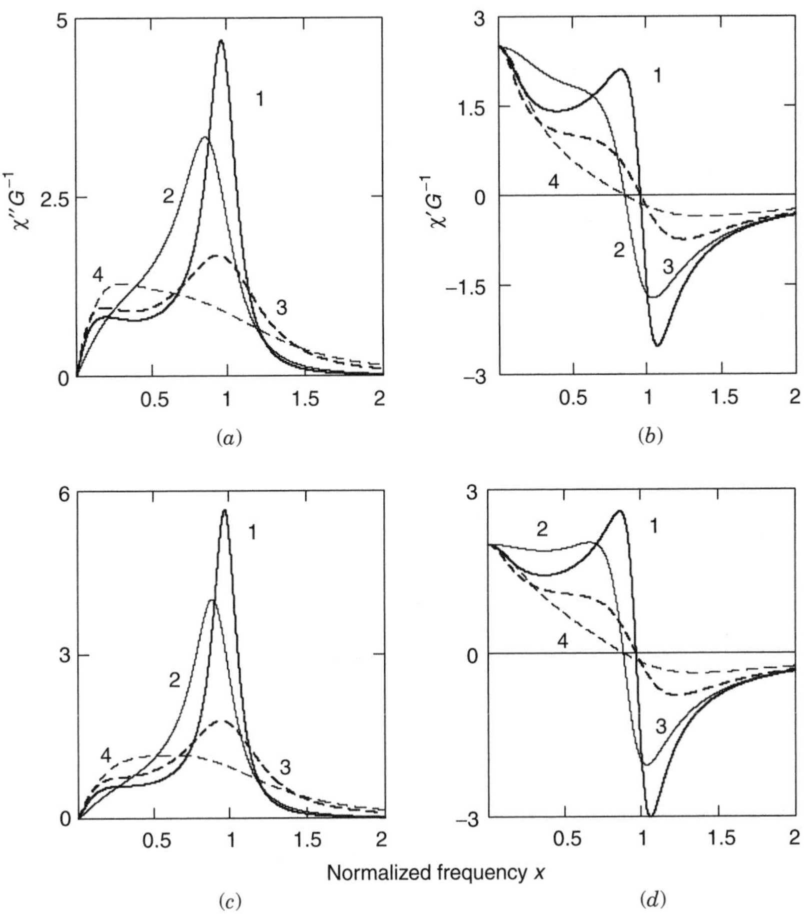

Figure 46. Evolution of frequency dependencies of loss (a, c) and of real part of the susceptibility (b, d) stipulated by change of the correlation factor g (nondimensional quantities). In Figs. (a, b) $g = 2.5$ and in Figs. (c, d) $g = 2$. Isothermal line (solid curves) and Gross line (dashed curves). Curves 1, 3 for $y = 0.4$ and curves 2, 4 for $y = 0.8$.

Note once more that the integrated absorption is the same for all three curves; it does not depend on y and g values.

If collisions become more *frequent,* the resonance lines widen and the resonance loss peak placed in the vicinity of $x = 1$ shifts to low frequencies.[57] The effect of collisions on the loss curves is more emphasized for the Gross line

[57]At F_B such a shift occurs to larger frequencies.

than for the isothermal line; see Figs. 46a,b, where curves 1, 3 refer to $y = 0.4$ and curves 2, 4 refer to $y = 0.8$. The frequency dependences of the real part of the susceptibility are illustrated by Figs. 46b,d.

If the lifetime τ substantially *increases* ($y \to 0$), then the loss line tends to $\delta(x - \omega/\Omega)$. This unphysical property of the *parabolic* potential well arises, since the oscillation frequency remains the same for *all* particles moving in the well.

When the *correlation factor g* decreases, both lines (371) and (373) become narrower[58] in the vicinity of the peak frequency (cf. Figs. 46a,b for $g = 2.5$ with Figs. 46c,d for $g = 2$).

We conclude: The submillimeter spectra calculated in terms of the harmonic oscillator model substantially differ from the spectra typical for the low-frequency Debye relaxation region. Such a fundamental difference of the spectra, calculated for water in microwave and submillimeter wavelength ranges, evidently reveals itself in the case of the composite hat-curved–harmonic oscillator model.

4. *About History of the Question*

The formulas for the susceptibility of a harmonic oscillator, presented above, were first derived in Ref. 18 with neglect of correlation between the particles' orientations and velocities. This derivation was based on an early version of the ACF method, in which the average perturbation theorem was not employed, so that the expression equivalent to Eq. (14c) was used. (The integrand of the latter involves the quantities perturbed by an a.c. field.) For a specific case of the parabolic potential, the above-mentioned theory is simple; however, it becomes extremely cumbersome for more realistic forms of the potential well.

In a classical resonance-absorption theory—given, for example, in Refs. 166 and 167—the complex susceptibility $\chi = \chi' - i\chi$ is found from reasoning based on calculation of the ratio

$$\chi = \hat{P}_E/\hat{E} \tag{383}$$

of the electric moment, comprised by a unit volume of the medium, to the amplitude of the a.c. electric field. Here $\hat{P}_E$ is assumed to be proportional to an ensemble average of the perturbed oscillation *amplitude*:

$$\hat{P}_E \propto \langle \hat{x}_E \rangle \tag{384}$$

[58]Change of g influences *insignificantly* the resonance line pertinent to the ensemble of *rigid dipoles* reorienting in a hat-like potential well, since the *resonance* absorption/loss peaks are usually placed at much higher frequencies than the Debye loss peak.

This amplitude is found from the equation of motion of a harmonic oscillator affected by an a.c. field $E(t)$. This approach yields the Lorentz and Van Vleck–Weisskopf lines, respectively, for a homogeneous and Boltzmann distributions of the initial a.c. displacements $x(t_0)$ established after instant t_0 of a strong collision. The susceptibility corresponding to the Van Vleck–Weisskopf line in terms of our parameters is given by [66]

$$\chi^*_{\mathrm{VVW}} = G(1 - yz)(1 - z^2)^{-1} \tag{385}$$

This line has an important advantages over the Lorentz line, since in Eq. (385): (i) the static susceptibility does not depend unlike that in Eq. (355a) on the collision frequency y and (ii) the loss curve is asymmetric. However, in contrast to the formula (382) the integral absorption corresponding to (385) diverges:

$$\Pi_{\mathrm{VVW}} = \infty \tag{386}$$

Thus, the lines generated by the autocorrelator (341b) and by the collision models described above *radically* differ from the lines generated by relations of the type (383) and (384). The difference between the two above-mentioned methods is summarized as follows.

Approach	**The ACF Method (GT, VIG)**	**The Method [66, 67]**
Boltzmann induced orientational distribution F_{B}	Lorentz line (355a)	Van Vleck–Weisskopf lines (385)
Homogeneous orientational distribution ($F = 1$)	Line (343)	Lorentz line (355a)

A classical resonance-absorption theory [66, 67] was aimed to obtain the formulas applicable for calculation of the complex permittivity and absorption recorded in *polar gases*. In the latter theory a *spurious* similarity is used between, (i) an almost harmonic perturbed law of motion of a charge affected by a parabolic potential (ii) and the law of motion of a *free* rotor, this law being expressed in terms of the projection of a dipole moment onto the direction of an a.c. electric field.

This analogy could indeed facilitate finding of the susceptibility like that given by Eq. (373) by the following reasoning [66, 67]. The law of molecular rotation could be represented in terms of two mutually perpendicular oscillations, so it seemed that the harmonic oscillator model could be applied. However, it turned out that development of a *rigorous* dielectric-relaxation theory applicable for two charged particles performing elastic oscillations along a straight line or for a reorienting rigid dipole does not present an easy task. Moreover, an assumption does not *rigorously* hold that the law of motion of a rigid dipole could be expressed *nearly* as an exponential dependence of the type $\exp(i\omega t)$. So, it would better to ground a modern linear-response theory on another evident idea: the law of motion of a particle starting its motion (and its interaction with an a.c. field) after the last strong collision *is repeated* after the time interval $2\pi/\omega$ equal to the period of an a.c. field (if other initial conditions are fixed for particles of a chosen sort). This idea actually presents a basis of the t_0 theorem proved in GT and VIG and explained in Section II.

The object to which the formulas derived here are applied is exactly characterized by oscillations of two charged particles.

Validity of our formulas for the resonance lines, which express the complex susceptibility through the spectral function, could be confirmed as follows. We have obtained an *exact* coincidence of the equations (353), (370), (371), which were (i) *directly* calculated here in terms of the harmonic oscillator model and (ii) derived in GT and VIG (see also Section II, A.6) by using a general linear-response theory.

VIII. DIELECTRIC RESPONSE OF AQUEOUS ELECTROLYTE SOLUTIONS

A. Schemes of Ion's Motion in Water Surroundings

If the concentration of an electrolyte is rather low, then *at least* three elementary types of motion stipulate dispersion of the complex permittivity in this polar fluid:

(i) Reorientation of single dipoles (water molecules) in an intermolecular potential well
(ii) Stretching vibrations of nonrigid dipoles (H-bonded water molecules)
(iii) Motion of a charged particles (ions) in dense surrounding of water molecules

Below we shall briefly consider only factors (i) and (iii) by taking into account the results of recent investigations [4, 5]. Factor (ii) so far was not studied with respect to water solutions. We shall emphasize its importance (and some other features of the problem, which is desirable to consider in the near future) at the end of the section. We think that the material below presents only a sort of a *preliminary studies* of several complicated phenomena.

Note that above we so far employed the absolute (CGS) system of units. Since in most publications deovted to electrolytes SI (MKS) system is usually used, we shall below duplicate the formula for both systems (if these formulas differ). Summing up the contributions to the complex dielectric permittivity due to the factors (i) and (iii), we write

$$\varepsilon^*(\omega) = \varepsilon^*_{\text{dip}}(\omega) + \begin{Bmatrix} 4\pi \\ 1 \end{Bmatrix} \chi^*_{\text{ion}}(\omega) \qquad \text{for} \quad \begin{Bmatrix} \text{CGS} \\ \text{MKS} \end{Bmatrix} \tag{387}$$

where the complex permittivity ε^* is a dimensionless parameter; it is the same in both systems.

By analogy with Eq. (278), we (a) suggest the contributions of dipoles and ions into the complex permittivity ε^* of the solution to be *additive* and (b) consider the ionic component of this permittivity to be proportional to the ionic susceptibility $\chi^*_{\text{ion}}(\omega)$. From Maxwell equations we have (see Appendix) the following relation between the ionic conductivity $\sigma = \sigma' + i\sigma''$ and susceptibility χ^*_{ion}:

$$\sigma(\omega) = -i\omega\chi^*_{\text{ion}} \begin{Bmatrix} 1 \\ \varepsilon_0 \end{Bmatrix} \qquad \text{for} \quad \begin{Bmatrix} \text{CGS} \\ \text{MKS} \end{Bmatrix} \tag{388}$$

In the MKS system of units, ε_0 is the dielectric constant of vacuum given by $\varepsilon_0 = \frac{1}{120\pi[\text{ohm}]\,c}$.

The second (ionic) term in Eq. (387) is assumed to vanish in the limit $\omega \to \infty$, just as does the term $\Delta\varepsilon^*(\nu)$ in Eq. (278) stipulated by oscillating *charges* of a nonrigid dipole. The first term in Eq. (387) will be calculated below in terms of the *hybrid model*, which was briefly described in Section IV.E. For the limit $\omega \to \infty$ we set this term to be equal to optical permittivity n^2_∞, *the same as in pure water.*

The existence of a hydration sheath surrounding an ion is modeled, just as in Section III for our basic *protomodel*, by an infinitely deep rectangular potential well. Thus we neglect contribution to ionic conductivity of charges, which penetrate through the hydration sheath. A fraction of such charges is shown to be very small [68].

The trajectory of an ion moving in such a potential presents a sequence of rectilinear sections placed between the points of elastic reflections of an ion from the "walls" of the well. We consider two variants of such a model related to one-dimensional and spatial motion of ion, depicted, respectively, in Figs. 47a and 47b. In the *first variant* the ion's motion during its lifetime[59] presents periodic oscillations on the rectilinear section $2\,l_c$ between two reflection points. In the second variant we consider a spherically symmetric potential well, to which a spherical hollow cavity corresponds with the radius l_c.

[59]We denote the mean value of this lifetime by τ_{ion}.

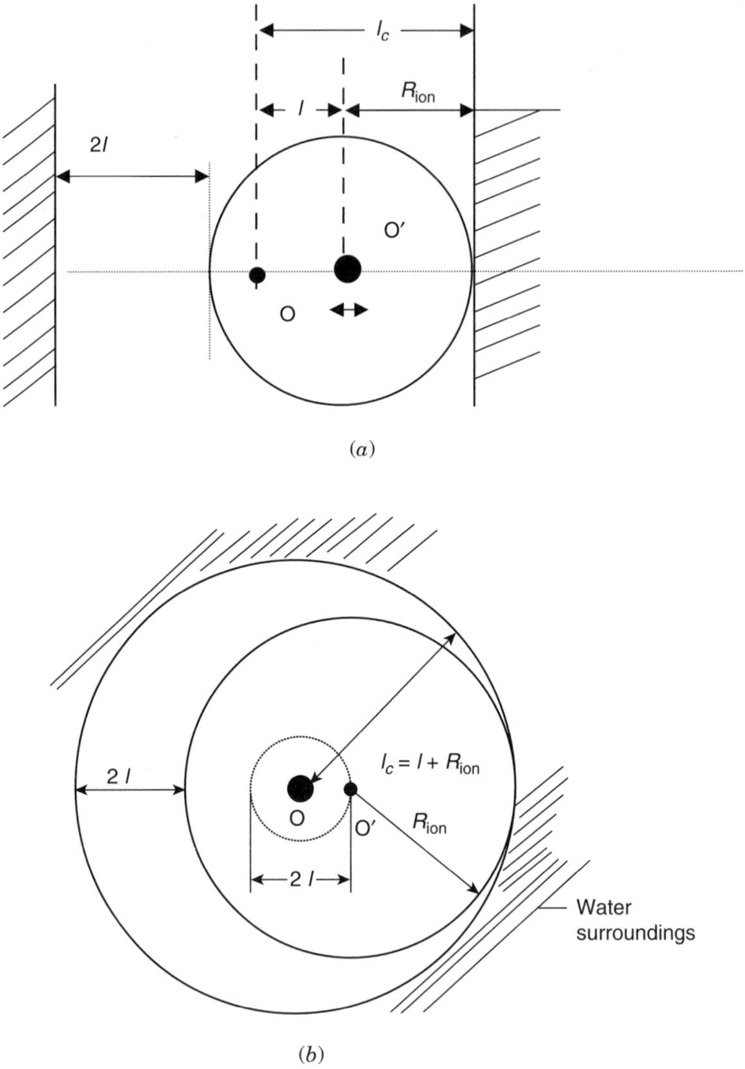

Figure 47. An ion inside a hydration cell. In the variant (a) an ion oscillates along a rectilinear section and in variant (b) inside a hollow reflecting sphere. (c) Schematic of the motion of an ion inside a spherical sheath. Dashed areas denote the hydration sheath.

We consider motion of an ion's *center of mass*. For a reasonable set of the parameters the length $2\,l_c$ is only *slightly* greater than the diameter $2\,R_{\text{ion}}$ of an ion [4, 5], so that the maximal deflection l of this center of mass O' about the center O of the case is very small, the absolute value of the current deflection

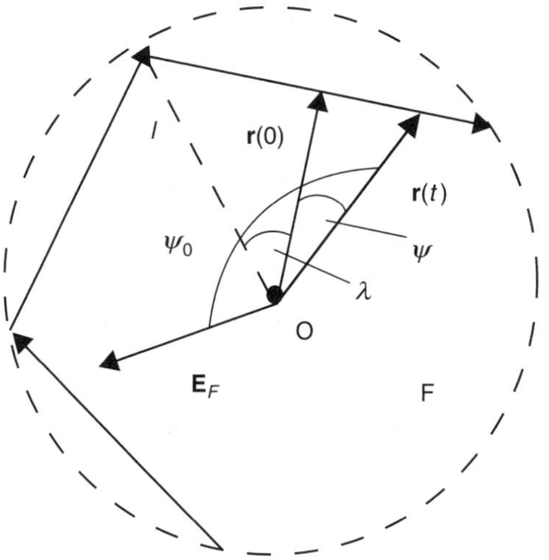

(c)

Figure 47 *(Continued)*

changes in the interval [0, 1]. It follows from Figs. 47a and 47b that $l = l_c - R_{\text{ion}}$.

During the lifetime the absolute value $|v|$ of the ion's velocity remains constant, since it is determined by a constant energy H of an ion:

$$|v| = \sqrt{2H/m} = v_{\text{T}}\sqrt{2h} \qquad \text{at} \quad |r(t)| \leq l \tag{389}$$

where m is mass of an ion, $r(t)$ is a current translational shift (we set $r = 0$ in the middle of the well), $v_T = \sqrt{(k_{\text{B}}T)/m}$ is the averaged ion's velocity, and $h = H/(k_{\text{B}}T)$ is the dimensionless energy.

1. *Motion Along a Line*

In the case of one-dimensional motion the time dependence of oscillating velocity of an ion can be expressed by a series, containing only odd harmonics:

$$v(t) = \sum_{n=1}^{\infty} v_{2n-1} \cos\left[\frac{2\pi}{\Im}(2n-1)t\right] \tag{390a}$$

where

$$\Im(h) \equiv \frac{4l}{v_T\sqrt{2h}} \tag{390b}$$

is the oscillation period. The Fourier amplitudes v_{2n-1} can be determined by using the standard formulas, if we write down a periodic law of ions motion with constant velocity from one reflection point to another one, taking into account the change of the *sign* of this velocity after the immediate reflection occurs.

2. *Motion Inside a Sphere*

A charged particle moves under effect of a *central force*. The latter is zero when a particle is found *inside* a sheath. At an instant, when the particle reaches the surface of a sphere, a strong force arises directed towards its center. A trajectory of a particle is placed in a diametric plane F of the sphere, since the angular momentum of the particle relative to the center O remains constant. Such a trajectory, being generally nonclosed, represents a sequence of equal-length chords shown in Fig. 47c; their lengths depend on energy and angular momentum of a particle.

Absolute value of the radius-vector $\mathbf{r}(t)$ of the center of mass (see Fig. 47c) varies in time periodically, from the minimum value $l\cos\lambda$ to l and conversely. Along any of the chords—that is, in the interval between two reflections—the angle λ is given by $\lambda = \psi(\Im/2)$ where $\Im$ is the period of the function $r(t)$ and $\psi(t)$ is the current turn of $\mathbf{r}(t)$ about the center O. In the above-mentioned figure, this angle is counted off from any "initial" direction $\mathbf{r}(0)$; the latter is set to be $\mathbf{r}_{\min}$. Let us express the projection r_E of the vector $\mathbf{r}(t)$ on the direction of an a.c. field $\mathbf{E}$. By analogy with Eq. (40) we write

$$r_E = r\sin\Theta\,\cos(\psi + \psi_0) \tag{391}$$

Θ is the angle between the normal to the plane F and the a.c. field vector $\mathbf{E}$; ψ_0 is the angle between the projection $\mathbf{E}_F$ of $\mathbf{E}$ onto the plane F and initial direction $\mathbf{r}_{\min}$. It follows from Eq. (389) and Fig. 47c that

$$r\cos\psi = l\cos\lambda = r_{\min} \tag{392a}$$

$$r\sin\psi = vt = v_T t\sqrt{2h} \tag{392b}$$

Then the period $\Im$ is given by

$$\Im(h) \equiv \frac{2l\,\sin\lambda}{v_T\sqrt{2h}} \tag{393}$$

B. The Frequency-Dependent Ionic Conductivity

We first consider the case of *one-dimensional* motion. Application of the ACF method to an ionic ensemble leads to the autocorrelator (341b) derived in

Section VII for a homogeneous induced distribution, characterized by $F = 1$. Substituting this formula into Eq. (388), we have the following expression for the complex conductivity:

$$\sigma(\omega) = \frac{a}{r_D^2}\int_\Gamma \exp(-h)v_E(0)\,d\Gamma \int_0^\infty v_E(t)\exp(i\hat{\omega}t)\,dt\left[\int_\Gamma \exp(-h)\,d\Gamma\right]^{-1} \tag{394}$$

$$a = \begin{cases} (4\pi)^{-1} & \text{for CGS} \\ \varepsilon_0 & \text{for SI} \end{cases} \tag{395}$$

The ionic susceptibility/conductivity is a function of the trajectories of the charges at *equilibrium*; that is, $\chi^*_{\text{ion}}(\omega)$ is proportional to the ACF spectrum of the E-projection of the steady-state velocity. One may regard Eq. (394) as a *convenient* (for numerical calculations) form of the Kubo formula [69] for the diagonal component of the conductivity tensor

$$\sigma_{xx}(\omega) = (k_B TN)^{-1}\text{Re}\int_0^\infty \langle J_x(t)J_x(0)\rangle \exp(-i\omega t)\,\text{d}t \tag{396}$$

In Refs. 4 and 5 was shown that χ^*_{ion}, found from Eq. (394), satisfies the sum rule for the integrated absorption for both one-dimensional and spatial ensembles:

$$\Pi = \int_{-\infty}^{\infty} \omega\chi''(\omega)d\omega \equiv \int_{-\infty}^{\infty} \sigma'(\omega)\text{d}\omega = \pi\omega_p^2 \tag{397}$$

where the plasma frequency is proportional to the charge of electron e squared[60]:

$$\omega_p^2 = \frac{e^2 N}{ma} \tag{398}$$

Substituting the series (390a) and (390b) into integral (394) and introducing a nondimensional function S of the reduced complex frequency Z, one can get the result for $S(Z)$ in an analytical form [5]:

$$\sigma = b\omega_p S(Z) \tag{399a}$$

$$b = \begin{cases} 1 & \text{for CGS} \\ 4\pi\varepsilon_0 & \text{for MKS} \end{cases} \tag{399b}$$

[60]We apply our theory for 1–1 electrolytes, so the charge of an ion is assumed to present the charge of electron.

where the series for S comprising the Fourier amplitudes v_{2n-1} could be summed up to transform to *simple* integral of analytical function:

$$S(Z,d) = -\frac{8iZ}{\pi^3\sqrt{\pi}}\int_0^\infty \sum_{n=1}^{\infty} \frac{1}{(2n-1)^2} \frac{\exp(-\xi^2)\xi^2 d\xi}{\left[\frac{\pi}{2d}(2n-1)\xi\right]^2 - Z^2}$$

$$\equiv \frac{i}{4\pi Z}\left[1 - \frac{4}{\sqrt{\pi}Zd}\int_0^\infty \tan\left(\frac{Zd}{\xi}\right)\exp(-\xi^2)\xi^3\, d\xi\right] \qquad (400)$$

Here d is the dimensionless parameter of the model

$$d \equiv l/\left(\sqrt{2}r_D\right) \qquad (401)$$

(the length l was defined above; see Fig. 47), r_D is the Debye radius for vacuum

$$r_D = \sqrt{\frac{a\, k_B T}{e^2 N_{ion}}} \qquad (402)$$

and the complex frequency Z is given by

$$Z = X + iY = \hat{\omega}/\omega, \qquad X \equiv \omega/\omega_p, \qquad Y \equiv (\omega_p \tau_{ion})^{-1} \qquad (403)$$

From Eq. (400) we have the following expression for the normalised static conductivity:

$$S_s(Y,d) = \frac{1}{4\pi Y}\left[1 - \frac{4}{\sqrt{\pi}Yd}\int_0^\infty \tanh\left(\frac{Yd}{\xi}\right)\exp(-\xi^2)\xi^3\, d\xi\right] \qquad (404)$$

For $d \to \infty$ the static conductivity tends to infinity, if $Y \to 0$, but if the half-width d is finite, then $S_s \to 0$ for $Y \to 0$. This is the principal distinction of our model from that corresponding to rarefied plasma, where d tends to infinity. Let now Y be finite. If $Y = 0.005$ and $d = 1$, then we find $S_{s\ plasma} \approx 15.9$ and $S_s(Y,d) \approx 2.64 \cdot 10^{-4}$. For the same Y and $d = 0.5$, $S_s(Y,d) \approx 6.6 \times 10^{-5}$. Consequently, the static conductivity decreases *by several orders of magnitude*, if translations in rarefied plasma (which occur without restrictions) are replaced by small-amplitude oscillations inside a local potential well introduced for a liquid state.

For rather *rare ionic collisions* ($Y \ll 1$), Eq. (400) yields simple formulas [5]:

(a) If the width d is *small*, then

$$S(Z,d) \approx -\frac{iZd^2}{6\pi} \quad \text{for } (Zd^2) \ll 1 \tag{405}$$

(b) If $Y \to \infty$ but the width d is *arbitrary*, then

$$\lim_{Y\to 0} \mathrm{Re}\{S(Z,d)\} \equiv S'(X,d) = \frac{16X^2d^3}{\pi^5\sqrt{\pi}} \exp\left[-\left(\frac{2Xd}{\pi}\right)^2\right] \tag{406}$$

For the case of *spatial* motion, Figs. 47b and 47c, it is convenient to start from the following expression for the autocorrelator of the E-projection $r_E(t)$ of the radius vector $r(t)$ [147]:

$$\sigma(\omega) = \frac{a(\hat{\omega})^2}{r_{\mathrm{D}}^2}\left(r_E(0)\int_0^\infty [r_E(t) - r_E(0)]\exp(i\hat{\omega}t)\,dt\right) \tag{407}$$

The ionic conductivity (407), analogously with that given by Eq. (394), is determined by equilibrium trajectories. If $r_E(t)$ is expressed through Eqs. (391)–(393), then (407) can be transformed to *double* integral from analytical function [4]:

$$S(Z,d) = \frac{i}{4\pi Z}\left(1 - \sqrt{\frac{1}{\pi}\frac{4}{Zd}}\int_0^\infty \exp(-t^2)t^5\,dh \int_0^1 \frac{f^3\sin[2\varsigma(f)]df}{f^2 - \sin^2[\varsigma(f)]}\right) \tag{408a}$$

$$\equiv \frac{i}{\pi\sqrt{\pi}Z^2d}\int_0^\infty \exp(-t^2)t^4dt \int_0^1 \left\{2Zd - \frac{tf\sin[2\varsigma(f)]}{f^2 - \sin^2[\varsigma(f)]}\right\} f^2\,df \tag{408b}$$

where

$$\zeta(f) \equiv Zfd/t \tag{408c}$$

Setting $X = 0$, we obtain an expression for the normalized static conductivity S_s as a function of Y and d:

$$S_s(Y,d) \equiv S(iY,d) = \frac{1}{4\pi Y}\left(1 - \sqrt{\frac{1}{\pi}\frac{2}{Yd}}\int_0^\infty e^{-h}h^2\,dh \int_0^1 \frac{\sinh[2\psi(f)]f^3\,df}{f^2 + \sinh^2[2\psi(f)]}\right)$$

$$\equiv \frac{1}{\pi\sqrt{\pi}Y^2d}\int_0^\infty \exp(-t^2)t^4dt \int_0^1 \left[2Yd - \frac{ft\sinh[2\psi(f)]}{f^2 + \sinh^2[2\psi(f)]}\right] f^2df \tag{409a}$$

$$\psi(f) \equiv \frac{fYd}{t} \tag{409b}$$

At small complex frequency Z, such that $|Zd| \ll 1$, Eq. (408a) could be reduced to a simple formula for the complex conductivity, analogous to Eq. (405):

$$S(Z, d) \approx -\frac{iZd^2}{10\pi}\left(1 + \frac{32}{21}d^2Z^2\right) \tag{410}$$

from which the expression

$$S_s(Y, d) = \frac{Yd^2}{10\pi}\left(1 - \frac{32}{21}d^2Y^2\right) \qquad \text{for } Yd \ll 1 \tag{411}$$

for the static conductivity is obtained. It follows from Eqs. (410) and (411), just as from Eq. (405), that for $Y \to 0$ the static conductivity becomes very small.

Figures 48a and 48b demonstrate, respectively, the frequency dependences of the real and imaginary parts of the normalized conductivity [Eq. (400)]. For the chosen set of parameters, $Y = 0.05$ and 0.1 and $d = 0.5$ and 1, which are more or less typical for electrolytes under consideration:

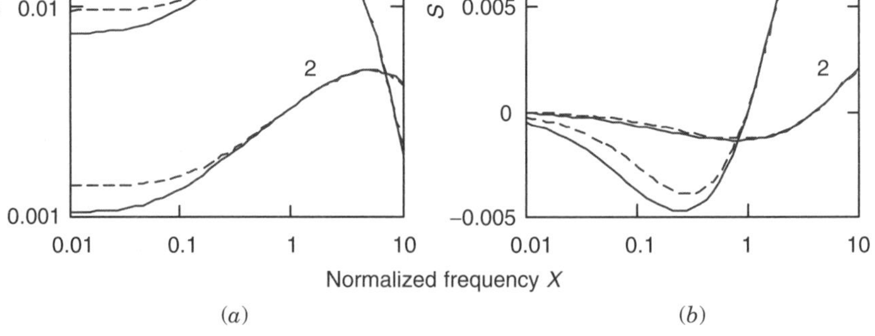

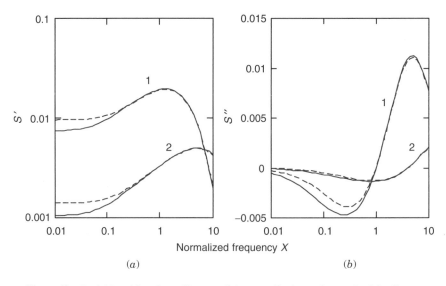

Figure 48. Real (a) and imaginary (b) parts of the normalized complex conductivity S versus normalized frequency X calculated from Eq. (400) for one-dimensional ensemble. Curves 1 are for $d = 1$; curves 2 are for $d = 0.5$. Normalized collision frequency $Y = 0.05$ (solid lines) and 0.1 (dashed lines).

(a) In the low-frequency region the real part S' is almost independent of frequency X, while the imaginary part S'' decreases from zero value with the rise of X, so that it is negative; the noticeable change of S' with X occurs for $X \approx 0.1$—that is, for ω about 1/10 of the plasma frequency ω_p.

(b) In the high-frequency region, S' goes through a maximum, while the imaginary part S'' goes through a minimum and then a maximum.

(c) These quasi-resonance curves shift to higher frequencies, if the width of the well decreases.

(d) The increase of the collision frequency Y results in the change of only low-frequency part of these X-dependences.

Formula (408a) derived for the case of spatial motion gives the frequency dependence of the normalized conductivity very close to that given by Eq. (400) for a one-dimensional ensemble. This statement is illustrated in Fig. 49. In the latter the maxima and minima of $S''(\nu)$ and $S''(\nu)$ are shifted to *lower frequencies*; moreover, the static susceptibility, expressed in terms of S_s, is a little *greater*. The dependences of the dimensionless ionic conductivity on frequency have a form of *fading oscillations*. This result looks reasonable, but it is desirable to use some particular example to show the importance of this phenomenon. We shall consider this question in terms of a one-dimensional ensemble, for which the formulas are simpler. This model will be applied to important aqueous solutions of NaCl and KCl, which in particular determine the biologic functions in the living organisms at the basic level of cells.

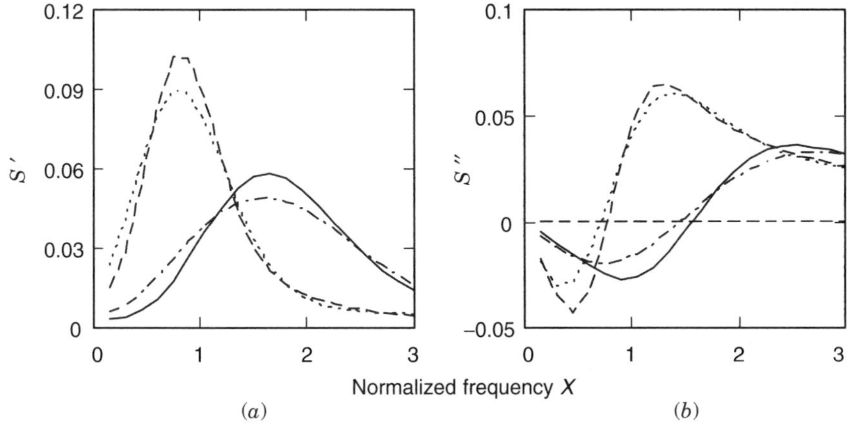

Figure 49. Frequency dependencies of the real (a) and imaginary (b) parts of the normalized complex conductivity S. Normalized collision frequency, Y, is 0.1. Spatial motion: Solid curves for $d = 1$, dashed curves for $d = 2$. One-dimensional motion: Dashed-and-dotted curves for $d = 1$, dotted curves for $d = 2$.

C. Effect of Ions on Wideband Spectra

Now we apply the *additivity approximation* corresponding to Eq. (387). Namely, we sum up the contributions of three sources of dielectric loss due to: (a) reorienting dipoles, (b) oscillating anions, and (c) oscillating cations. Combining Eqs. (387), (388), and (399a), we write the formula for the *total* complex permittivity:

$$\varepsilon^* = \varepsilon^*_{\text{dip}} + 2i\frac{\sigma}{c\nu b} \tag{412a}$$

The contributions of cations (Na^+ or K^+) and of the anions (Cl^-) to conductivity σ are added up:

$$\sigma = b\left(S^+\omega_p^+ + S^-\omega_p^-\right) \tag{412b}$$

where b is defined by Eq. (399b). The expression of the complex permittivity through the dimensionless quantities is given by

$$\varepsilon^* = \varepsilon^*_{\text{dip}} + 4\pi i\left[\frac{S(X^+)}{X^+} + \frac{S(X^-)}{X^-}\right] \tag{413a}$$

$$X^+ = \frac{\omega}{\omega_p^+} \quad \text{and} \quad X^- = \frac{\omega}{\omega_p^-} \tag{413b}$$

the signs + and − refer to cations and anions, respectively. Further calculations are made for the one-dimensional variant, in which the function $S(X)$ is given by Eqs. (400) or (405).

In our case of 1–1 electrolytes we have the following relations for the concentrations: $N^+ = N^- = N_{\text{ion}}$. The first term in Eq. (413a) is calculated by using the formulas presented in Section IV. However, generally speaking, the Kirkwood correlation factor now should be determined with account of ionic static permittivity as

$$g = \frac{[\varepsilon_s - \Delta\varepsilon^*_{\text{ion}}(0) - n^2_\infty][2\varepsilon_s - 2\Delta\varepsilon^*_{\text{ion}}(0) + n^2_\infty]}{12\pi G \cdot [\varepsilon_s - \Delta\varepsilon^*_{\text{ion}}(0)]} \tag{414a}$$

where ε_s is the *experimental* static permittivity,

$$\Delta\varepsilon^*_{\text{ion}}(0) \equiv \chi^*(0)\begin{Bmatrix} 4\pi \\ 1 \end{Bmatrix} \quad \text{for} \quad \begin{Bmatrix} \text{CGS} \\ \text{MKS} \end{Bmatrix} \tag{414b}$$

$$G = \frac{\mu^2 N_{\text{dip}}}{3k_B T}\begin{Bmatrix} 1 \\ \varepsilon_0^{-1} \end{Bmatrix} \quad \text{for} \quad \begin{Bmatrix} \text{CGS} \\ \text{MKS} \end{Bmatrix} \tag{414c}$$

and $\varepsilon_s - \Delta\varepsilon^*_{ion}(0)$ is the contribution to static permittivity due to *reorienting dipoles*. However, it will be shown that at rather small electrolyte concentration the usual formula holds:

$$g \approx \frac{(\varepsilon_s - n_\infty^2)(2\varepsilon_s + n_\infty^2)}{12\pi G\varepsilon_s} \tag{414d}$$

In further estimations we:

(i) Neglect distinction between the hydrated and bulk water.

(ii) Take the same lifetimes τ_{ion} for anions and cations, namely for Cl^- and NA^+ or Cl^- and K^+.
Consequently, the normalized collision frequency Y is related to τ_{ion} as

$$Y^+ = (\omega_p^+ \tau_{ion})^{-1} \quad \text{and} \quad Y^- = (\omega_p^- \tau_{ion})^{-1}$$

(iii) Set the mean ionic lifetime equal or much longer than that (τ), fitted for water molecules:

$$\text{variant 1:} \quad \tau_{ion} = \tau \quad \text{and} \quad \text{variant 2:} \quad \tau_{ion} = 10\tau$$

(iv) Assume that the length l is proportional to the sum of ion's and water molecule's radii:

$$l^\pm = (R^\pm + R_w)\gamma, \quad \text{so that} \quad d^\pm(\gamma) = (R^\pm + R_w)\gamma/(\sqrt{2}r_D) \tag{415}$$

where the coefficient γ is determined from the condition

Calculated static conductivity of the solution is equal to the experimental value σ_s (416)

After we put $\omega = 0$ in Eq. (415), the condition (416) yields the following equation for γ:

$$\frac{1}{b}\sigma_s = S_s^+[Y^+, d^+(\gamma)]\omega_p^+ + S_s^-[Y^-, d^-(\gamma)]\omega_p^- \tag{417}$$

where $S_s(Y, d)$ is determined by Eq. (404).

The number density N_{dip} of water molecules is found from the formulas

$$N_{dip} = \frac{N_A \rho_{eff}}{M_{ion}}, \qquad \rho_{eff} = \rho - 1000 C_M M_{ion} \tag{418}$$

TABLE XVI
Experimental/Fitted Parameters for NaCl–Water and KCl-Water Solutions[a]

	NaCl–water						KCl–Water					
C_M	ρ_{eff}	σ_s	ε_s	τ_D	τ	m_{dip}	ρ_{eff}	σ_s	ε_s	τ_D	τ	m_{dip}
0.0	0.998	0	80	9.8	0.42	7.6	0.998	0	80	9.8	0.42	7.6
0.5	0.988	3.89	71	9.2	0.44	8.0	0.984	5.11	74	9.0	0.41	7.4
1	0.980	7.22	63	8.8	0.47	8.5	0.969	9.78	69	8.3	0.40	7.2

[a]Temperature 20°C, concentration C_M in mol liter^{-1}; density ρ_{eff} of water in the solution in g · cm^{-3}; the static conductivity σ_s in (ohm · m)$^{-1}$; the relaxation time τ_D and the mean lifetime τ in ps.

Notes: $k_\mu = 1.12, I = 1.483 \times 10^{-47}$ kg · m^2, $n_\infty^2 = 1.7, \varepsilon_\infty \approx 5, u = 5.575, \mu_0 = 1.48$ D, $\beta = 19.4°$, $M_{NaCl} = 58.44, M_{KCl} = 74.55, M_w = 18$.

1 D $= 3.336 \times 10^{-30}$ C · m; in Absolute system of units $[\sigma] = c^{-1}$ and $\sigma_{[SI]} = 1/9 \times 10^{-9} \sigma_{[CGS]}$.

where N_A is Avogadro's number, M_{ion} is the molecular mass of an ion (M_{NaCl} or M_{KCl} in our case), ρ_{eff} [g/cm^3] is the effective density of water in the electrolyte solution, and ρ is the density of the solution. Experimental [70, 71] values ε_s, σ_s, τ_D, ε_∞, and ρ_{eff} and the fitted parameters of the hybrid model are presented in Table XVI for two concentrations C_M [mol liter^{-1}]. For liquid water—that is, at $C_M = 0$—we used experimental data given in VIG and in Ref. 17.

If the electrolyte concentration C_M varies, the wideband spectra are controlled only by one parameter (τ) of the hybrid model. Other parameters of this model—the normalized well depth u, the libration amplitude β, and the μ-correcting coefficient k_μ—can be set independent of C_M and therefore could be fit by comparison of the calculated and recorded [70, 71] spectra of water (see Table XVI).

In Tables XVII–XIX we present the fitted/calculated parameters of our model for ions: plasma frequency $\omega_p^\pm$, static ionic conductivity $\sigma_s^\pm$, fitted parameters

TABLE XVII
Experimental and Fitted Parameters for the Molecular Model of the Static Conductivity[a] σ_s

C_M (mol/liter)	NaCl–Water ω_p^-	ω_p^+	σ_s^-	σ_s^+	KCl–Water ω_p^-	ω_p^+	σ_s^-	σ_s^+
0.5	3.84	4.77	9.0	3.0	3.65	4.77	1.6	3.0
1	5.43	6.74	17	5.6	5.17	6.74	3.3	5.8

[a]Plasma frequency $\omega_p^\pm$ in rad · s$^{-1} \times 10^{-12}$, σ_s in (ohm · m)$^{-1}$.

Notes: For NaCl–water, $R^- = 0.95$ Å; for KCl–water, $R^- = 1.33$ Å; $R^+ = 1.8$ Å, $R_w = 1.5$ Å, $R^+ =$ 1.8 Å, $R_w = 1.5$ Å.

TABLE XVIII
Fitted Parameters of the Model[a]

C_M (mol liter^{-1})	NaCl–Water $\tau_{ion} = 10\tau$ d^+	d^-	$\tau_{ion} = \tau$ d^+	d^-	KCl–Water $\tau_{ion} = 10\tau$ d^+	d^-	$\tau_{ion} = \tau$ d^+	d^-
0.5	1.51	0.82	0.48	0.26	1.47	1.10	0.47	0.35
1	2.13	1.16	0.69	0.37	2.01	1.50	0.65	0.49
C_M (mol liter^{-1})	NaCl–Water $\tau_{ion} = 10\tau$ Y^+	Y^-	$\tau_{ion} = \tau$ Y^+	Y^-	KCl–Water $\tau_{ion} = 10\tau$ Y^+	Y^-	$\tau_{ion} = \tau$ Y^+	Y^-
0.5	0.0102	0.0082	0.102	0.082	0.0123	0.0130	0.123	0.130
1	0.0096	0.0077	0.096	0.077	0.0129	0.0136	0.129	0.136

[a]Lifetimes are given in ps.

$d^{\pm}$, and the corresponding *effective lengths* $l^{\pm}$. We see that $l^{\pm}$ has an order of a part of angstrom; this result looks reasonable. We remark that the mean distance ΔR_w between water molecules weakly depends on C_M, that is,

$$\Delta R_w = (N_A \rho_{eff}/M_w)^{-1/3} \approx 3.13\ \text{Å} \tag{419}$$

In Figs. 50 and 51 we illustrate the effect of σ-dispersion in terms of ionic contribution to the $\varepsilon^*(\nu)$ spectrum, calculated for the lowest concentration 0.5 mol liter^{-1}. In Figs. 50a,c and Figs. 51a,c we see that *additional loss* $\Delta\varepsilon''_{ion}(\nu)$ arises due to this dispersion. This loss increases (especially for NaCl) if ionic lifetime τ_{ion} becomes longer. For $\tau_{ion} = 10\tau$ the additional loss appears at millimeter waves, while for $\tau_{ion} = \tau$ it appears at submillimeter waves.

TABLE XIX
The Mean Distance l (in Å) Between the Surface of an Ion and the Hydrated Layer

C_M (mol liter^{-1})	NaCl–Water $\tau_{ion} = 10\tau$ l^+	l^-	$\tau_{ion} = \tau$ l^+	l^-	KCl–Water $\tau_{ion} = 10\tau$ l^+	l^-	$\tau_{ion} = \tau$ l^+	l^-
0.5	0.25	0.14	0.080	0.043	0.275	0.205	0.088	0.066
1	0.35	0.19	0.11	0.062	0.379	0.283	0.123	0.092

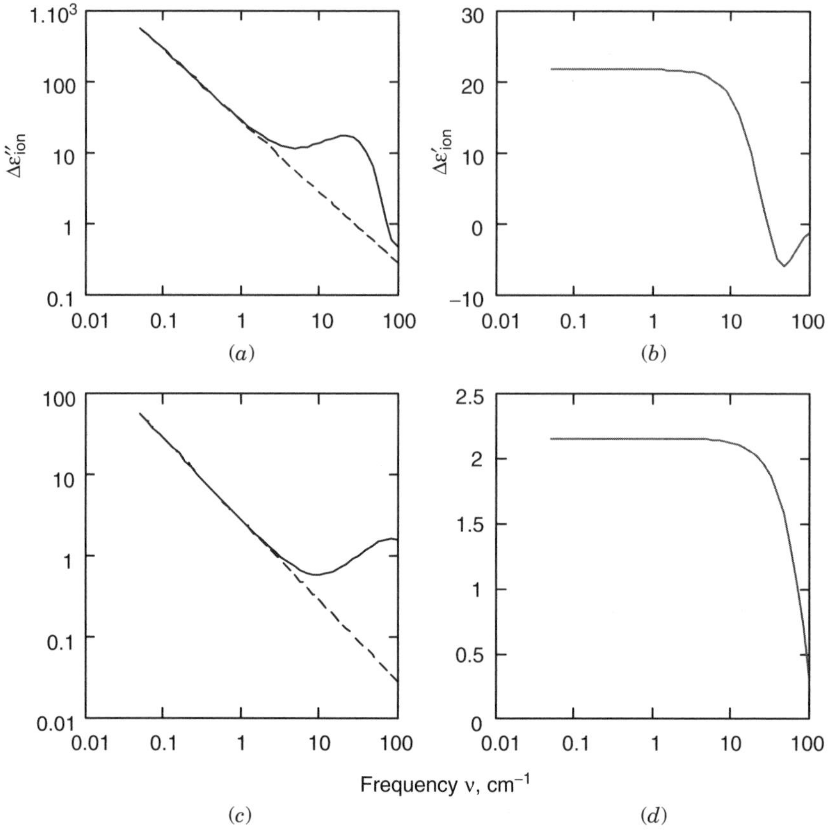

Figure 50. Calculated contributions of the ionic permittivity for the imaginary (a, c) and real (b, d) parts of the total complex permittivity (solid lines); dashed lines refer to the calculation, neglecting the ionic dispersion. (a, b) For NaCl–water solution; (c, d) for KCl–water solution. $C_M = 0.5\,\text{mol/liter}$, $\tau_{\text{ion}}/\tau = 10$.

In view of Figs. 51b,d, the ionic zero-frequency contribution $\Delta\varepsilon'_{\text{ion}}(0)$ to the static permittivity ε_s of the total permittivity is small at $\tau_{\text{ion}} = \tau$, so that we may indeed neglect the ionic term $\Delta\varepsilon'_{\text{ion}}(0)$ in Eq. (414a) in the comparison with the static permittivity ε_s. However, for $\tau_{\text{ion}} = 10\tau$, the value $\Delta\varepsilon'_{\text{ion}}(0)$ is large for NaCl and noticeable for KCl, as shown, respectively, in Figs. 50b and 50d. If such long ionic lifetimes indeed occur, one should use expression (414a) for the Kirkwood correlation factor, which is more accurate than Eq. (414d).

The calculated loss $\varepsilon''(\nu)$ and absorption $\alpha''(\nu)$ spectra are depicted for NaCl–water solution in Fig. 52 for $\tau_{\text{ion}} = 10\tau$. Again we take $C_M =$

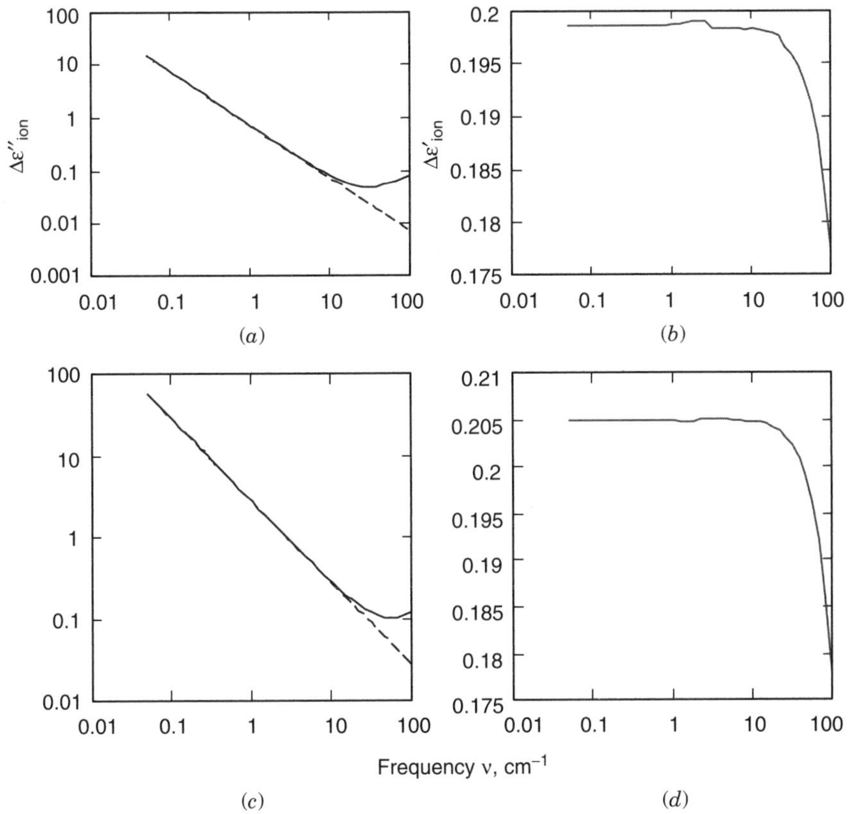

Figure 51. Same as in Fig. VIII.12 but $\tau_{ion}/\tau = 1$.

0.5 mol liter^{-1}. At microwaves the $\varepsilon''(\nu)$ curve (see Fig. 52a) comprises a shoulder, while in the submillimeter range it comprises the two maxima. They correspond to the R- (near 200 cm^{-1}) and librational (near 700 cm^{-1}) peaks of the absorption frequency dependence $\alpha(\nu)$, shown in Fig. 52b. At low frequencies up to $\nu \approx 2\,\text{cm}^{-1}$ for $\tau_{ion}/\tau = 10$ the rigorous $\varepsilon''(\nu)$ solid line curve is well approximated (cf. with the dotted curve) by the formula composed of the Debye term and of the ionic term for $\sigma \equiv \sigma_s$:

$$\varepsilon^*(\nu) = \varepsilon_\infty + \frac{\varepsilon_s - \varepsilon_\infty}{1 - 2\pi i c \nu \tau_D} + \frac{i\sigma_s}{2\pi a c \nu} \tag{420}$$

However, for $\nu > 10$ cm^{-1}, Eq. (420) does not hold.

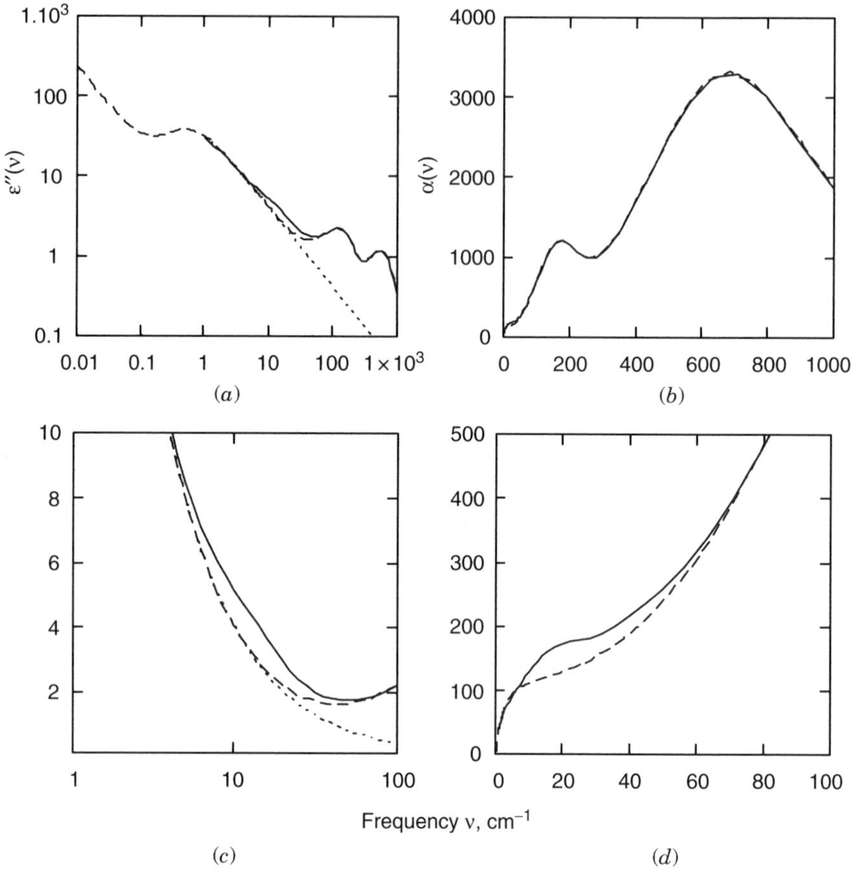

Figure 52. Frequency dependencies of dielectric loss (a, c) and of absorption coefficient (b, d) calculated for NaCl–water solution at concentration 0.5 mol/liter, temperature 20°C. Other explanations are given in the text.

It is also interesting to compare the dependence (413a) with

$$\varepsilon^* = \varepsilon^*_{\text{dip}} + 2i\frac{\sigma_s}{c\nu b} \tag{421}$$

(cf. dashed and solid curves in Fig. 52), in which the first dipolar term is rigorous, while the ionic dispersion is neglected in the second term. A noticeable distinction between the spectra calculated with or without account of the $\sigma(\nu)$

dispersion may arise in the case of NaCl–water solution at millimeter/submillimeter wavelengths, if the ratio τ_{ion}/τ is large (e.g., 10 in Fig. 52); if $\tau_{\text{ion}} \approx \tau$, the effect of the $\sigma(\nu)$ dispersion on the $\varepsilon^*(\nu), \alpha(\nu)$ dependences vanishes [5]. The same is true for KCl–water solution at both above-studied lifetimes τ_{ion}.

D. Discussion

It is useful to characterize the molecular dynamics/structure of a solution by the mean number $\breve{m}$ of reorientation cycles performed by a water molecule in the solution during its mean lifetime τ. Using for estimation Eq. (128), we see in Table XX that $\breve{m}$ is about 15. This means that the structure of each liquid is *rather rigid*. For NaCl–H_2O, $\breve{m}$ increases with concentration C_M, while for NaCl–H_2O it decreases, so that the first solute *strengthens* and the second one *weakens* the water structure. These results agree with the concepts of a "positive" (for NaCl) and a "negative" (for KCl) hydration [72]. Using Eq. (390b), we estimate the mean numbers $m^{\pm}$ of the ionic translation cycles, occurring in the well during the lifetime τ_{ion}:

$$m^{\pm} \approx \frac{\langle \Omega^{\pm} \rangle}{\tau_{\text{ion}}} = \frac{\sqrt{\pi}}{2d}\, \omega_p^{\pm} \tau_{\text{ion}} = \sqrt{\frac{\pi}{2}}\, \frac{v_T^{\pm} \tau_{\text{ion}}}{l^{\pm}} \tag{422}$$

where

$$\Omega = 2\pi/\Im \tag{423}$$

and $\Im$ is given by Eq. (390b). The numbers $m^{\pm}$ depend on the ionic lifetime τ_{ion}. In view of Table XX, for $\tau_{\text{ion}} = 10\tau$ the numbers $m^{\pm}$ are much greater than $\breve{m}$ but for $\tau_{\text{ion}} = \tau$ they are commensurable with $\breve{m}$.

The ionic dispersion $\sigma(\nu)$ has been discussed in Refs. 73 and 74. Namely, in Ref. 73 it was concluded on the basis of a mean-field approximation that the

TABLE XX
Mean Number of Ion's and Dipole's Reflections[a]

	NaCl–Water					KCl–Water				
	$\tau_{\text{ion}}\ 10\,\tau$		$\tau_{\text{ion}}\tau$			$\tau_{\text{ion}} = 10\,\tau$		$\tau_{\text{ion}} = \tau$		
C_M mol liter^{-1}	m^+	m^-	m^+	m^-	m_{dip}	m^+	m^-	m^+	m^-	m_{dip}
0.5	58	132	18	41	16	49	62	19	19	15
1	44	99	13	31	17	34	44	13	13	14

[a]Performed during their lifetimes in corresponding potential wells.

dispersion of the ionic conductivity is negligible in electrolytes. However, later (Ref. 74) such a dispersion was discovered. Yet the problem remained unsolved, since specific properties of the solvent (water) were *not* taken into account in Ref. 74. Using formulas (400) and (408a) and the approximations for them given above, we have shown that the dispersion $\sigma(\nu)$ should appear, if the normalized frequency $X = \omega/\omega_p$ exceeds ≈ 0.1 (see Fig. 48). In view of Figs. 49–52, the conductivity dispersion $\sigma(\nu)$ may cause an *additional* dielectric loss/absorption in the ranges of millimeter and submillimeter wavelengths, *if* the aqueous sheath around an ion lives much longer than a local-order cell in bulk water (i.e., if $\tau_{\text{ion}} \gg \tau$). From Table XX we see that in this case an ion should perform at $C_M = 0.5$ mol liter^{-1} about 50 or more cycles of translational motion inside the sheath. The question is whether such a situation is possible. On the other hand, if $\tau_{\text{ion}} \approx \tau$, then the *ionic-dispersion* phenomenon should *not* affect the calculated spectra. Then the approximation (421) holds, where $\sigma \equiv \sigma_s$ is set. It would be interesting to find *experimentally* which of two controversial inequalities ($\tau_{\text{ion}} \gg \tau$ or $\tau_{\text{ion}} \approx \tau$) is more realistic.

Let us consider in this respect the simulation result [75], which yields some additional information about the residence times of bulk-water molecules and about those in the hydration sheaths arising around the cations Na^+ and the anions Ca^-. In Ref. 75, Table 7, the following ratios were given:

$$\frac{\tau(H_2O - Na)}{\tau(H_2O - H_2O)} \approx \frac{21\,\text{ps}}{4.4\,\text{ps}} \approx 4.8 \tag{424}$$

$$\frac{\tau(H_2O - Cl)}{\tau(H_2O - H_2O)} \approx \frac{10.4\,\text{ps}}{4.7\,\text{ps}} \approx 2.2 \tag{425}$$

These data show that a *hydrated layer indeed lives longer than the bulk water molecules*. If we identify the ratios (424–425) with τ_{ion}/τ, then we conclude that the above-predicted additional loss/absorption due to movement of ions inside the hydration sheath should be *less pronounced* for NaCl, than is shown in Fig. 52. One may also expect that for KCl this effect should be *negligible*.

In Ref. 76, self-consistent expressions were derived for the *frequency-dependent* electrolyte friction and the conductivity. Unlike our approach, the effect of the surrounding solvent (water) was described using the phenomenological coefficients. However, no oscillatory component of $\sigma'(\omega)$ or $\sigma''(\omega)$ were discovered in Ref. 157, while our Figs. 48 and 49 show typical *damped oscillations*. A reservation should be made that *no* damped oscillations are seen in the case of a *solution* (see Fig. 50), if we employ the additivity approximation Eq. (387).

The key aspect of our models is that the motion of particles occur in rather deep and narrow potential wells.

The concept [72] of the positive and negative hydration is inherently confirmed in our calculations by the fitted $\tau(C_M)$ dependences which are *increasing* for NaCl and *decreasing* for KCl (Table XVI).

In view of the studies [75] of aqueous NaCl solutions by using molecular-dynamics simulation, one may suggest that our models are applicable[61] *only at low* salt concentrations C_M, for which the Kirkwood correlation factor g could be calculated from Eq. (414d), where the ionic contribution $\Delta\varepsilon^*_{ion}(0)$ is not involved.

However, the latter formula is not more applicable, *if* τ_{ion} is rather long and/or C_M is rather high, so that the zero-frequency ionic contribution $\Delta\varepsilon^*_{ion}(0)$ to permittivity is noticeable in comparison with the static permittivity ε_s of the solution. We note that the Kirkwood correlation factor g is used for calculation of the component ε^*_{dip} in Eq. (387). Thus, even in our additivity approximation the solvent permittivity ε^*_{dip} is determined in this case by concentrations of both solution components. This complication leads to a new self-consistent calculation scheme.

In the studies described in this section we have neglected (1) the ion–ion interactions, (2) the cross-correlation between the ionic and dipolar subensembles, and (3) the finiteness of the potential well depth *for ions*. It appears that in future it would be desirable to account for:

(i) Formation of contact ion pairs, which is substantial for $C_M > 1$ mol liter^{-1}
(ii) Specific lifetimes' values obtained from *independent* data, similar to the data [Eq. (423)]
(iii) Distinction between the properties of bulk water and water in hydration sheaths
(iv) Distinction between the lifetimes of the cations and anions, which was neglected above
(v) Development of a self-consistent calculation scheme, if the inequality $\Delta\varepsilon^*_{ion}(0) \ll \varepsilon_s$ is no longer valid
(vi) Finiteness of the potential well depth for an ionic subensemble

We suggest that the results of calculations described above for a small ($C_M \leq 1$ mol liter^{-1}) salt concentration give a correct *qualitative picture and relevant time/space scales* of molecular events.

In conclusion we shall briefly discuss the results of some new experiments.

(A) The studies [77] of interaction between ions in concentrated electrolyte solutions in the FIR (10–40 cm^{-1}) range show that the absorption coefficient is *small* for high concentration ($\sim$13 mol liter^{-1}).

[61] We remark that our theory is not applicable at very low frequencies, when the effect of the ionic atmosphere may cause a *very small* increase in the real part σ' of the conductivity.

(B) Recent experiments [78] concerning the 5- to 100-cm^{-1} band, where the absorption coefficient of liquid water was compared with that of LiCl and NaCl water solutions, show a number of interesting phenomena interpreted in terms of network breaking and restricted H_2O molecule motion.

(C) In Ref. 79 the phenomenon of "ion rattling" motions was supposed to influence the recorded absorption band at $\nu > 200\,\text{cm}^{-1}$. It seems that without a theory capable of describing the whole 0- to 1000-cm^{-1} band it is hardly possible to assign the measured alteration of absorption due to a number of specific physical factors. The model, presented above, can be used as a basis for this purpose. However, to achieve this aim, a few important factors should additionally be accounted for. (a) It appears that first of all it is reasonable to introduce an *additional fraction* containing water molecules of a hydration sheath around an ion. The dielectric response of this fraction should differ from that of bulk water. (b) A model of an "ion-dipole" system suggested by the authors in GT, p. 318, could possibly be employed for this studies. (c) Another important task concerns development of a model describing the specific loss mechanism due to vibration of H-bonded molecules.

(D) The authors of Ref. 80 arrive at a conclusion that "the plasma oscillation is not excited in the highly concentrated electrolyte solutions" in the 10- to 50-cm^{-1} experimentally investigated range. This *negative* conclusion could be foreseen in view of the experiments previously described, for example, in Refs. 81 and 82, where *no signs* of specific plasma oscillation were detected in the frequency region under interest. It would be interesting to emphasize that in view of Figs. 2 and 3 of Ref. 80 the *imaginary* part of a refractive index strongly decreases, *unlike the real part*, if the LiCl concentration increases in aqueous solutions. The absorption coefficient also substantially changes in this case. We reproduce these dependences in Fig. 53. The process of making and breaking of the *hydrogen bonds in the solution* could possibly determine this difference. In Section VII we proposed the composite model, which reveals specific loss in the submillimeter wavelength range and accounts for *vibration of the H-bonds* characterized by a short relaxation time τ_{Dvib}. Thus it is natural to suggest that the loss $\varepsilon''(\nu)$, recorded in the frequency range from 10 to 100 cm^{-1}, could be essentially determined by interaction of *ions* of aqueous solution with a nonrigid dipole formed by *charges* "attached" to two water molecules (or to their fragments). This argument is supported by the experimental result [81] that addition of an electrolyte to water corrupts, *first of all*, the R-band (in the vicinity of 200 cm^{-1}) and in the nearby submillimeter range, where the spectra are basically determined by the specific interactions in water related to the hydrogen bonds.

(E) The *difference* of the plasma phenomena [80] recorded experimentally at *low frequencies and in the FIR region* appears to be very important. This

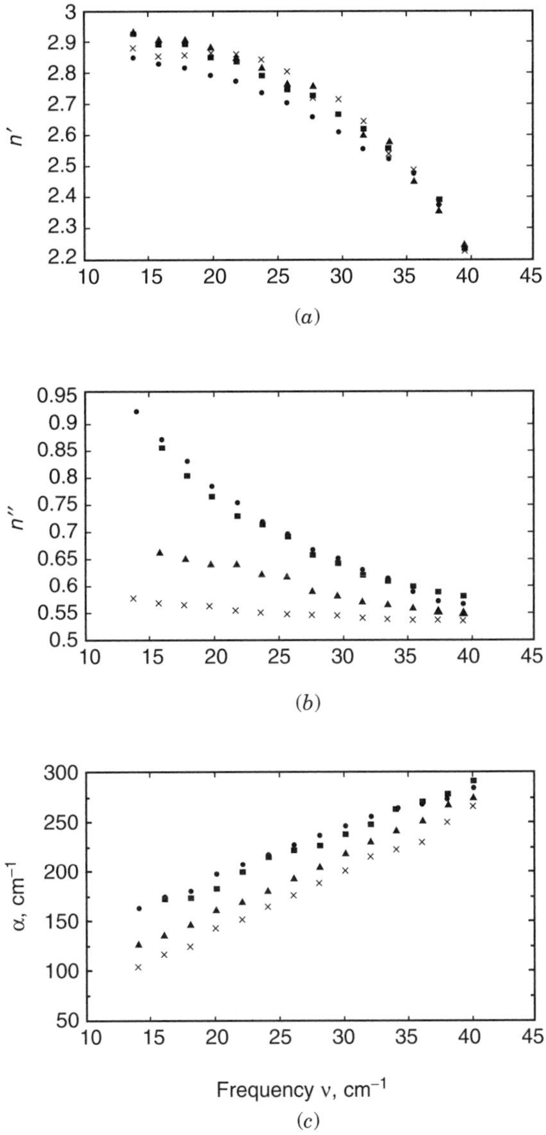

Figure 53. Frequency dependence of real (a) and imaginary (b) components of refractive index and of absorption coefficient (c) measured in LiCl electrolyte solutions. For curves from top to bottom the concentration of the solution comprises 0 (pure water), 2, 8, and 12 mol/liter. (Reproduced from Ref. 80).

property of aqueous solutions suggests that quite *different* molecular models should be worked out for low and FIR spectral ranges.

APPENDIX. RELATION OF IONIC SUSCEPTIBILITY TO CONDUCTIVITY

We start from Maxwell equations. Using the MKS system of units, we have for wave vector k perpendicular to electric field $\mathbf{E}$ the equation [see Eq. (3)]

$$k^{*2} - \frac{\omega^2}{c^2} = \frac{\omega^2}{c^2}\left(n^{*2} - 1\right)\hat{\mathbf{E}}^* = \frac{i\omega}{\varepsilon_0 c^2}\hat{\mathbf{J}}^* \tag{A1}$$

where n^* is complex refractive index, $n*^2 = \varepsilon^*$. Projecting this equation on direction of electric field and averaging over ensemble we have

$$n^{*2} - 1 = \frac{i}{\omega\varepsilon_0}\frac{\langle \hat{J}_E^* \rangle}{\hat{E}^*}$$

For ensemble of ions in vacuum we have the following, in accordance with the definition:

$$n^{*2} = \Delta\varepsilon_{\text{ion}}^* = 1 + \chi_{\text{ion}}^*$$

Then

$$\chi_{\text{ion}}^* = \frac{i}{\omega\varepsilon_0}\frac{\langle \hat{J}_E^* \rangle}{\hat{E}^*} \tag{A2}$$

If the conductivity does not depend on frequency, then in accord with Ohm's law the mean current density is: $\langle J_E \rangle = \sigma E$. Generalization for $\sigma = \sigma(\omega)$ yields

$$\langle \hat{J}_E^* \rangle = \sigma(\omega)\hat{E}^*$$

Therefore, it follows from Eq. (A2) that

$$\chi_{\text{ion}}^* = \frac{i\sigma(\omega)}{\varepsilon_0\omega} = \frac{i\sigma(\nu)}{2\pi\varepsilon_0 c\nu} \tag{A3}$$

$$\Delta\varepsilon_{\text{ion}} = \frac{\sigma'(\nu)}{2\pi\varepsilon_0 c\nu} \tag{A4}$$

Thus we have obtained the relations (388) and the second term of Eq. (412a).

IX. STRUCTURAL-DYNAMICAL (SD) MODEL[62]

A. Problem of Elastic Interactions

In Section VII we have applied a *phenomenological* approach in point of the forms of the potential wells characteristic for water. Now we shall consider a *principally* another way of modelling of intermolecular interactions pertinent to vibration of the H-bonded molecules. Recently [10, 12, 12a], a preliminary studies of the molecular dynamics was undertaken based on some picture (although very crude) of the *molecular structure* of water. We shall here briefly represent these results, namely:

(a) The "self-consistent" forms of the potential wells and the solution of 1D equation of motion will be found on the basis of the concept of *elastic interactions*. The distributions of various dynamic quantities will be found—particularly the frequency distribution, which resembles the absorption spectrum of water near 200 cm^{-1}.

(b) The spectrum of the dipolar ACF will be calculated for the same *structural-dynamical* (SM) *model.*

(c) A composite HC–SM model will be applied for calculation of wideband spectra of water.

As is well known, "there is a tendency for water to engage in tetrahedral configuration, and in ice Ih unquestionably involves tetrahedral configuration. Further, the C_{2V} point group ... arises naturally for the C_{2V} water molecule in a tetrahedral field of next-nearest oxygen neighbors" [16, p. 171]. Such molecular configuration is illustrated in Fig. 54, which we reproduce from Ref. 16. This idealized pattern is however pertinent to a very low temperature. It does not take into consideration that "Distortions from C_{2V} symmetry involving angles and distances are present in those molecules of the real liquid that are involved in the fully hydrogen-bonded structures, and the various distortions give rise to the breadth observed for the individual intermolecular components from liquid H_2O and D_2O. (Distortions are also present in the partially hydrogen-bonded structures.)" [16, p. 173]. We shall roughly consider, in terms of an analytical theory, such a "breadth" on example of a still simpler structure comprising *three* water molecules (Fig. 55). Here oxygen and hydrogen atoms are depicted as open and filled circles, respectively. The direction of a dipole moment is depicted by arrow. The hydrogen atoms B and B$''$, which belong to two neighboring molecules, deeply penetrate into a space-charge cloud formed by lone electron pair of the oxygen A$'$. Because of this, the dipole moment μ differs *in a liquid* with respect to the μ-value in an *isolated* water molecule. Molecules ABB and A$''$B$''$B$''$ exert various motions with respect to molecule A$'$B$'$B$'$.

[62] This section was written in collaboration with Ch. M. Briskina [12, 12a].

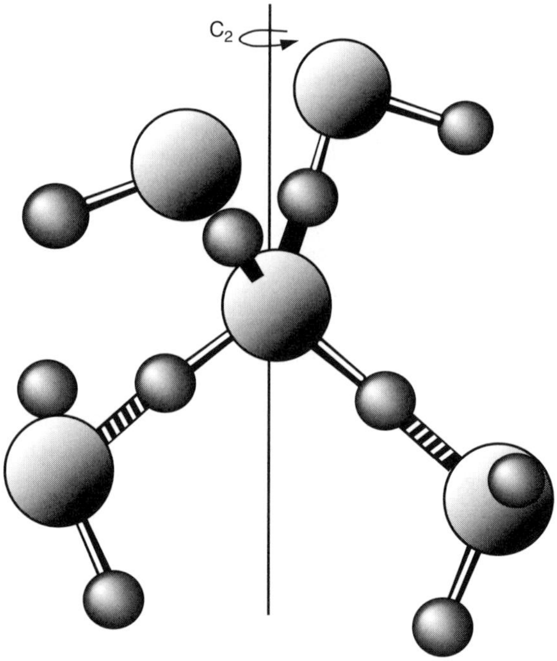

Figure 54. Five-molecule, fully hydrogen-bonded structure having intermolecular C_{2V} symmetry. Small spheres, H atoms; large spheres, O atoms. Disks refer to hydrogen bonds. The O–O distances are equal, and the oxygen atoms from a regular tetrahedron. The hydrogen bonds are equal in length, and in liquid H_2O they are about 1.8–9 Å. The OH bonds (rods) are also equal in length, and for the liquid they are about 1.0 Å. (Reproduced from Ref. 16, Fig. 9.)

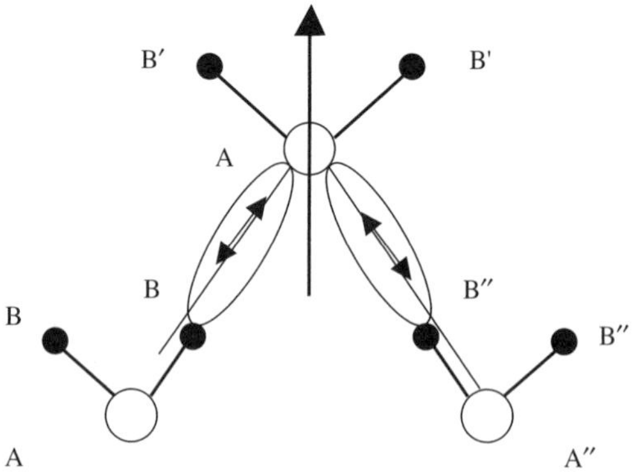

Figure 55. Simplified configuration pertinent to three H-bonded water molecules. Explanations are given in the text.

A mathematical treatment will be made for a still simpler *dimer configuration* shown in Fig. 56, which illustrates possible motions of specific parts of this structure due to thermal excitation. Figure 56a refers to a "pure" *transverse deflection* of the left water molecule perpendicular to an equilibrium (horizontal) position of an H-bonded molecule. The right molecule is assumed

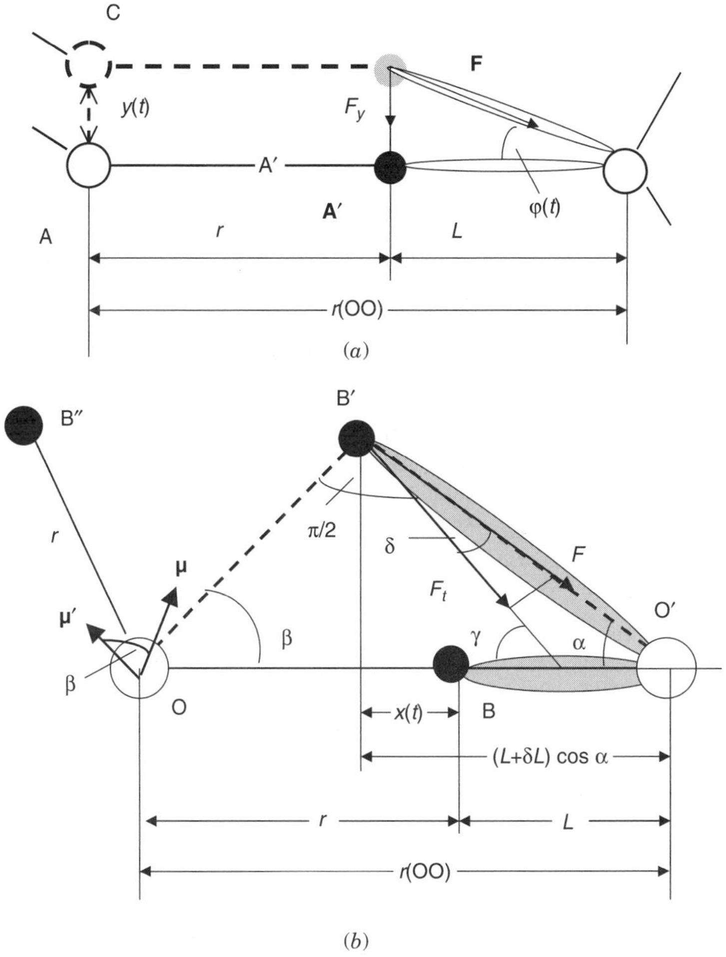

Figure 56. Scheme pertaining to dynamics of a water molecule, which suffers "pure" translational vibrations normal to the H-bond directions (a) or "pure" roations (b). Explanations are given in the text.

to be at rest. On the other hand, Fig. 56b depicts a "pure" *rotational deflection* of the left molecule, such that its center of mass is at rest. The right molecule again is immobile. In both cases the *H-bond expands*; the elastic force tends to restore an equilibrium position; therefore oscillations (vibrations) of various types should arise. For consideration of this problem such an important property [15, 16, 19, 83–85] is involved as the *elasticity of the H-bond.* Neglecting decaying of an "electronic string," we write down the law of motion

$$\bar{m}d^2x/dt^2 + kx = 0 \tag{426}$$

which describes harmonic oscillation of two water molecules *along* the H-bond. Here k is the *elasticity* constant of the latter, and $\bar{m}$ is the reduced mass. If both oscillating molecules are identical, then $\bar{m} = m/2$, where $m = Mm_H$ is molecular mass of water and m_H is mass of proton. It is usually assumed [16] that an experimentally observed R-absorption band around to 200 cm^{-1} is stipulated by stretching vibrations of water molecules along the H-bond. If we take the intrinsic frequency of such vibrations equal to $\nu_k = 180\,\text{cm}^{-1}$, then our estimation gives for water (H_2O) the value $k = 1.73 \cdot 10^4\,\text{dyn/cm}$, to which the following angular velocity corresponds: $\Omega = 2\pi\nu_k c = \sqrt{k/\bar{m}}$. The estimated k value falls into the range from 6.4×10^3 to 1.9×10^4 dyn/cm given in Ref. 15, p. 43.

In Section VII.A a strongly idealized picture was described. The *dielectric response* of an oscillating nonrigid dipole was found in terms of collective vibrations of two *charged* particles. Now a *more specific picture pertinent to an idealized water structure will be considered.* Namely, we shall briefly consider thermal motions of a *dipole* as (i) pure rotations in Fig. 56b and (ii) pure translations in Fig. 58a. Item (i) presents the major interest for us, since we would like to roughly estimate on the basis of a molecular dynamics form of the absorption band stipulated by *rotation of a dipole.* Of course, even in terms of a simplified scheme, the internal rotations of a molecule *should* also be accompanied by its translations, so the Figs. 56a and 56b should somehow interfere. However, in Section IX.B.1 we for simplicity will *neglect* this interference. This assumption approximately holds, since, as will be shown in Section IX.B.2, the mean frequencies of these two types of motion substantially differ.

B. Dynamics of Elastic Interactions (Account of Stretching Force)

1. *Equation of Motion and Its Solution*

Let us consider Fig. 56b, where F_t is the force, arising due to the H-bond *expanded* on δL. The expansion δL is stipulated by the turn of the H-bond; the

OO′ distance and the covalent OH bond length, denoted r, are assumed to be *fixed*. We also assume that a dipole moment $\boldsymbol{\mu}$ is rigidly connected with a molecule, so its turn on the angle β is accompanied by the same turn of the dipole-moment vector from the position $\boldsymbol{\mu}$ to $\boldsymbol{\mu}'$ (see Fig. 56b). For simplicity we consider rotation of a molecule OBB′ *in a plane*. Then the equation of motion under the torque due to stipulated by this force is given by

$$I\, d^2\beta/dt^2 = rF_t = -rk\delta L \cos\delta \tag{427}$$

The length r presents an arm of force F_t; the angle δ is shown in Fig. 56b. The fitted k value accounts for some that the molecule O exerts torque also from another H-bond (due to the second proton). The projection of the expanded H-bond on its "initial" direction, corresponding to a minimum length L, is given by

$$(L + \delta L)\cos\alpha = L + x, \qquad \text{where } x = r - r\cos\beta \tag{428}$$

Denoting $\lambda \equiv r/L$, after elementary but lengthy transformations we transform Eq. (427) to

$$I\frac{d^2\beta}{dt^2} = -krL(1+\lambda)\sin\beta\left\{1 - \frac{1}{\sqrt{1 + 2\lambda(1+\lambda)(1-\cos\beta)}}\right\} \tag{429}$$

This equation describes the law $\beta(t)$ of *periodic librations*. Let β_0 be their amplitude, which depends on the energy H of a molecule. Further transformation of Eq. (429) gives for $\beta_0 \ll \sqrt{12}$ the following simple formulas for the angular velocity $d\beta/dt$, period $\mathring{T}$, and rotational frequency $\mathring{\nu} = (\mathring{T}c)^{-1}$:

$$\frac{d\beta}{dt} = -\sqrt{\frac{k}{2I}}(r+L)\lambda\sqrt{\beta_0^4 - \beta^4} \tag{430}$$

$$\mathring{T} = (\beta_0 c d)^{-1} \tag{431}$$

$$\mathring{\nu} = \beta_0 d \tag{432a}$$

where

$$d \equiv \sqrt{\frac{k}{I}}\frac{1}{4cK(1/\sqrt{2})}\frac{r(r+L)}{L} \tag{432b}$$

and $K(\cdot)$ is full elliptic integral of the first kind; $K(1/\sqrt{2}) \approx 1.8541$.

The change dU of the potential energy produced by work of the force F_t along the arc $rd\beta$, corresponding to a turn of a molecule on the angle $d\beta$, is given by

$$dU(\beta) = -F_t r\, d\beta = kr\, dL \cos\delta\, d\beta$$

Choosing the constant of integration such that $U = 0$ at the position $\beta = 0$, we obtain from Eq. (429) the following "rotational" potential function:

$$U(\beta) = krL(1+\lambda)\left\{1 - \cos\beta - \frac{1}{\lambda(1+\lambda)}\left[\sqrt{1 + 2\lambda(1+\lambda)(1-\cos\beta)} - 1\right]\right\} \tag{433}$$

We shall take further into account that the rotator's energy H is equal to its potential energy U at the point β_0 of *maximum* deflection (in which its kinetic energy is zero):

$$H(\beta_0) \equiv U(\beta_0) \tag{434}$$

In view of our assumption about rigid connection of the dipole moment $\boldsymbol{\mu}$ with the amplitude β_0 of rotational (librational) motion, the energy [Eq. (433)] could be identified with the *dipole's potential energy* $\boldsymbol{\mu}\mathbf{E}_{\text{int}}$ in effective internal conservative electric field.

2. *Dynamics of Transverse Vibrations*

A water molecule shown on the right-hand side of Fig. 56a is assumed to be at rest, while the molecule A on the left-hand side is assumed to perform *transverse vibrations.* The covalent O—H bond oscillates up and down in the y direction (normal to equilibrium position of the O···H bond) due to existence of elastic force. The elastic constant is evidently the *same* (k) as that involved in Eq. (427). The equation of such back-and-forth motion is

$$md^2y/dt^2 = F_y = -k\Delta L \sin\varphi \tag{435}$$

The elastic force F, applied to the proton A′, is directed along the H-bond. The length of the expanded H-bond is $L + \Delta L = L/\cos\varphi$. Denoting the normalized displacement $y(t)/L$ by $\xi(t)$, we have: $\xi = \text{tg}\varphi$ so that $\sin\varphi = \xi/\sqrt{1+\xi^2}$. We rewrite Eq. (435) in terms of variable φ and then in terms of variable ξ:

$$m\frac{d^2y}{dt^2} = -kL(\tan\varphi - \sin\varphi) \tag{436a}$$

$$\frac{d^2\xi}{dt^2} = -\frac{k}{m}\xi\left(1 - \frac{1}{\sqrt{1+\xi^2}}\right) \tag{436b}$$

The first integral of Eq. (436b) expressed through reduced vibration amplitude $\xi_0 \equiv y_0/L$ is given by

$$\frac{1}{2}\left(\frac{d\xi}{dt}\right)^2 = -\frac{k}{m}\left(\frac{\xi^2 - \xi_0^2}{2} - \sqrt{1+\xi^2} + \sqrt{1+\xi_0^2}\right) \tag{437}$$

where the initial velocity $d\xi/dt|_{t=0}$ is set to be zero. If the oscillation amplitude is small, such that $\xi_0 = y_0/L \ll \sqrt{2}$, Eq. (437) transforms to

$$\frac{d\xi}{dt} = -\frac{1}{2}\sqrt{\frac{k}{m}}\sqrt{\xi_0^4 - \xi^4} \tag{438}$$

The solution of this differential equation could be expressed in terms of elliptic functions. The transverse-vibration frequency $\tilde{\nu}$ is determined by the formula analogous to Eqs. (431)–(432b):

$$\tilde{\nu} = b\,\xi_0 \tag{439a}$$

where

$$b \equiv \frac{\sqrt{k/m}}{4\sqrt{2}\,c\mathrm{K}(1/\sqrt{2})} \tag{439b}$$

The potential energy $U(\varphi)$ referring to equation of motion is connected with force F_y as

$$F_y = -kL\left(\frac{1}{\cos\varphi} - 1\right)\sin\varphi = -\frac{dU}{dy} = -\frac{1}{L}\frac{dU}{d\tan(\varphi)} \tag{440}$$

Integration aided by the condition $U(0) = 0$ gives the potential function

$$U(\varphi) = kL^2\left(\frac{1}{2} - \frac{1}{\cos\varphi} + \frac{1}{2\cos^2\varphi}\right) \tag{441}$$

To rewrite this formula in terms of a reduced transverse displacement $\xi \equiv y/L \equiv \tan\varphi$, we integrate Eq. (440) over ξ aided by the condition $U(\xi = 0) = 0$. Then we have

$$U(\xi) = kL^2\left(1 + \frac{1}{2}\xi^2 - \sqrt{1+\xi^2}\right) \tag{442}$$

The full energy H of transverse oscillations can be expressed as a function of the amplitude $\xi = 0$, if we set here $\xi \equiv \xi_0$ (at this ξ value the potential energy is equal to the full energy):

$$H(\xi_0) = U(\xi_0) \tag{443}$$

3. *Mean Frequencies/Amplitudes and Corresponding Density Distributions. Form of Rotational Absorption Band*

The mean over-ensemble value $\langle \beta_0 \rangle$ is found by integration over phase region Γ:

$$\langle \beta_0 \rangle = C \int_\Gamma \beta_0 \exp\left(-\frac{H}{k_B T} \right) d\Gamma \tag{444}$$

where

$$C^{-1} = \int_\Gamma \exp\left(-\frac{H}{k_B T} \right) d\Gamma \tag{445a}$$

In view of Eq. (432a) the mean values of $\mathring{\nu}$ and β_0 are proportional one to another:

$$\langle \mathring{\nu} \rangle = \langle \beta_0 \rangle d \tag{445b}$$

where d is determined by Eq. (432b). In our case of a one-dimensional motion it is convenient to choose the energy H and the constant of integration t_0 as a pair of the phase variables Γ. While calculating the integral in Eq. (444) we first fix the energy H of a particle and carry down the integral over t_0 in the interval $[0, \mathring{T}(H)]$, where $\mathring{T}(H) = [c\,\mathring{\nu}(H)]^{-1}$. Thus we take into account the contributions of all points along the particle's trajectory. Integration over t_0 results in multiplication by $\mathring{T}(H)$, since the integrand in Eq. (444) does not depend on t_0. Next, we find the integral over H in a reasonable interval of β_0 values:

$$\langle \beta_0 \rangle = D \int_0^{\pi/2} \beta_0 e^{-\frac{H(\beta_0)}{k_B T}} \frac{dH(\beta_0)}{d\beta_0} \frac{d\beta_0}{\breve{\nu}(\beta_0)} \tag{446a}$$

where

$$D^{-1} = \int_0^{\pi/2} e^{-\frac{H(\beta_0)}{k_B T}} \frac{dH(\beta_0)}{d\beta_0} \frac{d\beta_0}{\nu(\beta_0)} \tag{446b}$$

The integrand in Eq. (446a) is proportional to the normalized *distribution* of rotation amplitudes:

$$\overset{\circ}{w}(\beta_0) = Be^{-\frac{H(\beta_0)}{k_B T}} \frac{dH(\beta_0)}{d\beta_0} \left[\overset{\circ}{\nu}(\beta_0)\right]^{-1}, \qquad B \equiv \left[\int_0^{\pi/2} e^{-\frac{H(\beta_0)}{k_B T}} \frac{dH(\beta_0)}{d\beta_0} \frac{d\beta_0}{\overset{\circ}{\nu}(\beta_0)}\right]^{-1} \tag{447}$$

The constant B is found here from the condition $\int_0^{\pi/2} \overset{\circ}{w}(\beta_0) d(\beta_0) = 1$. In view of Eq. (445b) we estimate the distribution $\overset{\circ}{g}(\overset{\circ}{\nu})$ of the rotation frequencies as a quantity, proportional to Eq. (447):

$$\overset{\circ}{g}\left(\overset{\circ}{\nu}\right) \approx \frac{\overset{\circ}{w}}{d}\left(\frac{\overset{\circ}{\nu}}{d}\right) \tag{448}$$

The relationship (445b) is valid only at rather small β_0. However, since the function (447) tends to zero with β_0 so rapidly after $\overset{\circ}{w}(\beta_0)$ attains maximum (see Fig. 58a), no difficulty arises while *numerical* calculation of the quantities (447) and (448). Note that our calculations show that the maxima of the distributions, $\max\{\overset{\circ}{w}(\beta_0)\}$ and $\max\{\overset{\circ}{g}(\overset{\circ}{\nu})\}$, are attained just at the mean values of the corresponding variables, respectively, at $\beta_0 = \langle\beta_0\rangle$ and $\overset{\circ}{\nu} \equiv \langle\overset{\circ}{\nu}\rangle$.

Similar formulas hold also for *transverse vibrations*. The mean displacement is given by

$$\langle\xi_0\rangle = C\int_0^{\infty} \xi_0 e^{-\frac{H(\xi_0)}{k_B T}} \frac{dH(\xi_0)}{d\xi_0} \frac{d\xi_0}{\tilde{\nu}(\xi_0)} \tag{449a}$$

where

$$C^{-1} = \int_0^{\infty} \exp\left[-\frac{H(\xi_0)}{k_B T}\right] \frac{dH(\xi_0)}{d\xi_0} \frac{d\xi_0}{\tilde{\nu}(\xi_0)} \tag{449b}$$

In view of the relation

$$\langle\tilde{\nu}\rangle = b\langle\xi_0\rangle \tag{449c}$$

where b is given by Eq. (439b), the displacement and frequency distributions are, respectively, given by

$$\tilde{w}(\xi_0) = e^{-\frac{H(\xi_0)}{k_B T}} \frac{dH}{d\xi_0} \frac{1}{\tilde{\nu}(\xi_0)} \left[\int_0^{\infty} e^{-\frac{H(\xi_0)}{k_B T}} \frac{dH}{d\xi_0} \frac{d\xi_0}{\tilde{\nu}(\xi_0)}\right]^{-1} \tag{450}$$

$$\tilde{g}(\tilde{\nu}) \approx \frac{1}{b}\tilde{w}\left(\frac{\tilde{\nu}}{b}\right) \tag{451}$$

4. *Numerical Estimations*

To give an example pertinent to liquid water, we note that the distance $R(\mathrm{OO})$ between the oxygen's centers of mass in liquid water is about 2.85 Å [186]. The covalent-bond length $r \equiv r(\mathrm{OH})$ is *not* definitely known in the case of *liquid water*, unlike in isolated H_2O molecule, where it is equal to $r_{\mathrm{isol}} \approx 0.96$ Å. For instance, in accord with Ref. 57, r considerably exceeds the value r_{isol}, while in accord with Ref. 58 the bond length increases in water from 0.96 to only 1.02 Å. Thus, there is some uncertainty regarding the H-bond length $L = R(\mathrm{OO}) - r(\mathrm{OH})$. For our estimations we take several distances $r = r(\mathrm{OH})$. The results of these estimations are summarized by Table XXI.

The solid and dashed curves, shown in Fig. 57a for $L = 1.0$ and 1.42 Å, respectively, demonstrate that the form of the *self-consistent* potential well, found here, resembles, generally speaking, the form of the *hat-curved well* (see Fig. 20). The potential function $U(\beta)$ (433) widens near its bottom, if the H-bond length L increases (cf. solid and dashed curves calculated, respectively, for $L = 1.85$ and 1.54 Å). Note that the cosine-squared (CS) potential well[63] is substantially more *concave* than the self-consistent potential given by Eq. (433) (see dashed-and-dotted curve in Fig. 57a).

The dependences (447), (448), (450), and (451) are illustrated in Fig. 58. Since rotational motion of a water molecule is assumed to be rigidly accompanied by a turn of a dipole-moment vector, it is reasonable to propose that the absorption band recorded in water will be similar to the frequency distribution (448), which is determined by Eqs. (432a)–(434), and (447). Note that the validity of such a likening was previously justified in the example pertinent

TABLE XXI
Estimation of Some Characteristics Pertinent to Rotations and Transverse Vibrations[a]

O–H bond length $r \equiv r(\mathrm{OH})$, Å:	1.43	1.31	1.0
H-bond length L, Å:	1.42	1.54	1.85
Mean rotation amplitude $\langle\beta_0\rangle$, deg:	15.7	17.1	21.4
Mean fluctuation path $\langle\delta h\rangle \equiv r\sin\langle\beta_0\rangle$, Å:	0.39	0.39	0.37
Mean reduced transverse amplitude $\langle\xi_0\rangle$:	0.47	0.45	0.41
Mean transverse amplitude $\langle y_0\rangle$, Å:	0.66	0.69	0.75
Mean rotation frequency $\check{\nu}$, cm^{-1}:	382	351	280
Mean transverse-vibration frequency $\tilde{\nu}$, cm^{-1}:	35.5	34.0	30.9

[a]Water H_2O at $T = 300$ K.
Fixed parameters: $I = 1.483 \times 10^{-40}\ \mathrm{g \cdot cm^2}$; $k = 1.73 \times 10^4\ \mathrm{dyn \cdot cm^{-1}}$; $M = 18$ g/mole.

[63]The CS well has the form $U_{CS} = (1 - \cos^2\theta)$, where θ is angular deflection of a dipole from the symmetry axis. The steepness $p = \sqrt{U_{CS}/k_B T)}$ of this potential, just as in Ref. 190, was chosen equal to 1.7.

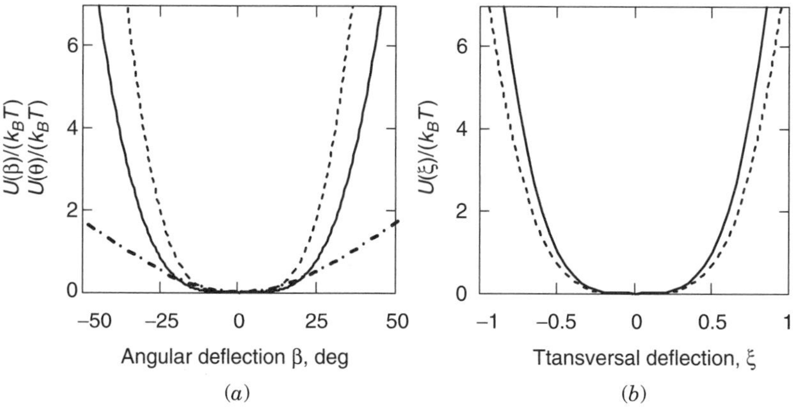

Figure 57. Forms of the potential well pertaining to the scheme shown in Fig. 56 for pure librations (a) and pure transverse translations (b). Solid line refers to the H-bond length $L = 1.0\,\text{Å}$, and dashed line refers to $L = 1.42\,\text{Å}$. Calculation for water H_2O at 27°C. In Fig. (a), dashed-and-dotted curve refers to the cosine-squared potential.

to the constant field model (cf. Figs. 18 and 19 in GT). In Fig. 59a we demonstrate by a dashed line the absorption spectrum calculated in Section VII for water H_2O at 27°C in terms of the composite hat-curved–harmonic oscillator model. The solid line shows a contribution to this spectrum due to stretching vibrations of the H-bonded molecules. Comparing Figs. 58b and 59a, we definitely conclude that the frequency dependence (448) is similar to the *translational* absorption band of water. This statement becomes more evident if we attempt to normalize the frequency dependence shown by the solid curve in Fig. 61a, so that it transforms to the curve plotted in Fig. 59b, and to juxtapose the latter with the normalized frequency dependences plotted in Fig. 58b. The best agreement between these two figures is attained for the lowest covalent OH length—that is, for $r = 1\,\text{Å}$ (see dashed-and-dotted line in Fig. 58b).

The distribution $\mathring{w}(\beta_0)$ of angles β_0 is shown in Fig. 58a. The mean angle $\langle\beta_0\rangle \approx 21.4°$ corresponds to the "the best" dashed-and-dotted line in Fig. 58b. This $\langle\beta_0\rangle$ value is very close to the well width fitted in Sections V–VII for the hat-curved model.

We see from Fig. 57b that if the *transverse* deflection ξ is small, the potential (20) has also a flat bottom similar to that calculated for the potential (443) at a small *angular* deflection β. As shown in Fig. 57b at $|\xi| > 0.25$, the function (443) rapidly increases. In view of Fig. 58d and Table XXI the estimated mean frequency $\langle\tilde{\nu}\rangle$ of transverse vibrations is about *order of magnitude less* than a mean rotational frequency. This result roughly justifies our neglect of the translational motion in derivations of the formulas for rotational motion.

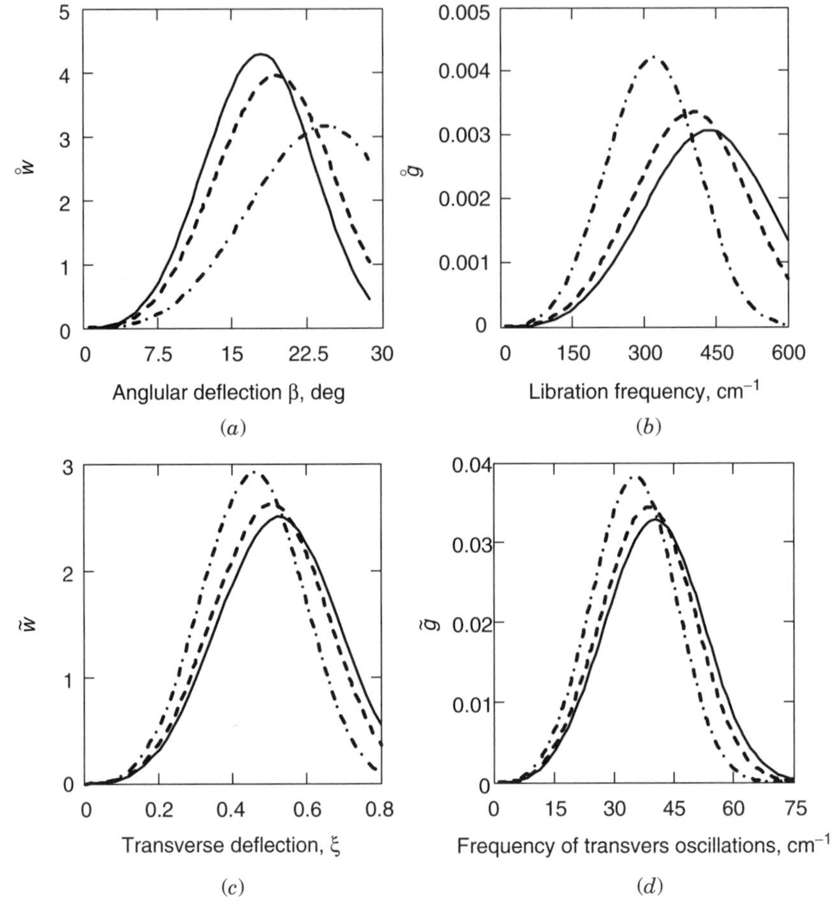

Figure 58. Density distributions of angular deflections β (a), internal-rotation frequencies (b), transverse librations (c), and their frequencies (d). Solid, dashed, and dashed-and-dotted curves refer to the H-bond length $L = 1.42, 1.54$, and 1.85 Å. Water H_2O at $T = 300$ K.

The estimated frequency $\langle\tilde{\nu}\rangle$ falls into the submillimeter region. Such behavior was predicted [16, 51] from experimental spectroscopic studies. The mean transverse-vibration amplitude is about 0.5 Å (see Table XXI). About the same is also the mean path $\langle\delta h\rangle \equiv r\sin\langle\beta_0\rangle$ of fluctuations stipulated by rotational motions. Note that the estimated amplitude of stretching vibrations is only approximately 0.2 Å [see Eq. (311)].

Returning to the rotational dynamics, we remark that the hat-curved model was applied in Sections V–VII for description of the *librational* band,

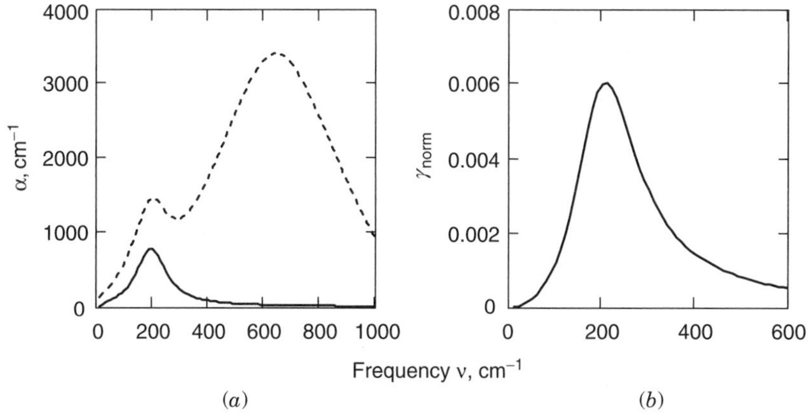

Figure 59. (a) Frequency dependences of the total absorption coefficient α_{tot} (dashed line) and of its part α_{vib} stipulated by stretching vibrations (solid line). (b) Frequency dependences of the normalized function γ_{LOR} calculated for the Lorentz line. The dependence $\gamma_{LOR}(\nu)$ approximately corresponds to the dashed-and-dotted curve in Fig. 60b. Liquid water H_2O at 27°C.

although Fig. 58b evidently concerns the R-band. Hence, the question arises whether it is possible to find by means of a "self-consistent" approach the potential well relevant to the librational band. We shall return to this question in Section X.

C. Dynamics of Elastic Interactions (Generalized Consideration)

Now we shall generalize previous consideration (1) by taking into account also the effect of the bending force constant K_α on rotational dynamics of an H-bonded molecule, (2) by rigorous solution of the 1D equation of motion, and (3) by application theory also for heavy water.

1. Calculation Scheme

We write down the expression for the potential energy of elastic interaction:

$$
\begin{aligned}
&U(\beta) = U_A(\beta) + U_B(\beta) \\
&U_A(\beta) = (k/2)[(L + \Delta L) - L]^2 \qquad U_B(\beta) = (K_\alpha/2)\alpha^2
\end{aligned}
\tag{452}
$$

Since dimensions of k and K_α differ on a quantity having dimension of a length squared, we introduce a dimensionless constant c_0 and express K_α through k and L as

$$K_\alpha = kL^2 c_0 \tag{453}$$

Using the scheme shown in Fig. 56b, we find from Eq. (452) after elementary transformations:

$$U_A(\beta) = (kL^2/2)[R_A(\beta) - 1]^2; \quad R_A(\beta) \equiv \sqrt{1 + 2\lambda(1+\lambda)(1 - \cos\beta)} \tag{454}$$

$$U_B(\beta) = (K_\alpha/2)[\alpha(\beta)]^2; \quad \alpha(\beta) \equiv \arctan\left(\frac{\lambda \sin\beta}{1 + \lambda(1 - \cos\beta)}\right) \tag{455}$$

If β_0 is an amplitude of angular deflection, then the full-energy H is found from Eqs. (454) and (455) at $\beta = \beta_0$ as

$$H = H(\beta_0) = U_A(\beta_0) + U_B(\beta_0) \tag{456}$$

Therefore, the kinetic energy, equal to the difference of the full and potential energies, is given by

$$\frac{I}{2}\left(\frac{d\beta}{dt}\right)^2 = E_{\text{kin}} = w^2(\beta, \beta_0) \equiv U_A(\beta_0) + U_B(\beta_0) - U_A(\beta) - U_B(\beta) \tag{457}$$

We find from here the restricted-rotation (RR) period $T_{\text{str}}(\beta_0)$ as a quadruplicated integral over β in the interval from 0 to β_0 and calculate frequency $\nu_{\text{str}} = (T_{\text{str}} c)^{-1}$ (we change notations $\mathring{T}$ and $\mathring{\nu}$ for T_{str} and ν_{str}):

$$\begin{aligned} T_{\text{str}}(\beta_0) &= \sqrt{8I}\, S_A(\beta_0) \\ \nu_{\text{str}}(\beta_0) &= \left(c\sqrt{8I}\right)^{-1} [S_A(\beta_0)]^{-1} \qquad S(\beta_0) \equiv \int_0^{\beta_0} \frac{d\beta_0}{w(\beta, \beta_0)} \end{aligned} \tag{458}$$

The dimensionless energy h equals $H(k_B T)^{-1}$, and its derivative of h over β_0 could be found as

$$\frac{\partial h(\beta_0)}{d\beta_0} = Q(\beta_0) \equiv Q_A(\beta_0) + Q_B(\beta_0) \tag{459}$$

$$\begin{aligned} Q_A(\beta_0) &= krL(1+\lambda)\sin\beta\left\{1 - [R_A(\beta)]^{-1}\right\} \\ Q_B(\beta) &\equiv \frac{krLc_0\alpha(\beta)[(1+\lambda)\cos\beta - \lambda]}{1 + 2\lambda(1+\lambda)(1 - \cos\beta)} \end{aligned} \tag{460}$$

The same arguments as were used in Section IX.B.3 express the mean values and the distributions over angular amplitudes and the RR frequencies in the form

$$\begin{aligned}\langle\beta_0\rangle &= D^{-1}\int_0^{\pi/2} e^{-h(\beta_0)}\beta_0 Q(\beta_0)S(\beta_0)\,d\beta_0 \\ \langle\nu_{\text{str}}\rangle &= (c\sqrt{8I})^{-1}D^{-1}\int_0^{\pi/2} e^{-h(\beta_0)}Q(\beta_0)\,d\beta_0\end{aligned} \tag{461}$$

$$\begin{aligned}g(\beta_0) &= D^{-1}\beta_0 e^{-h(\beta_0)}Q(\beta_0)S(\beta_0) \\ g(\nu_{\text{str}}) &= \frac{1}{c\sqrt{8I}}\left\{e^{-h(\beta_0)}Q(\beta_0)D^{-1}\left[\frac{d\nu_{\text{str}}(\beta_0)}{d\beta_0}\right]^{-1}\right\}_{\beta_0=\beta_{\text{str}}(\nu_{\text{str}})}\end{aligned} \tag{462}$$

where $D \equiv \int_0^{\pi/2} e^{-h(\beta_0)}Q(\beta_0)S(\beta_0)d\beta_0$ and $\beta_{\text{str}}(\nu_{\text{str}})$ is the function, inverse to $\nu_{\text{str}}(\beta_0)$ in (458).

2. *Results of Calculations*

a. Effect of H-Bond Bending. The used parameters of the SD model obtained from the solution of the equation of motion are given in Table XXII. In Fig. 60 the solid lines refer to an account of both force constants and the dashed lines refer to the disregard of the bending constant K_α. The chosen dimensionless coefficient c_0 (we set $c_0 = 0.1$) falls into the proper range of the k/K_α ratios [15, p. 42] and provides a reasonable $g(\nu_{\text{str}})$ dependence. The account of the effect of K_α (i) stipulates an increase of the potential energy u (Fig. 60a), (ii) does *not* substantially change the mean angles $\langle\beta_0\rangle$ and $\langle\alpha\rangle$, (iii) widens the angular distributions $g(\beta)$ and $g(\alpha)$ (Fig. 60b and Fig. 60c), (iv) narrows the *frequency dependence* $g(\nu_{\text{str}})$ and shifts its maximum to higher frequencies (Fig. 60d), and (v) cardinally changes the form of the $\nu_{\text{str}}(\beta_0)$ dependence. Indeed, some threshold RR frequency $\nu_{\min}$, equal in our example to $\sim$100 cm^{-1}, arises (see solid line in Fig. 60e), so that $\nu_{\text{str}}(\beta_0) > \nu_{\min}$.

The origin of this threshold could easily be explained. If $\beta^2 \ll 1$, then the potential $u(\beta)$ could be represented as a sum of the terms, proportional to β^2 and β^4. The first term involves the force constant K_α and the second term involves both constants K_α and k. At a very small β^2 the first parabolic-potential term predominates, to which the nonzero threshold resonance frequency ν_{str} corresponds to $\beta_0 \to 0$. This frequency is determined completely by the force constant K_α. Since the second term is positive, the potential curve $u(\beta)$ becomes steeper for larger β; this leads to the increase of the RR frequency ν_{str}.

TABLE XXII
Estimation of Parameters Pertinent to Restricted Rotations and Transverse Translations. $T = 300\,\text{K}$

Torque	r	1.43	1.31	1.15	1.02
	L	1.42	1.54	1.70	1.83
	Water H_2O				
A	$\langle\beta_0\rangle$	25.8	28.0	31.4	34.6
	$\langle\nu_{str}\rangle$	113	113	113	113
	$\langle\delta h\rangle$	0.62	0.62	0.6	0.58
	$\langle\alpha\rangle$	20.8	19.1	16.9	15.3
B	$\langle\beta_0\rangle$				35.3
	$\langle\nu_{str}\rangle$				192
	$\langle\delta h\rangle$				0.59
	$\langle\alpha\rangle$				15.5
C	$\langle\xi_0\rangle$	0.44	0.43	0.41	0.39
	$\langle y_\perp\rangle$	0.63	0.66	0.69	0.72
	$\langle\nu_\perp\rangle$	33.5	32.2	30.7	29.7
	Water D_2O				
	$\langle\beta_0\rangle$	23.8	25.8	28.9	31.9
	$\langle\nu_{str}\rangle$	81.4	81.4	81.4	81.4
	$\langle\delta h\rangle$	0.58	0.57	0.56	0.54
	$\langle\alpha\rangle$	19.7	18.1	16.1	14.5
D	$\langle\beta_0\rangle$	24.1			14.7
	$\langle\nu_{str}\rangle$	211			155
	$\langle\delta h\rangle$	0.58			0.54
	$\langle\alpha\rangle$	19.9			14.7
Fixed parameters:	For H_2O: $I = 1.483 \times 10^{-40}\,\text{g} \times \text{cm}^2; k = 6000\,\text{dyn} \cdot \text{cm}^{-1}$; $c_0 = 0.1, M = 18\,\text{g/mole}, r(\text{OO}) = 2.85\,\text{Å}$				
	For D_2O: $I = 2.765 \times 10^{-40}\,\text{g} \times \text{cm}^2; k = 8000\,\text{dyn} \cdot \text{cm}^{-1}$; $c_0 = 0.1, M = 20\,\text{g/mole}, r(\text{OO}) = 2.85\,\text{Å}$				

b. Description of FIR Isotopic Effect. We demonstrate in Fig. 61 a possibility of describing a small isotope shift of the R-band (notations are the same as in Fig. 60). In the case of heavy water, we take a greater force constant k, such a change agrees with the assertion [59, 60] that *the H-bond is stronger in D_2O than in H_2O*. A similar situation is met also in terms of the random network model of ice [19], in which approximately the same ratio $k(D_2O)/k(H_2O) \approx 1.3$ of the fitted force constants was used. It follows from the comparison of Figs. 61 and 60 that (i) for a given deflection β the potential energy $u(D_2O)$ is greater than $u(H_2O)$; (ii) the widths of the angular and of the frequency distributions remain approximately the same for both water isotopes; (iii) the effect of the bending force constant K_α is more emphasized for heavy water than for ordinary water.

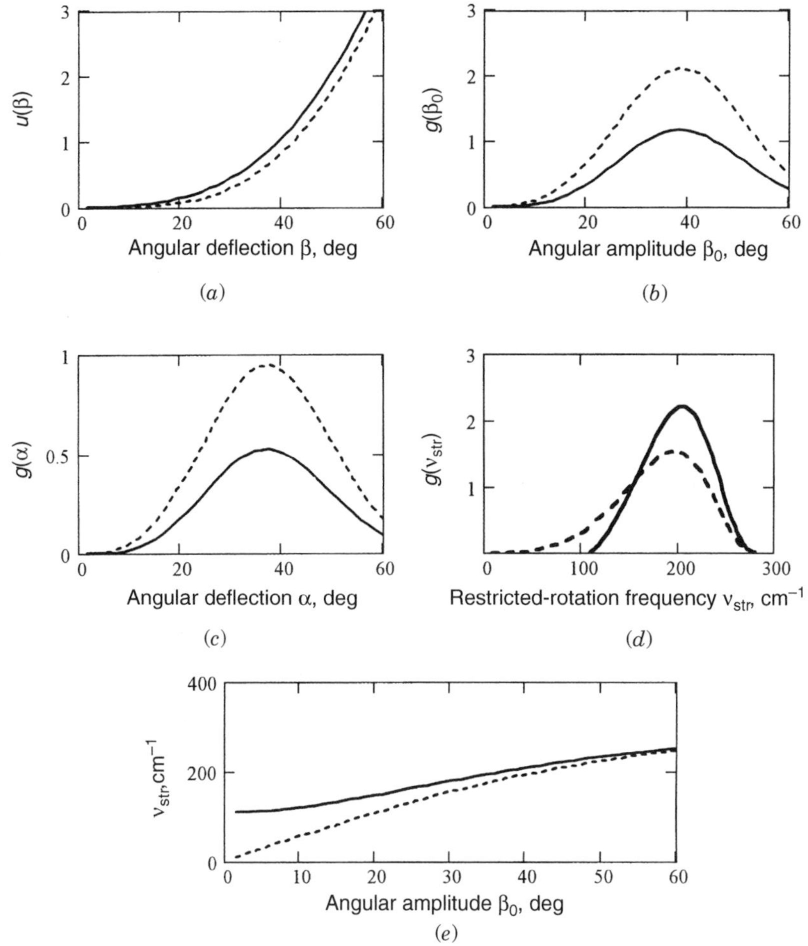

Figure 60. Effect of stretching and bending of H-bond on rotational dynamics of H_2O molecule. Solid lines account for the effect of the full torque, and dashed lines account for the effect of only the stretching. (a) Reduced potential u versus angular deflection β. (b,c) Distributions of amplitudes β_0 (b) and α (c). (d) Distribution of restricted-rotation frequencies ν_{str}. (e) The dependence of the RR frequency ν_{str} on amplitude β_0. $k = 6000\,\text{dyn}\,\text{cm}^{-1}, T = 300\,\text{K}, c_0 = 0.1, r = 1.02\,\text{Å}$.

Angles are expressed in degrees, distances in Å, and frequencies in cm^{-1}. Sections A and C of the table refer to calculation with the torque proportional to only the stretching force constant k, and sections B and D refer to the total torque.

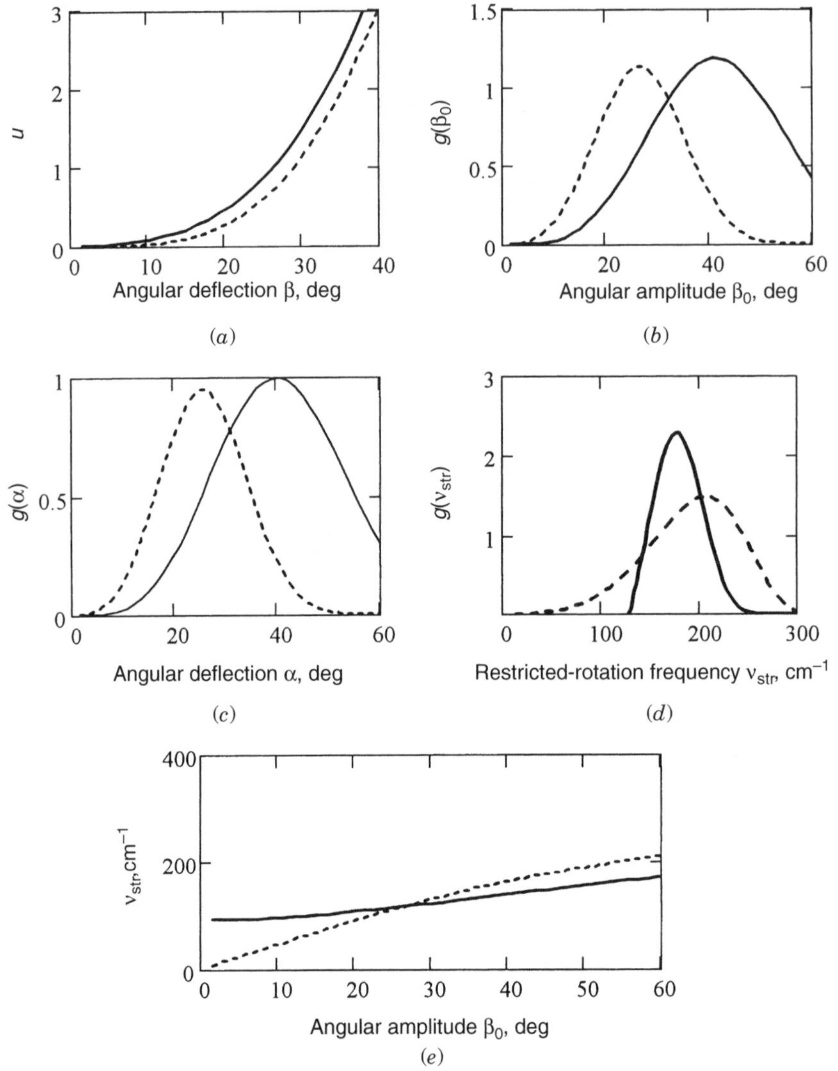

Figure 61. The same dependences and parameters as in Fig. IX.7 but calculated for liquid D_2O with $k = 8000\,\mathrm{dyn\,cm^{-1}}$.

D. Application of the ACF Method

We account for only the torque proportional to the string's expansion ΔL, which produces the main effect considered in this work. For calculation we employ the spectral function (SF) $L_{str}(Z)$, which is linearly connected with the spectrum of the *dipolar* ACF (see Section II), with $Z = x + iY$ being the reduced complex frequency. Its imaginary part Y is in inverse proportion to the "lifetime" τ_{str} of the dipoles exerting restricted rotation. The dimensionless absorption A_{str} is related to the SF L_{str} as

$$A_{str}(\nu) = x(\nu)\text{Im}\{L_{str}[Z(\nu)]\}$$
$$Z(\nu) \equiv x(\nu) + iY, \quad x(\nu) \equiv 2\pi c\eta\nu, \quad Y \equiv \eta/\tau_{str} \tag{463}$$

As usual, η is determined by the moment of inertia I of an effective *linear* molecule, $\eta \equiv \sqrt{I/(2k_B T)}$. Starting from the second equation in (452) and assuming that the angular deflection β is rather small ($\beta^2 \ll 1$), we shall give below, Eqs. (468)–(471), an approximate analytical expression for the SF $L_{str}(Z)$. Further calculations will be performed for the structural–dynamical model alone and for the composite HC–SD model.

1. *Spectral Function of Restricted Rotators*

To calculate $L(Z)$ in terms of the structural–dynamical model of water, we introduce the *longitudinal* and *transverse* dimensionless projections, $q_{\|} = \mu_{\|}/\mu$ and $q_{\perp} = \mu_{\perp}/\mu$, of a dipole-moment vector $\boldsymbol{\mu}$. These projections are directed, respectively, along and across to the local symmetry axis. In our case (see Fig. 56b), the latter coincides with an equilibrium direction of the H-bond. Next, we introduce the *longitudinal and transverse spectral functions* as

$$K_{\|}(Z) = iZ\left\langle q_{\|}(0)\int_0^{\infty}[q_{\|}(\varphi) - q_{\|}(0)]\exp(iZ\varphi)\,d\varphi\right\rangle$$
$$K_{\perp}(Z) = iZ\left\langle q_{\perp}(0)\int_0^{\infty}[q_{\perp}(\varphi) - q_{\perp}(0)]\exp(iZ\varphi)\,d\varphi\right\rangle \tag{464}$$

where $\varphi = t/\eta$ is the dimensionless time and $\langle F\rangle$ denotes an ensemble averaging of a dynamical variable F. For an isotropic *spatial* molecular ensemble the SF $L(Z)$ is determined as $L(Z) = 3[k_{\|}(Z) + 2K_{\perp}(Z)]$. Below (i) we consider a simple 1D ensemble (rotation in a plane) but (ii) in (464) perform statistical averaging pertinent to a *spatial* configuration. For an isotropic polar medium, this "quasi-spatial approximation" [GT, Eqs. (3.79a) and (3.80)] yields an *exact* square under the absorption curve and gives the following relationship:

$$L(Z) = L_{str}(Z) = 2[K_{\|}(Z) + K_{\perp}(Z)] \tag{465}$$

Assuming that the dipole-moment vector $\boldsymbol{\mu}$ and the molecule rotate in the same plane, we have in view of Fig. 56b:

$$q_{\parallel} = \cos(\Theta + \beta), \qquad q_{\perp} = \sin(\Theta + \beta) \tag{466}$$

It can be shown that the *sum* of the components $K_{\parallel}(Z)$ and $K_{\perp}(Z)$, given by Eqs. (464) and (465), is independent of Θ. Combining expressions (464)–(466) we replace Eq. (465) by

$$L_{\text{str}}(Z) = L_1(Z) + L_2(Z) \tag{467a}$$

$$L_1(Z) = 2\langle q_0 \psi_1 \rangle, \qquad L_2(Z) = \frac{1}{2}\langle q_0^2 \psi_2 \rangle \tag{467b}$$

$$\psi_n(Z) \equiv iZ \int_0^{\infty} (q^n - q_0^n) \exp(iZ\varphi)\, d\varphi, \qquad q \equiv \sin\beta, \qquad q_0 \equiv \sin\beta_0 \tag{467c}$$

In order to derive an analytic expression for $L_{\text{str}}(Z)$ from Eqs. (467), we generate the recurrence relations for the dynamical quantity ψ_n. Assuming small angular deflections β, we omit the terms comprising $\sin^m \beta$ with $m > 2$. Then using equations of motion for the angle $\beta(t)$, we derive

$$\begin{aligned} L_1(z) &= 2C \int_0^{\infty} \frac{[F_1(h) - F_2(h)] h^{3/4} \exp(-h)\, dh}{(h - Z^2)(9h - Z^2) + 3p^2 h} \\ L_2(Z) &= C \int_0^{\infty} \frac{h^{5/4}(G_1\sqrt{h} - G_2) \exp(-h)\, dh}{4h - Z^2} \end{aligned} \tag{468}$$

where

$$p^2 = (1/2)k(r/L)^2 r(\text{OO})^2 (k_B T)^{-1} \tag{469}$$

$$\begin{aligned} F_1(h) &= [2\mathbf{E} - \mathbf{K}] p\sqrt{h}(9h + 3p^2 - Z^2) \\ F_2(h) &= \frac{1}{9}\mathbf{K}(18h^2 + 6p^2 h - 2Z^2 h + 3p^2 Z^2) \end{aligned} \tag{470}$$

$$C = 4\sqrt{\frac{2}{\pi}} \frac{1}{p^2 \Gamma(1/4)}, \qquad G_1 = \frac{14}{9}\mathbf{K}, \qquad G_2 = [2\mathbf{E} - \mathbf{K}]p \tag{471}$$

$\mathbf{K}$ and $\mathbf{E}$ are the full elliptic integrals, respectively, of the first and second kind with the modulus $1/\sqrt{2}$; $\mathbf{K}$ was defined above, $2\mathbf{E} - \mathbf{K} = 0.847$.

The spectral function (467) and (468) is determined by the same set of parameters of the model, as was involved in Section IX.B, combined now in the form of a *single* parameter p defined by the formula (469). This parameter characterizes an inhomogeneity of the self-consistent potential $u(\beta)$, since at

$p = 0, u \equiv 0$. *Additionally* the mean lifetime τ_{str}, which is an intrinsic parameter in the ACF method, is involved in the formula for $L_{str}(Z)$. Note that now we describe spectra in terms of the a.c. field frequency ν, not in terms of the resonance RR frequency ν_{str} determined by the period of a rotational motion of a molecule. If β is small, then the dimensionless potential $u \equiv U/(k_B T)$ reduces to $u(q) \approx (1/4)p^2 q^4 \approx (1/4)p^2 \beta^4$. Hence, the larger the value of p, the more our self-consistent potential differs from the homogeneous one, namely from $u \equiv 0$. In the calculations described below, the employed value 0.31 of $\lambda = r/L$ is less than that (0.56) used in Section 1D. The distinction could be ascribed to the assumption $\beta^2 \ll 1$, used in derivation of Eqs. (468)–(471).

2. *Dielectric Response of Restricted Rotators*

We shall show now that: (i) The R-band arises due to rotational motion of a polar H-bonded molecule determined by the elastic force constant k. (ii) The dimensionless absorption $A_{str}(\nu)$ (463) agrees qualitatively with the $g(\nu_{str})$ frequency dependence found in Section IX.B. (iii) The used SD model describes a *small isotopic shift* of the R-band also in terms of the ACF method.

In Fig. 62a we demonstrate by solid lines the absorption dependence (463) in the R-band calculated for liquid H_2O (a) and liquid D_2O (b) at temperature 22.2°C; the ordinates are fitted in such a way that the magnitude of the $A_{str}(\nu)$ peak is equal to 1. The chosen dimensionless collision frequency Y is chosen 0.6 for H_2O and 1.32 for D_2O, with correspondingly the lifetimes $\tau_{str} \approx 0.071$ and 0.036 ps. These curves exhibit a *damped resonance*, since the fitted lifetime τ_{str} is very short; they resemble the dashed curves $g(\nu_{str})$ in Fig. 60d and 61d.

3. *Composite HC–SD Model: Frequency Dependence of Wideband FIR Absorption*

The structural–dynamical (SD) model can be applied for calculation of the *absorption coefficient* $\alpha(\nu)$ of liquid water in the FIR frequency range, provided that we *combine* it with the hat-curved (HC) model (described in Section V), to which the hat-like potential corresponds. This model was successfully employed in Section VII in combination with the harmonic–oscillator (HO) model or with the cosine-squared (CS) potential model. Replacing the HO model by the SD model allows us to understand the origin of the R-band *in terms of physical arguments*, while introducing the parabolic potential in the HO model is based on the assumption that two *charged* molecules vibrating *along* the H-bond direction exist. Now there is no need to accept such an assumption, since a rotating water molecule is *polar* with a known value of its *dipole moment*. On the other hand, replacing the description of dielectric response of two *effective vibrating charges* $\pm\delta q_{vib}$ *of unknown origin* by a clearer dielectric-response mechanism, stipulated by a restricted rotation of a polar H-bonded water molecule, yields similar final results (the calculated spectra).

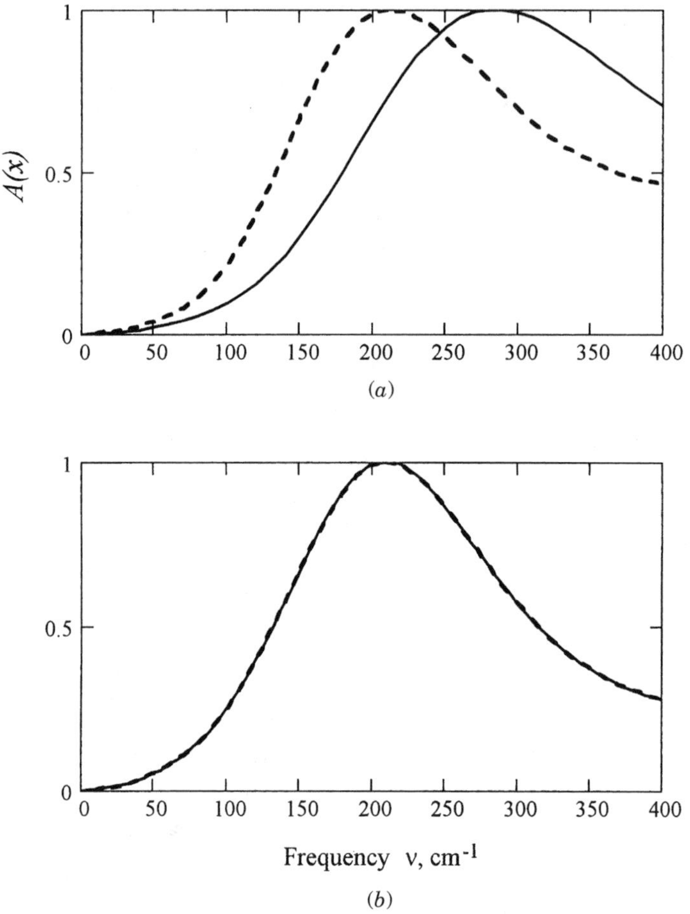

Figure 62. Solid lines: contribution of the structural–dynamical model to the dimensionless absorption $A_{str}(\nu)$ calculated in the R-band by the ACF method. (a) Calculation for H_2O, the dimensionless collision frequency $Y = 0.6, r/L = 0.27, p = 2.07$. (b) Calculation for D_2O with $Y = 1.3, r/L = 0.4, p = 3.54$. Dotted lines: total absorption calculated for the composite HC–SD model. Temperature 22.2°C. The peak ordinates are set equal to 1.

Referring to Section VII.B, we note that the formulae used in this section are applicable also for the HC–SD model, if we replace the spectral function (315) of the CS model as follows:

$$L_{vib}(Z)|_{\text{From Eq. (315)}} \rightarrow = L_{str}(Z) \tag{472}$$

We remind the reader that the following free parameters are employed in the HC model: (a) the reduced potential will depth $u = U_0/(k_B T)$; (b) the angular half-width β of the well; (c) the mean lifetime τ, during which a near-order state exists in a liquid; and (d) the form-factor f defined as follows: $f =$ (flat part of the well's bottom)/(total well's width 2β). The SD *model* is characterized by (e) the inhomogeneity-potential parameter p of our "self-consistent" well, (f) the lifetime τ_{str}, of restricted rotation, and (g) the fraction r_{vib} of dipoles performing RR with respect to their total concentration N.

The following *molecular constants* are used in further calculations: density ρ of a liquid; the static (ε_s) and optical (n_∞^2) permittivity; moment of inertia I, which determine the dimensionless frequency x in both HC and SD models; the dipole moment μ; the molecular mass M; and the static permittivity ε_1 referring to an ensemble of the restricted rotators. The results of calculations are summarized in Table XXIII. In Fig. 62 the dimensionless absorption around frequency 200 cm^{-1}, obtained for the *composite* model, is depicted by dots in the same units as the absorption A_{str} described in Section B. Fig. 62a refers to H_2O and Fig. 62b to D_2O. It is clearly seen in Fig. 62b that the total absorption calculated in terms of the composite model decreases more slowly in the right wing of the R-band than that given by Eq. (460). Indeed, the absorption curve due to dipoles reorienting in the HC well overlaps with the curve generated by the SD model, which is determined by the restricted rotators.

TABLE XXIII

Free and Estimated Parameters Pertinent to the Composite HC–DYN model. Temperature 22.2°C

Liquid	Ordinary Water	Heavy Water
The Dynamic Model (Application of the ACF Method)		
$p = \frac{r(OO)}{L}\sqrt{\frac{k}{k_B T}}$	2.4	2.775
$\lambda = r/L$	0.31	0.31
Proportion r_{vib} of molecules performing restricted rotation, %	3.5	8
Lifetime τ_{str} of restricted rotors, ps	0.061	0.036
Contribution $\Delta\varepsilon_s$ of this ensemble to static permittivity ε_s	1.92	1.8
The Hat-Curved Model		
Optical permittivity n_∞^2	1.7	1.7
Angular width of intermolecular potential β, deg	23	25
Lifetime τ, ps	0.25	0.32
Well depth u of intermolecular potential (in units $k_B T$)	5.8	6.2
Form factor of intermolecular potential well, f	0.8	0.72
Mean number of librations, m_{lib}	4.9	3.6
Proportion of the over-barrier particles, %	3.9	2.6

Note: $\Delta\varepsilon_s$ could be estimated from the experimental data as the difference $\Delta\varepsilon_s = \varepsilon_1 - \varepsilon_\infty$ of the "end points" of the principal Debye region and the so-called second Debye relaxation region.

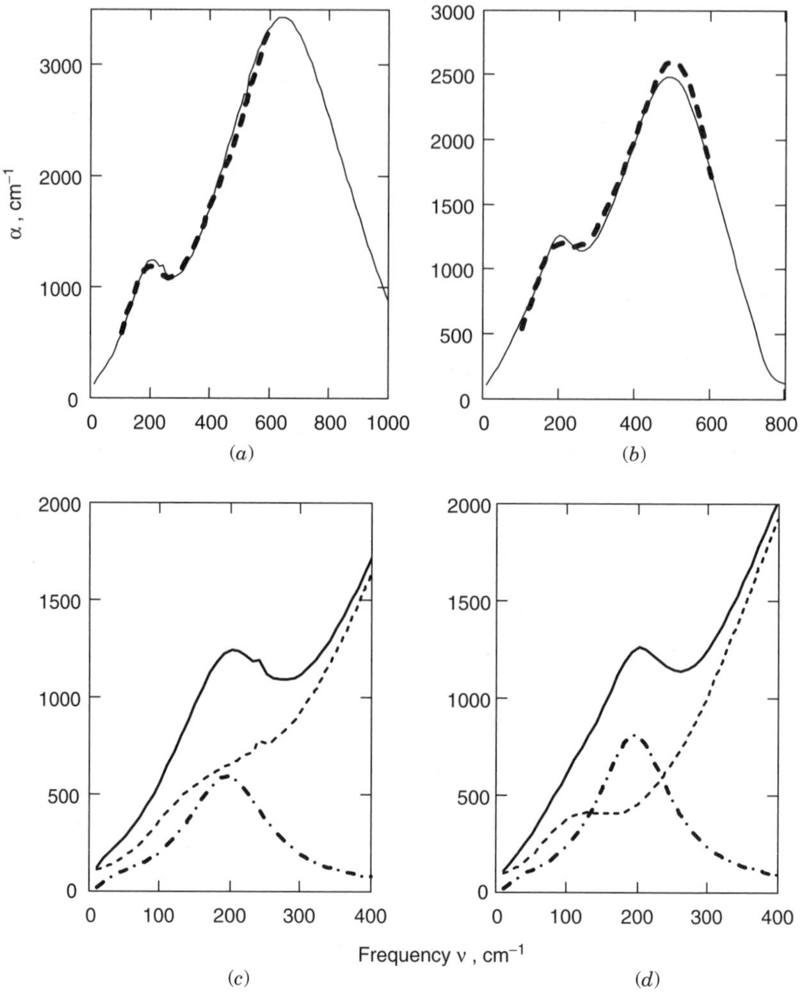

Figure 63. (a,b) Solid lines: absorption frequency dependences calculated for liquid H_2O (a) and D_2O (b) in terms of the composite HC–SD model. Dashed lines refer to the experimental data [51]. (c,d) Contributions to the absorption in the R-band due to molecules reorienting in the hat-curved potential (dashed lines) and performing restricted rotation (dash-and-dotted lines); solid lines refer to the calculated total absorption. Temperature 22.2°C.

The theoretical wideband absorption spectra of water shown by solid lines in Figs. 63a and 63b, respectively for water H_2O and D_2O, agree with the experimental spectra [51] drawn by dashed lines. Figures 63c and 63d show the contributions of librators reorienting in the HC well (dashed lines) or of restricted rotators (dash-and-dotted lines) to the total absorption (solid lines). For both isotopes the absorption feature at $\sim$200 cm^{-1} is contributed mainly by the molecules' exerted restricted rotations. If we employ the terms used in Refs. 54 and 86, then it appears that a small fraction of the RR molecules, which determine the 200-cm^{-1} absorption feature, actually belongs to a "high-density liquid" (HDL) water. A more populated fraction of dipoles, reorienting in the hat-curved well, with high local tetrahedral ordering mostly belongs to the "low-density liquid" (LDL) water.

A success of this composite model allows us to clarify a physical picture of molecular motions in water and to recognize a physical meaning of the hat-curved model itself. We shall return to this question in the final section.

X. CONCLUSIONS AND PERSPECTIVES

A. Effect of Temperature on Wideband Dielectric Spectra

Above (in Sections V–VII) the effect of the temperature T on wideband spectra of water was briefly studied (only for two temperatures). The following important properties of our models were then found:

(i) The well depth/width of intermolecular potential, in which a single water molecule *reorients*, and the parameters pertinent to *cooperative* motions of the H-bonded water molecules in water show a very small T-dependence.

(ii) The main factors that determine the effect of the temperature on spectra come in our theory from (a) *explicit* T-dependences of the Boltzmann factor $\exp[-H/(k_B T)]$ and of the time-scale parameter $\eta = \sqrt{I/(2k_B T)}$ and (b) *implicit* dependences on T of the Debye relaxation time τ_D and other molecular parameters, which we take from the experimental data.

The properties (i)–(ii) allow us to conclude that (a) the water *structure* exerts (unlike statistical distributions and characteristic time scales) only *small changes* in a restricted (but very important) range of T including room temperatures and (b) to promise advantage while describing the effect of temperature on the wideband spectra by means of analytical formulas presented in our review. It appears that some analogy exists between these properties of our models and the "fluctuation concept" [87] of hydrogen bonding. According to this concept, very broad bands in the vibrational spectra of water OH oscillators reflect a statistical distribution of geometrical configurations of the O–H$\cdots$O hydrogen bridge inherent in liquids (unlike crystals) and caused by fluctuations of the local environment of different H_2O molecules. The formalism of such fluctuation theory is based on the hypothesis by Zhukovsky [88], who had suggested that we should describe the frequency distribution in the statistical ensemble, $P(\nu, T)$—that is, the statistical contour related to the experimental IR and Raman spectra—as

$$P(\nu, T) = Q^{-1}(T)W(\nu)\exp\left(-\frac{E(\nu)}{k_B T}\right)$$

The central point of this hypothesis that functions $W(\nu)$ and $E(\nu)$ are *temperature-independent* is actually based on apparent constancy of water density over the temperature range under study and this statement gives an important ability to *predict* spectrum shape at any temperature from the *experimental* IR or Raman spectra recorded at the only one temperature.

Returning to our problem, we remark that the temperature dependences of such parameters as, for example, the Debye relaxation time $\tau_D(T)$ (which determines the *low-frequency dielectric spectra*), or the static permittivity ε_s are fortunately known, at least for ordinary water [17]. As for the reorientation time dependence $\tau(T)$, it should probably correlate with $\tau_D(T)$, since the following relation (based on the Debye relaxation theory) was suggested in GT, p. 360, and in VIG, p. 512:

$$\tau(T) \approx a_D[\beta(T)]^2 \tau_D(T) \tag{473}$$

where the constant a_D is independent on T and β is a characteristic angular width of an intermolecular potential well, where a single molecule reorients. The validity of this relation can be confirmed by the example considered in Section VII. Inserting into Eq. (425) the fitted values $\tau = \tau_{or}$, β, and the experimental τ_D values (given in Tables XII and XIVB), we find that a_D is the same (0.174) for both temperatures (22.2 and 27°C) studied in this example.

Further studies based on our *molecular* models will possibly allow us to convincingly explain the well-known experimental FIR spectra of water recorded [51] in a wide temperature range. We hope to propose in the future a *weak*, physically reasonable and analytically described temperature dependence of the model parameters. It is hoped that initiation of such a theory will allow us to predict evolution of the wideband water spectra affected by the temperature (this is important for engineering studies) and to connect the experimental spectra with other molecular quantities—for example, with the barriers corresponding to various elementary processes.

B. Far-Infrared Dielectric Response of Ice

The hat-curved model aided by a reasonable account of cooperative (collective) effects arising due to vibration of the H-bonded water molecules could hopefully provide a good perspective for spectroscopic investigation of ice I (more precisely, ice Ih), which is formed near the freezing point (0°) at a rather low

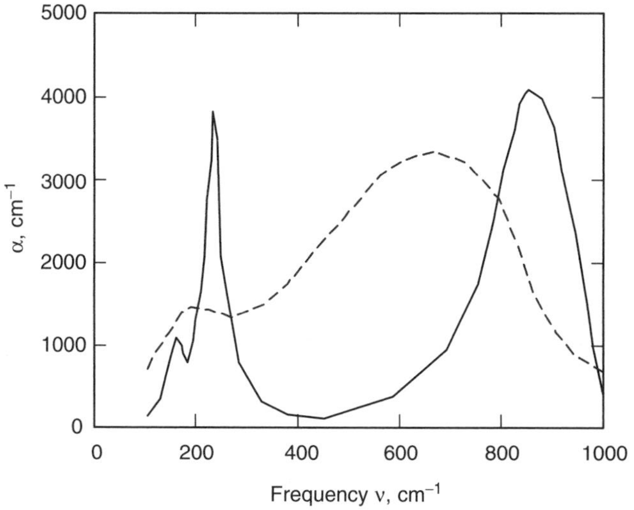

Figure 64. The absorption coefficient vs. frequency for ice Ih at 100 K (solid line) and for liquid water at 300 K (dashed line). Comparison of the experimental data.

pressure. The molecular structure of ice I evidently resembles the water structure. At lower temperature or/and higher pressure other ices, II–VIII, may exist [57, 89]. Ice Ih is a polar substance, characterized by high static permittivity ε_s ($\varepsilon_s = 95$) exceeding that of water. The main absorption band, centered at $\sim 840\,\text{cm}^{-1}$, presents an analogue of the rotation band of water. In the case of ice, this band is shifted to *higher frequencies* on $\sim$200 cm^{-1} and is substantially narrower than in water. The R-band in ice is also narrower and is also split. The main absorption peak ν_R is centered at 225 cm^{-1}, while $\nu_R(H_2O) \approx 200\,\text{cm}^{-1}$ [90; see also VIG, p. 57].

The great distinction between the *relaxation* spectra of water and ice originates due to the phase transition occurring near 0°C. Because of the latter the *translational mobility* is extremely low in ice, unlike in water. However, an evident resemblance of the FIR spectra of water and ice, demonstrated in Fig. 64, suggests an idea that the *rotational mobility* does not differ so much in these two phase states of H_2O. If we apply the hat-curved model also for ice I, the fitted form factor f obviously should be less than $f(H_2O)$ to give narrower rotational band. The intermolecular potential should be spread in ice at longer distances than in water, just as the steepness of such potential is less in water than in a nonassociated liquid (see Section VII).

Similar reasoning concerns also the *R-band*, if we shall apply for calculations the harmonic oscillator model (we mean now a rough qualitative description).

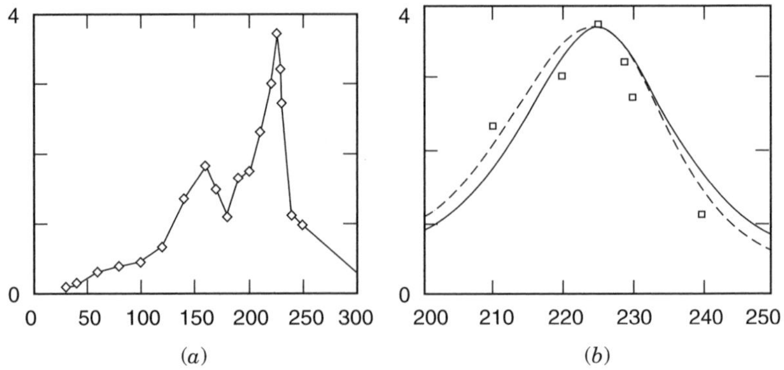

Figure 65. Frequency dependence of dielectric loss in ice Ih. (a) Recorded [171] spectrum in the translational band, $T' = 100\,\mathrm{K}$. (b) The main part of the loss line shown in Fig. (a), calculated for the constant field (solid line) and for the cosine-squared potential (dashed line); points represent experimental data from Ref. 171.

We remark that previously [90–93] rather successful attempts were undertaken by using the so-called "field models" for describing the main part of the loss $\varepsilon''(\nu)$ curve (near frequency 225 cm^{-1}). Namely, the constant field (CF) and the cosine-squared (CS) potential models were applied to polar *reorienting* molecules. The result of such modeling is illustrated by Fig. 65, to which the fitted lifetimes τ_{vib} 0.2 and 0.49 ps correspond, respectively, for CF and CS models. These lifetimes are much longer than the vibration lifetimes estimated for water in terms of the harmonic oscillator model, possibly, since vibrations in ice appear to be more regular than in water. Correspondingly, much larger (about 10) than in water is an estimated number of the vibration cycles performed during this time τ_{vib}. An approach based on the harmonic oscillator model or on its variants may suggest another (possibly closer to reality) molecular mechanism stipulating the R-band—that is, connected with stretching vibrations of H-bonded molecules. In further studies it would also be interesting to explore what a molecular mechanism could invoke the splitting of the R-band.

It should be noted that a number of the experimental data [93–98] is available concerning the wideband spectra of ice, which are sufficient at least for the start of above mentioned theoretical research. Useful empiric formulas were also suggested [99]. The hypothesis that a substantial contribution to the R-band is given *in ice* by stretching vibrations of the H-bonded molecules appears to be still more plausible than in the case of water, for which the experimental data [62, 63] confirm this viewpoint (see also Section VII).

C. About Evolution of Molecular Models

Our review is mostly based on a phenomenological approach. The employed potential forms, being more or less reasonable, yield rather simple expressions in terms of the ACF method and provide at least satisfactory agreement between the calculated and recorded spectra. Moving sequentially from Section III to Section VII.A, we gradually complicated the forms of employed intermolecular potential wells and also used calculation schemes based on our linear response theory (the ACF method). Using the hat-curved–harmonic oscillator (HC–HO) composite model, we finally have reached a very good description of the experimental wideband dielectric/FIR spectra of liquid H_2O and D_2O in the range from 0 to 1000 cm^{-1}. So far, no information is available that like coincidence between the calculated and recorded $\varepsilon^*(\nu)$ and $\alpha(\nu)$ spectra have been obtained in any of the published works. Incidentally, our analytical theory is rather simple, and the fitting parameterization procedure does not present much troubles. The hat-curved model accounts for *unspecific* interactions (which determine in water the Debye relaxation and the librational band in the region from 300 to 1000 cm^{-1}), while a primitive harmonic oscillator model accounts for *specific* interactions in water, stipulating the submillimeter-wavelength spectrum of complex permittivity (from 10 to 100 cm^{-1}) and the R-band around frequency 200 cm^{-1}. However, as was shown in Section VII.B, a good coincidence of the theoretical and experimental wideband spectra could also be obtained by using the HC–CS composite model (hat-curved–cosine-squared potential). In the latter, two vibrating *charged* H-bonded molecules, pertinent to the HO model, are replaced by a *dipole*, performing restricted rotation. Hence, good description of spectra based on molecular theory could be achieved by various ways. We see that the phenomenological (heuristic) modeling of intermolecular potentials has an inherent drawback. Using this approach, it is hardly possible (and probably impossible) to recognize a *physical nature* of the employed potential well. For instance, (i) it is not clear what physical factor causes the hat-curved potential-well-bottom to be rounded and (ii) in the case of a most successful approach, in which *two potential wells* are employed, one actually can *equally effectively* describe the FIR spectra of water using quite different models and interpretations accounting for specific interactions.

In Section IX we have outlined a *nonheuristic* way of modeling of specific interactions. This way is based on studies of dynamics of a reasonable but very crude picture of water structure. In view of the obtained results, the absorption R-band arises in water due to *elastic* interactions. In terms of the structural dynamic (SD) model, these are characterized by two force constants: one (k) refers to an *expansion* ΔL and another (K_α) refers to a *turn* α of the H-bond. The success of the composite hat-curved–structural-dynamical (HC–SD) model suggests an idea that this model actually describes *two states of water*, which coexist like two components of a solution. The first state (A-state) is pertinent to a rather long molecular rotation of a dipole in a deep (with depth about 6 $k_B T$) intermolecular well. The short-lived B state is pertinent to a restricted rotation of a dipole in a shallow (with depth about 1 $k_B T$ or less) potential well. One may suggest that *at least one H-bond is broken* in the A-state. Indeed, relevant molecules rotate rather freely due to existence of a broad flat part on the potential well's bottom. Our estimations show that about 95% of water molecules belong to the A-state. The B-state, comprising about 5% of molecules, is characterized by a tighter configuration and by a greater number of the H-bonds. It became clear now that "specific" interactions in water could be described in one or another way in terms of *elastic interactions* arising between the H-bonded molecules. As for the hat-curved potential, it represents a rather formal tool for describing of "unspecific" interactions. We intend to elaborate in the near future a new model of dielectric relaxation based, just as the structural-dynamical model, on the structure of water, but now with an *explicit* account of the effect of a broken H-bond. It appears in this context that "unspecific" interactions in water are actually also "specific," since for their description one should account for the influence of H-bonds on molecular rotation, but in another way than it was modeled in Section IX.

D. Hat-Curved–Elastic Bond (HC–EB) Model as a Simplest Composite Model

Having clarified a physical sense of the hat-curved model (as that describing the contribution to spectra of rather freely rotating molecules), finally we shall briefly consider a very simple composite model. The contribution to spectra of the vibrating H-bonded molecules is basically described in the elastic bond (EB) model by the *Lorentz-like line* [see GT (p. 258) and VIG (p. 363)]. We note that (i) the Lorentz line is also involved in the well-known *empirical* description of the R-band, given by Liebe et al. [17], and (ii) at *small* angular declinations of a dipole from an equilibrium position the self-consistent potential considered in Section IX, Figs. 60b,c and 61b,c, resembles that pertinent to the elastic bond model. It appears that the composite HC–EB model is almost as good as the composite hat-curved–structural-dynamical model. The difference between them is follows: The parameter p of the SD model [Eq. (469)] is actually such a characteristic of the model, which reflects the structural-dynamical properties of the H-bonded molecules, while the only parameter p_m of the EB model represents the steepness of the parabolic well introduced on an intuitive basis. In the case of an isotropic polar medium the spectral function of the EB model is determined by the following simple formula [VIG, p. 367]:

$$L_{\text{vib}}(Z) = \frac{1}{p_m^2 - Z^2} + \frac{1}{2p_m^2} \frac{1}{4p_m^2 - Z^2} \tag{474}$$

The first and second terms in the right-hand part are, respectively, the transverse and longitudinal components of the spectral function. In other words, these terms are stipulated by reorientation of the projections of a dipole moment, which are, respectively, normal and collinear to the potential symmetry axis. The potential under consideration comprises two wells with oppositely directed symmetry axes. Such is the cosine-squared potential

$$U(\theta)/(k_B T) = P_m^2(1 - \cos^2\theta) \tag{475}$$

It exhibits maxima at $\theta = 0$ and $\theta = \pi$, with p_m being the dimensionless resonance frequency of angular harmonic declinations θ performed near the bottom of the well. In the EB model, small declinations θ are assumed. Then Eq. (475) yields the parabolic dependence on θ:

$$U(\theta)/(k_B T) = p_m^2\theta^2 \tag{476}$$

to which formula (474) corresponds. The first term in Eq. (474) presents the Lorentz line with the resonance frequency $\nu_m = p_m(2\pi c\eta)^{-1}$, and the second term presents such a line with two times larger frequency.

We have used the calculation scheme described in Section VII.B for the HC–CS model. Now we employ the composite HC–EB model. The only difference from the above-mentioned calculations concerns the formula for the spectral function: Now we use Eq. (474) instead of Eq. (315). Setting ν_m to be close to $\nu_R \approx 200\ \text{cm}^{-1}$, we estimate the steepness of the parabolic well as

$$p_m \approx 2\pi \cdot 200c\eta \quad \text{namely,} \quad p_m(H_2O) \approx 1.61$$
$$\text{and} \quad p_m(D_2O) \approx 2.2 \quad \text{at } T = 195\,\text{K} \tag{477}$$

We see in Table XXIV that the fitted p_m-values are rather close to these estimations.

In Figs. 66 and 68 the calculated absorption and loss spectra are depicted for ordinary water at the temperatures 22.2°C and 27°C and for heavy water at 27°C. The solid curves refer to the composite model, and the dashed curves refer to the experimental spectra [42, 51]. For comparison of our theory with experiment at *low frequencies*, in the case of H_2O we use the empirical formula [17] comprising double Debye–double Lorentz frequency dependences. In the case of D_2O we use empirical relationship [54] aided by approximate formulae given in Appendix 3 of Section V. The employed molecular constants were presented in previous sections, and the fitted/estimated parameters are given in Table XXIV. The parameters of the composite model are chosen so that the calculated absorption-peak frequencies ν_{lib} and ν_R come close to the

TABLE XXIV
Fitted (A) and Estimated (B) Parameters Involved in the Hat-Curved–Elastic-Bond Model Pertaining to Liquid Water

A

		Reorientation				Vibration		
Liquid	T (°C)	τ_{or}	u	β	f	τ_{vib}	p_m	r_{vib}
H_2O	27	0.197	5.9	23	0.82	0.0471	1.4	0.035
H_2O	22.2	0.196	6.2	23	0.75	0.053	1.5	0.030
D_2O	22.2	0.228	6.2	23	0.65	0.053	2.1	0.07

B

Liquid	T (°C)	y	M_{or}	$\mathring{r}_{or}$	m_{lib}	m_{vib}	Y
H_2O	27	0.215	3.25	3.4%	3.69	0.27	0.9
H_2O	22.2	0.218	3.71	2.9%	3.71	0.32	0.8
D_2O	22.2	0.255	3.25	3.5%	3.25	0.32	1.1

Times are given in ps, frequency and absorption coefficient are given in cm^{-1}, and density is given in $g \times cm^{-3}$.

experimental values, and the form of the absorption curve resemble the recorded form. Comparing the solid and dashed curves in Figs. 66 and 68, we ascertain that the theoretical and experimental spectra agree well. Several parameters of the HC model (the fitted lifetime τ_{or}, the libration amplitude β, and the number m_{lib} of the librational cycles, performed during this time) remain almost

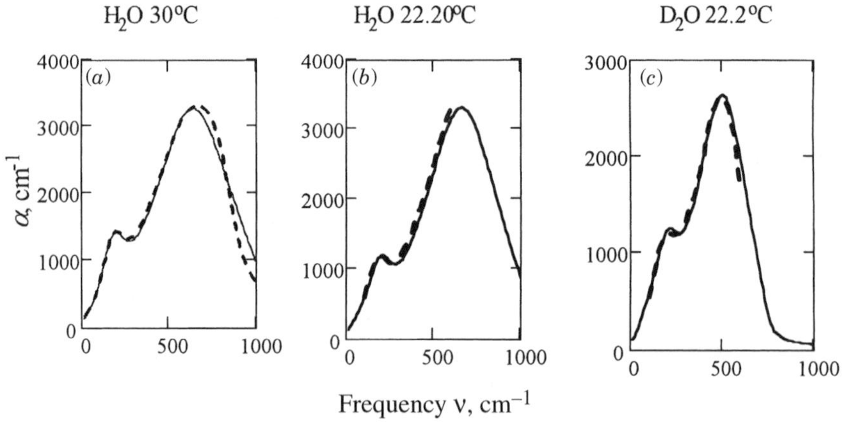

Figure 66. Absorption coefficient of liquid water versus frequency. Calculation for the composite model is depicted by solid lines, and experimental spectra [17, 51, 54] are depicted by dashed lines.

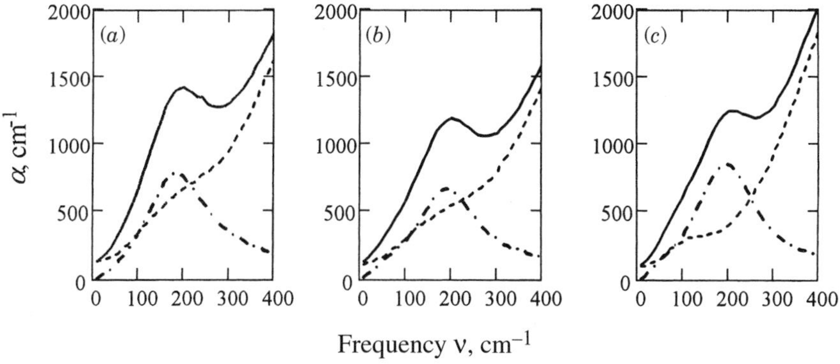

Figure 67. Calculated absorption in the R-band (solid lines), contribution to this absorption due to reorienting dipoles (dashed lines), and contribution due to vibrating particles (dash-dotted lines).

unaltered, when temperature T changes, while the *reduced* well depth u slightly and the fitted f value noticeably decrease with T. Therefore, the depth of HC well lowers and the well's bottom becomes *more rounded*, if the temperature rises. Hence, *weakening of the water structure* occurs with temperature.

In Fig. 67 we show by solid lines the calculated absorption $\alpha(\nu)$ in the R-band for the same cases as in Fig. 66; the dash and dash-dot lines show the contribution to α due to dipoles, rotating, respectively, in the hat-curved well (dashed lines) and the parabolic well (dash-dot lines). At $\nu \approx 200\ \text{cm}^{-1}$ the absorption peak α_{vib}, which could be attributed to vibration of the H-bond, is larger in the case of D_2O than H_2O. The loss maximum, stipulated by the vibrating dipoles, is placed in the submillimeter wavelength range and is very shallow (see Fig. 68). In view of Table XXIV the vibration lifetime (≈0.05 ps) comprises about one-fourth of the reorientation lifetime τ_{or}. The estimated number (≈3) of libration cycles, performed by a dipole in the hat-curved potential, is an order of magnitude larger than such number m_{vib} calculated for the parabolic potential. Note that m_{vib} is estimated from the formula

$$m_{\text{vib}} \approx \tau_{\text{vib}}/T_{\text{vib}} = c\nu_{\text{vib}}\tau_{\text{vib}} \approx 200c\tau_{\text{vib}} \tag{478}$$

Due to such small value of this parameter, the *loss curve* $\varepsilon''_{\text{vib}}(\nu)$ is very shallow (see dash-dot lines in Fig. 68b). Therefore, vibration of the H-bonded molecule is so damped ($m_{\text{vib}} \approx 1/3$) that resonance properties are not actually revealed for this type of molecular rotation. The *absorption spectra* obtained above in terms of the HC–EB model (Fig. 66b,c) evidently are close to spectra (Fig. 63a,b) calculated for the composite HC–SD model. The absorption/loss spectra resemble those found in Section VII.B.

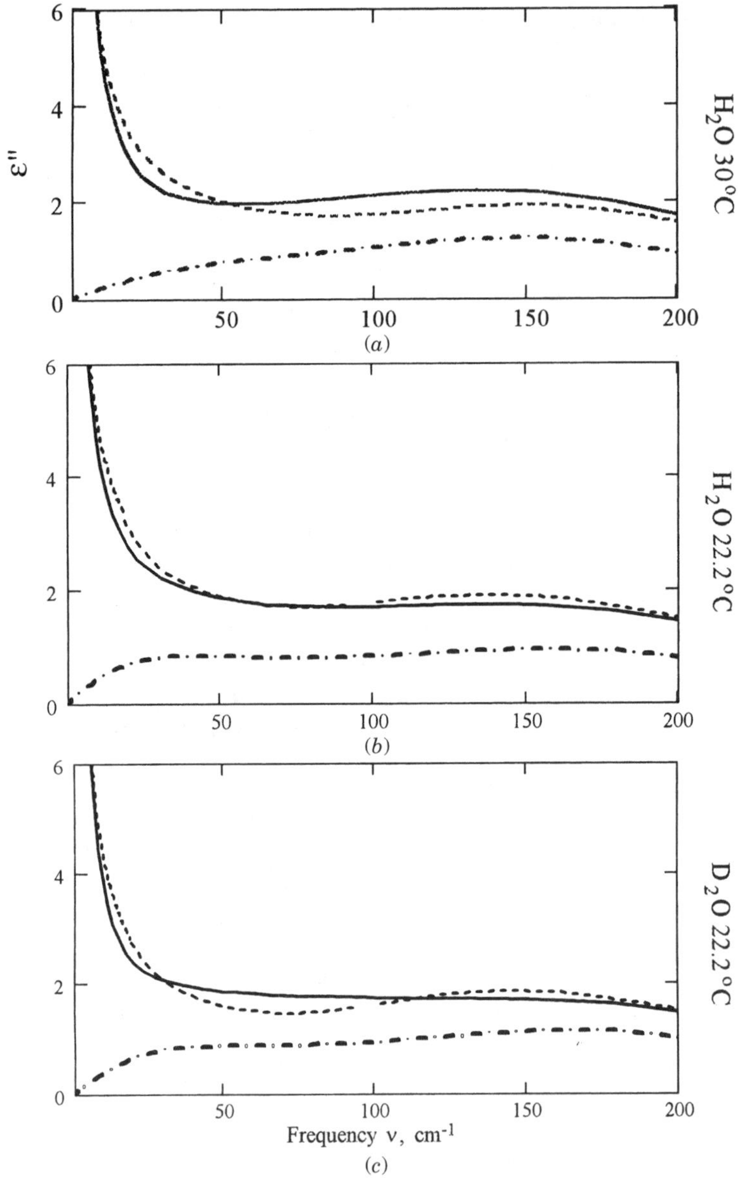

Figure 68. Dielectric loss calculated for the composite model (solid lines); experimental curves [17, 51, 54] (dashed lines); contribution to loss due to dipoles performing restricted rotation in the parabolic potential (dot-dashed lines).

However, the following important distinction exists between the estimated proportions r_{vib} of the H-bonded molecules: for the HC–SD model $r_{vib}(D_2O) \approx r_{vib}(H_2O)$, while for the HC–EB model $r_{vib}(D_2O)$ is about two times larger than $r_{vib}(H_2O)$. On the other hand, application of the CS well characterized by a *finite welldepth* yields the lifetimes τ_{vib} commensurable with the lifetimes τ_{or} of dipoles reorienting in the hat-curved well, while *much shorter* lifetimes τ_{vib} were obtained, if the elastic bond or the structural-dynamical models were applied. Correspondingly, in the first case (for the CS well) the partial-loss curve $\varepsilon''_{vib}(\nu)$ pertinent to vibrating dipoles is more emphasized (in the submillimeter wavelength region) than such curve pertinent to the EB or SD models. It seems that the approximation of a parabolic (for the EB model) or a parabolic-like (for the SD model) potential possibly yields *unreasonably short* lifetimes τ_{vib} or *unreasonably shallow* $\varepsilon''(\nu)$ frequency dependences, so that a purely elastic restoring force should be replaced in the future by a more adequate force versus deflection *quasi-elastic* dependence, for which the structural-dynamical model possibly would give the results close to those obtained in Section VII.B for the cosine-squared potential model.

We conclude as follows:

1. The composite hat-curved–elastic bond model presents a climax of this work regarding description of wideband spectra of liquid water generated by rotation of water molecules in two employed phenomenological intermolecular potentials characterized by simple profiles. The same is true for the composite hat-curved–cosine squared potential model. The latter possibly is preferable (as compared with the HC–EB model) in view of a longer fitted lifetime τ_{vib} and a more emphasized $\varepsilon''(\nu)$ dependence. Using these (or somehow improved) phenomenological models for detailed studies of evolution of wideband spectra with temperature, for modeling FIR spectra of ice Ih and of dielectric/FIR spectra of aqueous solutions (first of all, of electrolyte solutions), we hope to obtain a lot of new physical information and to determine reasonable limitations of our theory.

2. It seems that the structural–dynamical model pertinent to vibrating H-bonded molecules opens a new paradigm in calculating of wideband spectra of liquid water. It would be important to extend this model also for molecules performing rather free rotation in a structure characterized by a broken H-bond (or by more broken bonds). It appears that this new semimicroscopic approach gradually could allow us to avoid application of phenomenological models (such as the hat-curved model). It is hoped that the so extended structural–dynamical models could be later employed for calculating spectra of various aqueous fluids.

Acknowledgments

The authors highly appreciate collaboration with Professor D. S. F. Crothers, Professor J. K. Vij, and Dr. Ch. M. Briskina. The generalized results of this collaboration were employed in Sections V–X. The authors are thankful to Professor S. A. Rice for important information concerning the random network model of water and ice. The discussions with Professor M. N. Rodnikova and Professor I. N. Kochnev were useful regarding the role of elasticity of the spatial H-bond network and mechanisms of the so-called specific interactions in water. The authors acknowledge the help of Ms. Dee Gumbel and Ms. Kellsee Chu in the preparation of the manuscript for publication. One of the authros (V. I. G.) is also grateful to Professor W. T. Coffey for his hospitality in Dublin.

References

1. V. I. Gaiduk and B. M. Tseitlin, *Adv. Chem. Phys.* **87**, 124–378 (1994) (cited throughout as GK).
2. V. I. Gaiduk, *Dielectric Relaxation and Dynamics of Polar Molecules*, World Scientific, Singapore, 1999, XVIII + 365 pp. (cited throughout as VIG).
3. W. T. Coffey, V. I. Gaiduk, B. M. Tseitlin, and M. E. Walsh, *Physica A* **282**, 384 (2000).
4. V. I. Gaiduk, B. M. Tseitlin, and D. S. F. Crothers, *J. Mol. Liq.* **89**, 81 (2000).
5. V. I. Gaiduk, B. M. Tseitlin, and J. K. Vij, *Phys. Chem. Chem. Phys.* **3**, 523 (2001).
6. V. I. Gaiduk, B. M. Tseitlin, and Ch. M. Briskina, *Rep. Russ. Acad. Sci.* **379**, 609 (2001).
7. V. I. Gaiduk, O. F. Nielsen, and T. S. Perova, *J. Mol. Liq.* **95**, 1 (2002).
8. V. I. Gaiduk, B. M. Tseitlin, Ch. M. Briskina, and D. S. F. Crothers, *J. Mol. Struct.* **606**, 9 (2002).
9. V. I. Gaiduk, *Opt. Spectrosc.* **94**(2), 219 (2003).
10. V. I. Gaiduk, B. M. Tseitlin, and D. S. F. Crothers, *J. Mol. Liq.* (submitted).
11. V. I. Gaiduk and J. K. Vij, *Phys. Chem. Chem. Phys.* **4**, 5289 (2002).
12. Ch. M. Briskina and V. I. Gaiduk, *Russ. J. Chem. Phys.* (submitted).
12a. V. I. Gaiduk, D. S. R. Crothers, Ch. M. Briskina, and B. M. Tseitlin, *J. Mol. Struct.* (submitted).
12b. V. I. Gaiduk and J. K. Vij, *Phys. Chem. Chem. Phys.* **3**, 5173 (2001).
12c. V. I. Gaiduk and J. K. Vij, *J. Chem. Phys.* **118**(20), 9457 (2003).
13. M. W. Evans, G. J. Evans, W. T. Coffey, and P. Grigolini, *Molecular Dynamics and Theory of Broadband Spectroscopy*, Wiley-Interscience, New York, 1982.
14. W. T. Coffey, Yu. P. Kalmykov, and J. T. Waldron, *The Langevin Equation*, World Scientific, Singapore, 1996.
15. G. V. Yukhnevich, *Infrared Spectroscopy of Water*, Nauka, Moscow, 1973 (in Russian).
16. G. E. Walrafen, in: F. Franks, ed., *Water, A Comprehensive Treatise*, Vol. 1, Plenum, New York, 1972, p. 151.
17. H. J. Liebe, G. A. Hufford, and T. Manabe, *Int. J. of Infrared and Millimeter Waves* **12**, 659 (1991).
18. V. I. Gaiduk and Yu. P. Kalmykov, *J. Mol. Liq.* **34**, 1–222 (1987).
19. R. McGraw, W. G. Madden, M. S. Bergren, and S. A. Rice, *J. Chem. Phys.* **69**, 3483 (1978).
20. V. I. Gaiduk, Yu. P. Kalmykov, and B. M. Tseitlin, *Radiotekhn. Elektron.* **24**, 1170 (1979) [*Radio Eng. Electron. Phys.* **24**, 133 (1979)].
21. V. I. Gaiduk and Yu. P. Kalmykov, *J. Chem. Soc. Faraday Trans. II* **77**, 929 (1981).
22. E. P. Gross, *J. Chem. Phys.* **23**, 1415 (1955).
23. R. A. Sack, *Proc. Phys. Soc.* **70B**, 414 (1957).

24. A Kocot, S. Gandor, J. Sciesinski, T. Grochulski, L. Pszczolkowski, and K. Leibler, *Mol. Phys.* **53**(6), 1481 (1984).
25. T. Grochulski, Z. Kiziel, L. Pszczolkowski, and K. Leibler, *Mol. Phys.* **58**(3), 647 (1984).
26. K. Pasterny and A. Kocot, *J. Chem. Soc., Faraday Trans.* **82**(2), 485 (1986).
27. M. Godlewska, J. A. Janik, A. Kocot, X. P. Nguyen, J. Sciesinski, and W. Witko, *Liq. Crystals* **1**(6), 529 (1986).
28. K. Pasterny and A. Kocot, *J. Chem. Soc., Faraday Trans. 2* **83**(8), 1439 (1987).
29. J. K. Vij, T. Grochulski, A. Kocot, and F. Hufnagel, *Mol. Phys.* **72**(2), 353 (1991).
30. T. Grochulski and A. Gerschel, *Mol. Phys.* **77**(3), 539 (1992).
31. T. Grochulski and A. Kocot, *Mol. Phys.* **81**(3), 705 (1994).
32. V. I. Gaiduk, T. A. Novskova, and B. M. Tseitlin, *Chem. Phys. Rep.* **17**, 887 (1998).
33. V. I. Gaiduk and V. V. Gaiduk, *Mendeleev Commun.* (2), 76 (1997).
34. V. I. Gaiduk and V. V. Gaiduk, *Russ. J. Phys. Chem.* **71**, 1637 (1997).
35. N. Agmon, *J. Phys. Chem.* **100**, 1072 (1996).
36. S. Woutersen, U. Emmerichs, and H. J. Bakker, *Science* **278**, 658 (1997).
37. E. P. Gross, *Phys. Rev.* **97**, 395 (1955).
38. A. P. Prudnikov, Yu. A. Brychkov, O. I. Marichev, *Integrals and Series*, Nauka, Moskow, 1981.
39. B. M. Liberman and V. I. Gaiduk, *Radiotekhn. Electron.* **44**(1), 97 (1999).
40. H. Frohlich, *Theory of Dielectrics, Dielectric Constant and Dielectric Loss*, 2nd ed., Clarendon Press, Oxford, 1958.
41. A. R. von Hippel, *Dielectrics and Waves*, Wiley, New York, 1954 (Russian translation: Izdat. Inostr. Literat., Moscow, 1960, p. 64).
42. H. D. Downing and D. Williams, *J. Geophys. Res.* **80**, 1656 (1975).
43. A. Gerschel, I. Dimicoli, A. Jaffre, and A. Riou, *Mol. Phys.* **32**, 679 (1976).
44. W. T. Coffey, D. A. Garanin, D. J. McCarthy, *Adv. Chem. Phys.* **117**, 483 (2001).
45. K. Higasi, R. Minami, H. Takahashi, and A. Ohno, *J. Chem. Soc., Faraday Trans. 2*, **69**, 1598 (1973).
46. R. Minami and A. Ohno, *J. Phys. Soc. Jpn.* **35**, 1730 (1973).
47. V. I. Gaiduk, T. A. Novskova, and V. V. Brekhovskikh, *J. Chem. Soc. Faraday Trans.* **89**, 1975 (1993).
48. V. Gaiduk and M. N. Rodnikova, *J. Mol. Liq.* **82**, 47 (1999).
49. T. A. Novskova, A. K. Lyashchenko, and V. I. Gaiduk, *Chem. Phys. Rep.* **18**, 467 (1999).
50. V. I. Gaiduk, B. M. Tseitlin, and E. S. Kryachko, *Nuovo Cimento* **13D**, 1247 (1991).
51. H. R. Zelsmann, *J. Mol. Structure* **150**, 95 (1995).
52. P. Debye, *Polar Molecules*, Chemical Catalog Company, New York, 1929.
53. L. W. Pinkley, P. P. Sethna, and D. Williams, *J. Opt. Soc. Am.* **56**, 64 (1977).
54. C. Rønne, P.-O. Åstrand, and S. R. Keiding, *Phys. Rev. Lett.* **82**, 2888 (1999).
55. R. Buchner, J. Barthel, and J. Stauber, *Chem. Phys. Lett.* **306**, 57 (1999).
56. C. Rønne, L. Thrane, P.-O. Åstrand, A. Wallqvist, K. V. Mikkelsen, and S. R. Keiding, *J. Chem. Phys.* **107**, 5319 (1997).
57. G. N. Zatsepina, *Physical Properties and Structure of Water*, Moscow University Publishing House, Moscow, 1987 (in Russian).

58. J. O'M. Bockris and A. K. N. Reddy, *Modern Electrochemistry 1*, 2nd ed., Plenum, New York, 1994.
59. V. I. Lobyshev, *Thesis*, Moscow University, 1974.
60. G. E. Walrafen, W.-H. Yang, Y. C. Chu, and M. S. Hokmabadi, *J. Phys. Chem.* **100**, 1381 (1996).
61. V. I. Gaiduk, V. V. Gaiduk, T. A. Novskova, B. M. Tseitlin, and J. McConnell, *J. Chem. Phys.* **103**, 5246 (1995).
62. I. N. Kochnev, *J. Mol. Liq.* **101**(1–3), 169 (2002).
63. Y. Amo and Y. Tominaga, *Physica A* **276**, 401 (2000).
64. Yu. Amo and Ya Tominaga, *Phys. Rev. E* **60**, 1708 (1999).
65. P. L. Bhatnagar, E. P. Gross, and M. Krook, *Phys. Rev.* **94**, 511 (1954).
66. W. F. Brown, *Encyclopedia of Physics, Vol. 17, Dielectrics*, S. Flugge, ed., Springer-Verlag, Berlin, 1956 (Russian translation: W. Brown, *Dielectrics*).
67. Ch. H. Townes and A. L. Shawlow, *Microwave Spectroscopy*, Dover Publications, New York, 1975 (Russian translation: Inostran. Literat., Moscow, 1959).
68. V. I. Gaiduk, B. M. Tseitlin, and B. M. Liberman, *J. Commun. Technol. Electron.* **43**, 1297 (1998).
69. R. Kubo, *Statistical Mechanics*, North Holland, Amsterdam, 1965, Chapter 6.
70. J. A. Lane and J. A. Saxton, *Proc. R. Soc.* **214A**, 531 (1952).
71. Ya. Yu. Akhadov, *Dielectric Properties of Binary Solutions*, Nauka, Moscow, 1977. (in Russian).
72. O. Ya. Samoilov, *Structure of Aqueous Solutions of Electrolytes and Hydration of Ions*, Izdat. Akademii Nauk SSSR, Moscow, 1957 (in Russian).
73. D. Wei and G. N. Patey, *J. Chem. Phys.* **94**, 6795 (1991).
74. A. Chandra, D. Wei, and G. N. Patey, *J. Chem. Phys.* **99**, 2083 (1993).
75. A. P. Lyubartsev and A. Laakonen, *J. Phys. Chem.* **100** (1996) 16410. 7553 (1998).
76. A. Chandra and B. Bagchi, *J. Chem. Phys.* **112**, 1876 (2000).
77. T. Dodo, M. Sugawa, and E. Nonaka, *J. Chem. Phys.* **98**, 5310 (1993).
78. E. Zoidis, J. Yarwood, and M. Besnard, *J. Phys. Chem.* **A 103**, 220 (1999).
79. T. Dodo, M. Sugawa, E. Nonaka, H. Honda, and S. Ikawa, *J. Chem. Phys.* **102**, 6208 (1995).
80. T. Dodo, M. Sugava, and E. Nonaka, *J. Chem. Phys.* **116**, 5701 (2002).
81. M. Bennouna, H. Cachet, and J. C. Lestrade, *Chem. Phys.* **62**, 439 (1981).
82. Y. Amo and Ya. Tominaga, *Phys. Rev. E.* **58**, 7553 (1998).
83. S. A. Rice and M. G. Sceats, *J. Phys. Chem.* **85**, 1108 (1981).
84. M. G. Sceats and S. A. Rice, *J. Chem. Phys.* **72**, 3236 (1980).
85. M. N. Rodnikova, *Zhurn. Khim. Phys.* **67**, 275 (1993).
86. P. H. Poole, F. Sciortino, T. Grande, H. E. Staneley, and C. A. Angell, *Phys. Rev. Lett.* **73**, 1632 (1994).
87. Yu. Ya Efimov and Yu. I. Naberukhin, *Mol. Phys.* **30**, 1621 (1975); **33**, 759 (1977); **36**, 973 (1978).
88. A. P. Zhukovsky, *Zh. Struct. Khim.* **17**, 931 (1976).
89. D. Eisenberg and W. Kauzmann, *The Structure and Properties of Water*, Clarendon Press, Oxford, 1969.
90. V. I. Gaiduk, V. V. Gaiduk, T. A. Novskova, B. M. Tseitlin, and J. McConnell. *J. Chem. Phys.* **103**, 5246 (1995).

91. T. A. Novskova and V. I. Gaiduk, *J. Mol. Liq.* **69**, 143 (1996).
92. V. I. Gaiduk and T. A. Novskova, *Russ. J. Phys. Chem.* **71**, 646 (1997).
93. J. E. Bertie, H. J. Labbe, and E. Whalley, *J. Chem. Phys.* **50**, 4501 (1969).
94. V. M. Zolotaryev, *Opt. Spektrosk.* **29**, 1126 (1970).
95. E. Whalley, S. J. Jones, and L. W. Gold, eds., *Physics and Chemistry of Ice*, Royal Society of Canada, Ottawa, 1973.
96. B. Minceva-Sukarova et al., *J. Phys. C.* **175**, 833 (1984).
97. M. Marchi et al., *J. Chem. Phys.* **85**, 2414 (1986).
98. O. B. Toon, M. A. Tolbert, B. G. Koehler, A. M. Middlebrook, and J. Jordan, *J. Geophys. Res.-Atmos.* **99**(D12), 25631–25654 (1994).
99. G. Hufford, *Int. J. Infrared Millimeter Waves* **12**, 677 (1991).

AUTHOR INDEX

Numbers in parentheses are reference numbers and indicate that the author's work is referred to although his name is not mentioned in the text. Numbers in *italic* show the pages on which the complete references are listed.

Acharya, A. R., 38(71), *61*
Ackland, G. J., 28(48), 35(57), [29(48,56-57)], 34(48), 36(48), 38(71), 51(111), *60–61, 63*
Agmon, N., 84(35), 149(35), 178(35), 205(35), *329*
Agrawal, R., 45(90), 49(90), *62*
Akhadov, Ya. Yu., 282(71), *330*
Alder, B. J., 50(107), *63*
Alexandrowicz, Z., 51(115), *63*
Alfe, D., 50-51(109), *63*
Amo, Y., 238(63), 240(64), 290(82), 320(63), *330*
Angell, C. A., 302(86), 317(86), *330*
Åstrand, P.-O., 156(54), 174-175(56), 198(54), 201(54), 203(54), 214(54), [223(54,56)], 240(56), 244-247(54), 317(54), 323-324(54), 326(54), *329*

Bagchi, B., 288(76), *330*
Bakker, H. J., 84(36), *329*
Barthel, J., 223(55), 240(55), *329*
Bennett, C. H., 31(61), 50(107), *61, 63*
Bennouna, M., 290(81), *330*
Berg, B. A., 18(27), 26(27), 27(41), 37(27), 53(27), [54(27,124)], *60, 63*
Bergren, M. S., 82(19), 205(19), 296(19), 308(19), *328*
Bertie, J. E., 320(93), *331*
Besnard, M., 290(78), *330*
Bhatnagar, P. L., 263(65), *330*
Binder, K., 8(7), 40(75), 51(113), *59, 61, 63*
Binney, J. J., 47(98), *62*
Bockris, J. O'M., 224(58), *330*
Bolhuis, P. G., 45(91), *60*(35), *62*
Borgs, C., 47(94-95), *62*
Brangian, C., 51(117), *63*
Brekhovskikh, V. V., 156(47), *329*
Briskina, Ch. M., [72(6,8,12-12a)], 81(12-12a), [158(6,8)], 179(12-12a), [203(6,8)], [206(6,8,12-12a)], 215(6), [224(6,8)], 249(12-12a), 293(12-12a), *328*
Brown, W. F., 268-269(66), *330*
Bruce, A. D., 28(48), [29(48,56-57)], 30(58), 34(48), 35(57), [36(48,56-58)], 37(58), 38(71), 40(76), 47(58), 48(100), 54(122), 56-57(122), *60–63*
Brychkov, Yu. A., *329*
Buchner, R., 223(55), 240(55), *329*

Caccamo, C., 2(2), *59*
Cachet, H., 290(81), *330*
Ceperley, D. M., 51(113), *63*
Challa, M. S. S., 28(49), *60*
Chandra, A., 287(74), [288(74,76)], *330*
Chu, Y. C., 238(60), 308(60), *330*
Coffey, W. T., [72(3,13-14)], 78(13-14), [135(3,44)], [154(14,44)], 156(3), 161(3), *328–329*
Consta, S., 51(115), *63*
Cowley, R. A., 3(3), *59*
Crothers, D. S., [72(4,8,10)], 81(12a), 179(12a), 206(12a), 249(12a), 270(4), 272(4), 275(4), 277(4), 293(12a), *328*

Debye, P., 178(52), *329*
de Oliveira, P. M. C., 55(128), *63*
De Pablo, J. J., 23-25(38), 45(91), 55(126), *60, 62–63*

Advances in Chemical Physics, Volume 127, Edited by I. Prigogine and Stuart A. Rice
ISBN 0-471-23583-0

De Wijs, G. A., 50-51(109), *63*
Diep, H. T., 50(105), *62*
Dijkstra, M., 51(114), *63*
Dimicoli, I., 150-151(43), 177-178(43), *329*
Dobry, A., 50(105), *62*
Dodo, T., 289(77), 290(79-80), 291(80), *330*
Downing, H. D., 144-145(42), 148(42), 174-175(42), 194-195(42), 197(42), 203(42), 206(42), 211-212(42), 214(42), 233(42), 235(42), 323(42), *329*
Dowrick, N. J., 47(98), *62*

Efimov, Yu. Ya., 317(87), *330*
Eisenberg, D., 319(89), *330*
Emmerichs, U., 84(36), *329*
Escobedo, F. A., 45(91), *62*
Evans, G. J., 72(13), 78(13), *328*
Evans, M. G., 72(13), 78(13), *328*
Evans, R., 2(1), 51(114), *59, 63*

Falk, H., 41(81), 43(81), *61*
Faller, R., 55(126), *63*
Ferrenberg, A. M., 58-59(133), *64*
Fischer, J., 40(77), 41(79-80), *61*
Fisher, A. J., 47(98), *62*
Fisher, M. E., 51(116), *63*
Fosdick, L. D., 12(12), *59*
Frenkel, D., 20(28-29), 21(28), [22(28,31)], 23(34), 29(28), 38(34), 47(97), [50(97,108)], 51(115), *60, 62–63*
Frohlich, H., 95(40), 140(40), 160(40), 176(40), *329*

Gaiduk, V. I., 71(1), 72(1-12c), 77(1-2), 80(1-2), [81(12-12a,18)], [83(18,20-21)], [84(1-2,32-34)], 86(1-2), [90(1-2,18)], 92-94(1), 95(1-2), 100-101(1-2), 103(2), 104-105(1), 106(1-2), 110(1), 112(1-2), 114(1), [115(1-2,32,39)], 135(2-3), 136(2), 141(1-2), 150(1), 152(2), [154-155(1-2, 20-21)], [156(2-3,6-7,12c,32-34,39, 47-49)], [157(1-2,32,50)], 158(8), 159-160(1-2), [161(2-3,7)], 165(1-2), 168(1-2), [179(2,12-12b)], 180(32), 185-186(1-2), 188-189(1), 190-191(2), 199(10), [203(6-8,11)], [205(7-8,12b)], [206(6-8,12-12b)], 208(1-2), [214(7,12b,33-34)], 215(6-7), 216(9), 217(1), 218(1-2), [222(2,21)], [224(6,8)], 226(18), 229(1-2), [236(2,9)], 238(61), [240(2,18)], [242(1-2,12b)], 244(1-2), 246(1-2), [247(1-2,9,12b,18)], [249(10,12-12a)], 250(1-2), 254(1-2), 256(10-11), 260(1-2), a4262(1-2), [263(1-2,18)], 268(18), 270(4-5), 271(68), 272(4-5), 275(4-5), 276(5), 277(4), 282(2), 287(5), 290(1), [293(10,12-12a)], 303(1), 311(1), 318(1-2), [319(2,90)], 320(90-92), 322(1-2), *328–331*
Gaiduk, V. V., 84(33-34), 156(33-34), 214(33-34), 238(61), 319-320(90), *329–330*
Gandor, S., 83(24), *329*
Garanin, D. A., 154(44), *329*
Gerschel, A., 83(30), 150-151(43), 177-178(43), *329*
Geyer, C. J., 16(18), *59*
Gillan, M. J., 50-51(109), *63*
Godlewska, M., 83(27), *329*
Gold, L. W., 320(95), *331*
Gomez, L., 50(105), *62*
Grachev, M. E., 50(106), *63*
Grande, T., 302(86), 317(86), *330*
Grigolini, P., 72(13), 78(13), *328*
Grochulski. T., [83(24-25,29-31)], *329*
Gross, E. P., 83(22), 155(22), 222(37), 263(65), *328–330*
Grossmann, B., 27(42), *60*

Hansen, J. P., 22(30), *60*
Hansmann, U. H. E., 27(41), *60*
Heerman, D. W., 8(7-8), *59*
Herrmann, H. J., 55(128), *63*
Hetherington, J. H., 28(49), *60*
Higasi, K., 154(45), *329*
Hocken, R., 48(99), *62*
Hokmabadi, M. S., 238(60), 308(60), *330*
Honda, H., 290(79), *330*
Hoover, W. G., 22(32), *60*
Hufford, G. A., 81(17), 83(17), 144-145(17), 147-148(17), 175(17), 194(17), 197-198(17), 202-203(17), 206(17), 211-212(17), 214-216(17), 222-224(17), 236(17), 240(17), 245 -247(17), 282(17), 318(17), 320(99), 322-324(17), 326(17), *328, 331*
Hufnagel, F., 83(29), *329*
Hukushima, K., 16(19), *59*
Hunter, J. E. (III), 43(83-84), *62*
Huse, D. A., 43(87), *60*(36), *62*

Iba, Y., 16(17), *59*
Ikawa, S., 290(79), *330*

Jackson, A. N., 28(48), [29(48,57)], 34(48), 35(57), [36(48,57)], *60–61*
Jacucci, G., 31-32(63), 37(63), *61*
Jaffre, A., 150-151(43), 177-178(43), *329*
Janik, J. A., 83(27), *329*
Janke, W., 47(95), *62*
Jarzynski, C., 38(70), 42(82), *61*
Jones, S. J., 320(95), *331*
Jordan, J., 320(98), *331*

Kalmykov, Yu. P., 72(14), 81(18), 83(18), 78(14), [83(18,20-21)], 90(18), 154-155(20-21), 222(21), 226(18), 240(18), 247(18), 263(18), 268(18), *328*
Kashurnikov, V. A., 50(106), *63*
Kauzmann, W., 319(89), *330*
Keiding, S. R., 156(54), 174-175(56), 198(54), 201(54), 203(54), 214(54), [223(54,56)], 240(56), 244-247(54), 317(54), 323-324(54), 326(54), *329*
Khatchaturyan, A. G., 43(85), *62*
Kiziel, Z., 83(25), *329*
Kob, W., 51(117), *63*
Kochnev, I. N., 156(62), 238(62), 320(62), *330*
Kocot, A., [83(24,26-29,31)], *329*
Koehler, B. G., 320(98), *331*
Kofke, D. A., 45(90-91), 49(90), *62*
Kolesik, M., 51(118), *63*
Kotecky, R., 47(94), *62*
Kresse, G., 50-51(109), *63*
Kristof, T., 49(102), *62*
Krook, M., 263(65), *330*
Kryachko, E. S., 157(50), *329*
Kubo, R., 275(69), *330*

Laakonen, A., 288-289(75), *330*
Labbe, H. J., 320(93), *331*
Ladd, A. J. C., 20(28), 21-22(28), 29(28), *60*
Landau, D. P., 8(8), 51(112), 53-54(112), *59, 63*
Lane, J. A., 282(70), *330*
Laursen, M. L., 27(42), *60*
Lee, J., 52(119), *63*
Lee, L. W., 55(129), *63*
Leibler, K., 83(24-25), *329*
Lekkerkerker, H. N. W., 22(31), *60*
Lestrade, J. C., 290(81), *330*
Liberman, M., 115(39), 156(39), 271(68), *329–330*
Liebe, H. J., 81(17), 83(17), 144-145(17), 147-148(17), 175(17), 194(17), 197-198(17), 202-203(17), 206(17), 211-212(17), 214-216(17), 222-224(17), 236(17), 240(17), 245 -247(17), 282(17), 318(17), 322-324(17), 326(17), *328*
Lisal, M., 45(91), *62*
Liszi, J., 49(102), *62*
Lobyshev, V. I., 238(59), 308(59), *330*
Lotfi, A., 41(79), *61*
Lyaschenko, A. K., 156(49), *329*
Lyubartsev, A. P., 18(23), 37(23), *60* 288-289(75), *330*

Madden, W. G., 82(19), 205(19), 296(19), 308(19), *328*
Manabe, T., 81(17), 83(17), 144-145(17), 147-148(17), 175(17), 194(17), 197-198(17), 202-203(17), 206(17), 211-212(17), 214-216(17), 222-224(17), 236(17), 240(17), 245 -247(17), 282(17), 318(17), 322-324(17), 326(17), *328*
Marchi, M., 320(97), *331*
Marichev, O. I., *329*
Marinari, E., 16(20), 18(24), *59–60*
Martsinovski, A. A., 18(23), 37(23), *60*
Matsuda, H., 22(33), *60*
Mau, S.-C., 43(87), *60*(35), *62*
McCarthy, D. J., 154(44), *329*
McConnell, J., 328(61), 319-320(90), *330*
McGraw, R., 82(19), 205(19), 296(19), 308(19), *328*
Metropolis, N., 10(10), *59*
Mezei, M., 18(26), *60*
Middlebrook. A. M., 320(98), *331*
Mikkelsen, K. V., 174-175(56), 223(56), 240(56), *329*
Miller, M. A., 43(84), *62*
Minami, R., 154(45-46), *329*
Minceva-Sukarova, B., 320(96), *331*
Möller, D., 40(77), *61*
Mon, K. K., 40(75), *61*
Moody, M. C., 37(68), *61*
Mooij, G. C. A. M., 51(115), *63*
Muser, M. H., 51(113), *63*

Naberukhin, Yu. I., 317(87), *330*
Narayan, O., 8(9), *59*

Nemoto, K., 16(19), *59*
Neuhaus, T., 18(27), 26(27), 27(41), 37(27), 53-54(27), *60*
Newman, M. E. J., 47(98), *62*
Nguyen, X. P., 83(27), *329*
Nielaba, P., 51(113), *63*
Nielsen, O. F., 72(7), 156(7), 161(7), 203(7), 205-206(7), 214-215(7), *328*
Nikitendo, O. A., 50(106), *63*
Nishiyama, Z., 29(54), *61*
Nonaka, E., 289(77), 290(79-80), 291(80), *330*
Novoskova, T. A., 84(32), 115(32), [156(32,47,49)], 157(32), 180(32), 238(61), 319(90), 320(90-92), *329–331*
Novotny, M. A., 51(118), *63*

Ogawa, T., 22(33), *60*
Ogita, N., 22(33), *60*
Ogura, H., 22(33), *60*
Ohno, A., 154(45-46), *329*

Panagiotopoulos, A. Z., 39(72-73), 40(73), 51(116), *61, 63*
Pant, P. V., 49(103), *62*
Parisi, G., 18(24), *60*
Pasterny, K., [83(26,28)], *329*
Patey, G. N., 287(73-74), 288(74), *330*
Penna, T. J. P., 55(128), *63*
Perova, T. S., 72(7), 156(7), 161(7), 203(7), 205-206(7), 214-215(7), *328*
Pinkley, L. W., 156(53), 194(53), 196-197(53), *329*
Polson, J. M., 47(97), 50(97), *62*
Poole, P. H., 302(86), 317(86), *330*
Privman, V., 47(93), *62*
Pronk, S., 23(34), 38(34), 47(97), [50(97,108)], *60, 62–63*
Prudnikov, A. P., *329*
Pszczolkowski, L., 83(24-25), *329*

Quirke, N., 49(101), *62*

Rahman, A., 31-32(63), [37(63,68)], *61*
Ray, J. R., 37(68), *61*
Reddy, A. K. N., 224(58), *330*
Ree, F. H., 22(32), *60*
Reinhardt, W. P., 43(83), *62*
Rice, S. A., 82(19), 205(19), [296(19,83-84)], 308(19), *328, 330*
Rickwardt, C., 51(113), *63*
Riou, A., 150-151(43), 177-178(43), *329*
Ritvold, P. A., 51(118), *63*
Rodnikova, M. N., 156(48), 296(85), *329–330*
Rønne, C., 156(54), 174-175(56), 198(54), 201(54), 203(54), 214(54), [223(54,56)], 240(56), 244-247(54), 317(54), 323-324(54), 326(54), *329*
Rosenbluth, A. W., 10(10), *59*
Rosenbluth, M. N., 10(10), *59*
Rowlinson, J. S., 3(4), *59*
Rudnev, I. A., 50(106), *63*
Rushbrooke, G. S., 3(4), *59*

Sack, R. A., 83(23), 155(23), *328*
Samoilov, O. Ya., 287(72), 289(72), *330*
Saxton, J. A., 282(70), *330*
Sceats, M. G., 296(83-84), *330*
Sciesinski, J., [83(24,27)], *329*
Sciortino, F., 302(86), 317(86), *330*
Sengers, A. L., 48(99), *62*
Sengers, J. V., 48(99), *62*
Sethna, P. P., 156(53), 194(53), 196-197(53), *329*
Shawlow, A. L., 268-269(67), *330*
Shevkunov, S. V., 18(23), 37(23), *60*
Smit, B., 20(29), 51(115), *60, 63*
Smith, G. R., 29(55), 54(122), 56(122), [57(55,122)], *61, 63*
Sollich, P., 50(104), *62*
Staneley, H. E., 302(86), 317(86), *330*
Stapleton, M. R., 49(101), *62*
Stauber, J., 223(55), 240(55), *329*
Stoobants, A., 22(31), *60*
Stuhn, T., 51(117), *63*
Sugawa, M., 289(77), 290(79-80), 291(80), *330*
Swendsen, R. H., 55(130), 58-59(133), *63–64*

Takahashi, H., 154(45), *329*
Takayama, H., 16(19), *59*
Tay, T. K., 55(130), *63*
Teller, A. H., 10(10), *59*
Teller, E., 10(10), *59*
Temperley, H. N. V., 3(4), *59*
Theodoru, D. N., 49(103), *62*
Thompson, E. A., 16(18), *59*
Thrane, L., 174-175(56), 223(56), 240(56), *329*
Tildesley, D. J., 49(101), *62*
Tolbert, M. A., 320(98), *331*

Tominaga, Y., 238(63), 240(64), 290(82), 320(63), *330*
Toon, O. B., 320(98), *331*
Torrie, G. M., 18(22), 52(22), *60*
Townes, Ch. H., 268-269(67), *330*
Trappenberg, T., 27(42), *60*
Trizac, E., 47(97), 50(97), *62*
Tseitlin, B. M., 71(1), [72(1,3-12)], 77(1), 80(1), 81(12a), 83(20), [84(1,32)], 86(1), 90(1), 92-94(1), 95(1), 100-101(1), 104-105(1), 106(1), 110(1), 112(1), 114(1), [115(1,32)], 135(3), 141(1), 150(1), [154-155(1,20)], [156(3,32)], [157(1,32,50)], [158(6,8)], 159-160(1), 161(3), 165(1), 168(1), 179(12a), 180(32), 185-186(1), 188-189(1), 199(10), [203(6,8)], [206(6,8,12a)], 208(1), 215(6), 217(1), 218(1), [224(6,8)], 229(1), 238(61), 242(1), 244(1), 246-247(1), [249(10,12a)], 250(1), 254(1), 256(10), 260(1), 262-263(1), 270(4-5), 271(68), 272(4-5), 275(4-5), 276(5), 277(4), 287(5), 290(1), [293(10,12a)], 303(1), 311(1), 318(1), 329-310(90), 322(1), *328–330*

Ueda, A., 22(33), *60*

Vacek, V., 45(91), *62*
Valleau, J. P., [18(22,25)], 39(74), 52(22), *60–61*
van Roij, R., 51(114), *63*
Verlet, L., 22(30), *60*
Vij, J. K., 72(5), 83(29), 203(11), 156(12c), 179(12b), 205-206(12b), 214(12b), 242(12b), 247(12b), 256(11), 270(5), 272(5), 275-276(5), 287(5), *328*
von Hippel, A. R., 140-141(41), 160(41), *329*
Vorontsov-Velyaminov, P. N., 18(23), 37(23), *60*
Voter, A. F., 37(69), *61*
Vrabec, J., 41(79-80), *61*

Waldron, J. T., 72(14), 78(14), 154(14), *328*
Wallqvist, A., 174-175(56), 223(56), 240(56), *329*
Walrafen, G. E., 75(16), 82(16), 158(16), 203(16), 216(16), 222(16), 238(60), 239(16), 293(16), 296(16), 304(16), 308(60), *328, 330*
Walsh, M. E., 72(3), 135(3), 156(3), 161(3), *328*
Wang, F., 51(112), 53-54(112), *63*
Wang, J. S., 55(129-130), *63*
Wei, D., 287(73-74), 288(74), *330*
Whalley, E., [320(93,95)], *331*
Widom, B., 41(78), *61*
Wiese, U. J., *60*
Wilding, N. B., 3(5), 27(43-44), 29(56-57), 30(58), 35(57), 36(56-58), 37(58), 40(44), 47(58), [48(44,100)], 50(104), 51(115), *59–63*
Williams, D., 144-145(42), 148(42), 156(53), 174-175(42), [194(42,53)], 195(42), 196(53), [197(42,53)], 203(42), 206(42), 211-212(42), 214(42), 233(42), 235(42), 323(42), *329*
Witko, W., 83(27), *329*
Woodcock, L. V., *60*(36)
Woutersen, S., 84(36), *329*

Yamamoto, R., 51(117), *63*
Yan, Q., 23-25(38), 55(126), *60, 63*
Yang, W.-H., 238(60), 308(60), *330*
Yarwood, J., 290(78), *330*
Young, A. P., 8(9), *59*
Yukhnevich, G. V., 73(15), 82(15), 205(15), 296(15), 307(15), *328*

Zatsepina, G. N., 224(57), 288(57), 319(57), *329*
Zelsmann, H. R., 198-200(51), 205-206(51), 211-212(51), 230-232(51), 236(51), 245-246(51), 302(51), 304(51), 316(51), 318(51), 323-324(51), 326(51), *329*
Zhukovsky, A. P., 317(88), *330*
Zoidis, E., 290(78), *330*
Zolotaryev, V. M., 320(94), *331*
Zong, F. H., 51(113), *63*
Zwanzig, R. W., 31-33(62), 41(62), *61*

SUBJECT INDEX

Ab initio calculations, equilibrium phase diagrams, future issues, 51
Absorption bandwidth:
  hat-curved-elastic bond (HC-EB) model, 325–327
  hat-curved model, polar fluids, spectral function calculations, 179–182
  molecular model correlation, 104–105
  structural-dynamical model, rotational band formation, 300–301
Absorption coefficient:
  hat-curved-structural dynamical (HC-SD) model, wideband FIR absorption, 313–317
  molecular model correlation, dipolar ensemble localization, 102–103
Absorption frequency dependence:
  hat-curved-harmonic oscillator (HC-HO) model, heavy water (D_2O), 232–235
  rectangular potential well protomodel spectral function, 115
  specific interactions in liquid water, 211–214
Absorption-peak frequency estimation:
  hat-curved models, polar fluids, 169–173
  hat-flat (HF) models, mean number of reflections, 133–135
  hat-flat models, nonassociated fluids, 153
  molecular model correlation, 103–104
Absorption spectrum, hat-curved models, polar fluids, 169–173
Adaptive umbrella sampling, equilibrium phase diagrams, 18
Additivity approximation:
  aqueous electrolyte solutions, dielectric response, wideband spectra, ionic conductivity, 280–287
  defined, 72
  reorienting single dipole, 81
Analytical representation:
  hat-curved models, unspecific interactions, 154–158
  hat-flat models:
    librator spectral function, 126–127
    qualitative spectral dependencies, 137–140
  nonrigid oscillator composite models, spectral function (Boltzmann collision model), 256–260
  spectral function, rectangular potential well protomodel, 111–112
Anharmonicity:
  molecular model correlation, absorption bandwidth, 104–105
  rectangular potential well protomodel, 117–118
Anisotropic medium, dielectric response, 75
A priori probabilities, equilibrium phase diagrams, 4
  statistical mechanics and thermodynamics, 8
Aqueous electrolyte solutions:
  dielectric/far-infrared spectra, 270–287
    frequency-dependent ionic conductivity, 274–279
    ionic motion, 270–274
    wideband spectra, ionic effects, 280–287
  ionic conductivity susceptibility, 292
Autocorrelation function (ACF):
  aqueous electrolyte solutions, dielectric response, frequency-dependent ionic conductivity, 275–279
  dielectric/FIR spectra of water, 73, 75
    molecular model parameters:
      aborption bandwidth, 104–105
      absorption-peak frequency, 103–104
      dipolar ensemble localization, 102–103
      relaxation spectrum characteristics, 105–106

Advances in Chemical Physics, Volume 127, Edited by I. Prigogine and Stuart A. Rice
ISBN 0-471-23583-0

Autocorrelation function (ACF) (*Continued*)
reorienting dipole spectral function, local
axisymmetric potential, 96–102
ensemble averaging, 100
Fourier amplitudes, 100–102
spectral function and complex
susceptibility, 85–96
average perturbation (AP) theorem,
90–92
complex refraction index-current density
connection, 85–86
current density, spectral function
theorem, 86–88
dipole autocorrelator spectral function,
92–93
statistical distributions, 88–89
susceptibility component expression,
94–96
hat-curved models, polar fluids, linear
response theory, 159–160
molecular model evolution, 321–322
structural-dynamical (SD) model, 311–317
composite hat-curved (HC)-SD model
frequency dependence, 313–317
restricted rotators:
dielectric response, 313
spectral function, 311–313
undamped motion, 75
Average perturbation (AP) theorem:
dielectric/far-infrared spectra, liquid water,
spectral function (SF), complex
susceptibility, 90–92
nonrigid oscillator composite models,
equilibrium trajectories, 254–256
Avogadro's number:
aqueous electrolyte solutions, dielectric
response, wideband spectra, ionic
conductivity, 282–287
hat-flat models, polar fluids, 141–143

Bain switch operation, equilibrium phase
diagrams, Fast Growth (FG) methods,
43
Bending force constant, structural-dynamical
(SD) model:
elastic interactions, 305–307
hydrogen-bond bending, 307–308
Berg-Neuhaus strategy, extended sampling (ES)
distributions, macrostate visit statistics,
53–57
Bessel function, hat-curved-cosine-squared
potential composite model, heavy water
(D_2O)/liquid water spectral calculations,
243–246
Bhatnagar-Gross-Krook model, isothermal
equations, 263
Boltzmann collision model:
dielectric/FIR spectra, liquid water, complex
susceptibility expression through
spectral function, 94–96
nonrigid oscillator composite models, spectral
function analytic formula, 256–260
Boltzmann distribution:
dielectric/far-infrared spectra, liquid water:
average perturbation (AP) theorem, 91–92
spectral function (SF), complex
susceptibility, collision models, 94–96
hat-curved models, polar fluids:
linear response theory, 159–160
spectral function, norm *C* derivation,
182–184
isothermal collisional model, frequency
dependences, 261–263
rectangular potential well protomodel, phase
space configuration, 109–110
specific interactions in liquid water,
harmonically changing time-varying
derivation, 219–221
Boltzmann sampling technique, equilibrium
phase diagrams, 15
Bounded charges, nonrigid oscillator composite
models, equilibrium trajectories, 256
Broad-histogram relation, extended sampling
(ES) distributions, 57

Canonical distribution, equilibrium phase
diagrams:
Boltzmann sampling technique, 15
Extended Sampling Interface Traverse
(ESIT), 27–29
Monte Carlo techniques, 10–13
polydispersity, 49–50
Central Limit Theorem, equilibrium phase
diagrams, finite-size effects, 48
Charged particle motion, nonrigid oscillator
composite models, 251–260
Boltzmann collision model, 256–260
complex susceptibility:
trajectory equilibrium, 253–256
trajectory perturbation, 251–253

Clausius-Clapeyron equation, equilibrium phase diagrams, phase boundary tracking, 45–46
Cole-Cole diagrams, hat-curved-harmonic oscillator (HC-HO) model, heavy water (D_2O), 235
Collision models:
  dielectric/far-infrared spectra, liquid water, spectral function (SF), complex susceptibility, 75, 94–96
  frequency dependences, 260–263
  hat-curved models, polar fluids, linear response theory, 160
  nonrigid oscillator composite models, spectral function analytic formula, 256–260
  specific interactions in liquid water, 204–206
Complex permittivity:
  aqueous electrolyte solutions, dielectric response, 271–274
    wideband spectra, ionic conductivity, 280–287
  dielectric/FIR spectra for water, 73, 83
  hat-curved-cosine-squared potential composite model, 246–248
  hat-curved-harmonic oscillator (HC-HO) model, 223–225
    vibrational susceptibility, 228
  hat-flat models:
    empirical formula, 144
    polar fluids, 140–143
  heavy water (D_2O), double Debye approximation, 198–199
  specific interactions in liquid water, 204–206
    composite model, two relaxation times, 217
    unspecific interactions and, 216
Complex refraction index, dielectric/far-infrared spectra, liquid water, current density and, 85–86
Complex susceptibility:
  dielectric/far-infrared spectra, liquid water, spectral function (SF), 75, 85–96
    average perturbation (AP) theorem, 90–92
    complex refraction index and current density, 85–86
    dipole autocorrelator, 92–93
    prehistorical review, 83–84
    statistical distributions, 88–89
    susceptibility component expression, 94–96
    t_0 theorem, current density representation, 86–88
  nonrigid oscillator composite models:
    equilibrium trajectories, 253–256
    perturbed trajectories, 251–253
Composite models, liquid water, 221–270
  hat-curved-cosine-squared potential composite model, 241–250
  hat-curved-harmonic oscillator model, 221–241
  nonrigid oscillator, 250–270
    back-and-forth motion of charged particles, 251–260
    equation of motion, 250–251
    frequency dependences, 260–270
  specific interactions, 217
Computational strategies, equilibrium phase diagrams:
  extended sampling distributions, 51–57
  fast growth methhods, 41–43
  Gibbs Ensemble Monte Carlo, 39–40
  histogram reweighting, 57–59
  Monte Carlo techniques, 8–13
  NPT-TP method, 40–41
  path-based techniques, 13–18
    extended sampling, 26–29
    generic routes, 13–15
    meaning and specification, 13
    numerical integration and reference states, 19–23
    parallel tempering, 23–26
    sampling strategies, 13–18
    wormholes, phase switching, 29–38
  statistical mechanics and thermodynamics, 5–8
Cone angle β, rotator spectral function, 130
Cone confined rotator (CCR) model:
  molecular model correlation, absorption-peak frequency estimation, 103–104
  rectangular potential well protomodel, conical cavity dipole, 107–109
  spectral function series representation, 110–111
Configurational-integral ratios, equilibrium phase diagrams, path-based transition to wormholes, 31–38
Configuration space, equilibrium phase diagrams, statistical mechanics and thermodynamics, 29–38

Confined rotator-extended diffusion model, dielectric/FIR spectra, liquid water, 83
Conical cavity dipole:
dielectric/far-infrared spectra, liquid water:
potential wells, 77
rectangular potential well protomodel, 106–110
geometry and law of motion, 106–109
phase-space configuration, 109–110
hat-flat models:
axial and mirror symmetry, 121–125
interpolation approximation, 130
Correlation factor, frequency dependences, 268
Cosine squared (CS) potential:
composite liquid water model, 221
hat-curved models:
ordinary/heavy water comparisons, 242–246
polar fluids, "field" models, 157
specific interactions in liquid water, 203–206
Critical point systems, equilibrium phase diagrams, hyper parallel tempering (HPT), 23–26
Crystalline defects, equilibrium phase diagrams, 50
Current density, dielectric/far-infrared spectra, liquid water:
complex refraction index, 85–86
t_0 theorem, 86–87

Damped resonance:
aqueous electrolyte solutions, dielectric response, 288–292
structural-dynamical (SD) model, restricted rotator dielectric function, 313
Debye-like low-frequency dependence, nonrigid oscillator composite models, 260, 263–265
Debye relaxation. *See also* Double Debye relaxation; Second Debye term
dielectric/far-infrared spectra, liquid water, spectral function (SF), complex susceptibility, 96
hat-curved-elastic bond (HC-EB) model, 323–327
hat-curved-harmonic oscillator (HC-HO) model:
liquid water, 223
spectral function (SF) calculations, 226
hat-curved model, 79–80
polar fluids, spectral function calculations, 178–182
hat-flat models, 78–79
polar fluids, spectral characteristics, 143
simple water molecules, 144–146
hat-flat potential, 78–79
hybrid hat-flat (HF) models, 136–137
liquid water dielectric/FIR spectra, 177–180
liquid water optical constants, 194–200
molecular model correlation, relaxation spectrum, 105–106
specific interactions in liquid water:
multiple relaxation times, 205–206
spectral function modification, 209–210
Density distribution:
equilibrium phase diagrams, polydispersity, 49–50
structural-dynamical model elastic interaction, 300–301
Density of states, extended sampling (ES) distributions, macrostate visit statistics, 55–57
Derivatives, equilibrium phase diagrams, numerical integration, 19–23
Dielectric/far-infrared spectra:
ice, 318–320
liquid water:
aqueous electrolyte solutions, 270–287
frequency-dependent ionic conductivity, 274–279
ionic motion, 270–274
wideband spectra, ionic effects, 280–287
autocorrelation function (ACF), spectral function and complex susceptibility, 85–96
average perturbation (AP) theorem, 90–92
complex refraction index-current density connection, 85–86
current density, spectral function theorem, 86–88
dipole autocorrelator spectral function, 92–93
statistical distributions, 88–89
susceptibility component expression, 94–96
composite models, 221–270
hat-curved-cosine-squared potential composite model, 241–250

hat-curved-harmonic oscillator model, 221–241
nonrigid oscillator, 250–270
back-and-forth motion of charged particles, 251–260
equation of motion, 250–251
frequency dependences, 260–270
dynamic methods:
local axisymmetric potential dipole reorienting spectral function, 96–102
molecular model correlation, 102–106
hat-curved models:
hat-curved-cosine-squared potential composite model, 241–250
calculated spectra, water and heavy water, 241–246
hat-curved-harmonic oscillator model, 221–241
complex permittivity components, 223–225
dielectric relaxation mechanisms, 221–223
free and statistical parameters, 228–232
librational band narrowing, 236–234
molecular motion, 239–240
potential profile, 235–239
reorientation, 225–226
temperature dependence, wideband spectra, 235–236
vibration, 226–228
water/heavy water comparisons, 232–235
polar fluids, 153–199
librators, Fourier amplitudes, 192–193
linear response theory, 158–160
molecular subensembles, 160–173
optical constants, 194–200
precessors spectral function, integral transformation, 193–194
spectral calculations, 174–181
spectral function derivation, 182–192
unspecific interactions, 154–158
hat-flat model, 120–150
hybrid model, 135–137
librator spectral function, 125–128
polar fluid applications, 140–153
qualitative spectral dependencies, 137–140
rotator spectral function, 128–130
statistical parameters, 130–135
trajectory classification, 120–125
ice molecules, 318–320
ionic susceptibility and conductivity, 292
modified spectral function derivation, 217–221
optical constants, 194–199
complex permittivity, double Debye approximation, 198–199
refractive index, 199–200
temperature dependence, 194–197
rectangular potential well protomodel, 106–120
conical cavity dipole, 106–110
planar librations-regular precession (PL-RP) approximation, 112–115
rigorous spectral function, 110–112
series expression, spectral function, 118–120
statistical parameters, 115–118
specific interactions, 199, 201–206
calculated spectra, 210–214
modified spectral function, 206–210
unspecific interactions, 216
wideband spectra:
aqueous electrolyte solutions, 280–287
temperature effects, 317–318
molecular model evolution, 321–322
Dielectric relaxation mechanisms:
hat-curved-harmonic oscillator (HC-HO) model, 221–223
historical background, 83–84
Dimer configuration, structural-dynamical (SD) model, elastic interaction, 295–296
Dipolar ensemble:
hat-curved models, subensemble proportions, 163–166
molecular model correlation, localization, 102–103
specific interactions in liquid water, 206
Dipolar reorientation, dielectric/far-infrared spectra, liquid water, local axisymmetric potential, 96–102
ensemble averaging, 100
Fourier amplitudes, 100–102
Dipole autocorrelator, dielectric/far-infrared spectra, liquid water, spectral function (SF), complex susceptibility, 92–93

Dissipation mechanism, hat-curved models, polar fluids, linear response theory, 160
  defined, 72
Double Debye approximation:
  hat-curved-cosine-squared potential composite model, heavy water (D_2O)/ liquid water spectral calculations, 242–246
  heavy water permittivity, 198–199
Dynamic methods
  dielectric/far-infrared spectra, liquid water:
    local axisymmetric potential dipole reorienting spectral function, 96–102
    molecular model correlation, 102–106
    undamped motion, 72
  hat-curved models, polar fluids, spectral function derivation, 185

Effective potentials, hat-curved models, subensemble proportions, 162–166
Effective weight measurements, equilibrium phase diagrams, Monte Carlo techniques, 12–13
Einstein Solid Method (ESM), equilibrium phase diagrams, numerical integration to reference macrostates (NIRM), 20–23
Elastic interaction:
  hat-curved-elastic bond (HC-EB) model, 322–327
  structural-dynamical (SD) model, 293–296
    calculation protocol, 305–307
    equation of motion, 296–298
    FIR isotopic effect, 308–310
    hydrogen-bond bending, 307–308
    mean frequencies/amplitudes and density distributions, 300–301
    numerical estimations, 302–305
    transverse vibration dynamics, 298–300
Elasticity constants:
  dielectric response, liquid water, 82
  specific interactions in liquid water, 205–206
Empirical formula, hat-flat models, complex permittivity, 144
Ensemble averaging, dielectric/far-infrared spectra, liquid water, local axisymmetric potential, dipoar reorientation, 100
Entropy, numerical integration to reference macrostates (NIRM), 21–23
Equation of motion:
  elastic interaction, structural-dynamical (SD) model, 296–298
    bending force constant, 305–307
  harmonic oscillator, parabolic potential, nonrigid oscillator composite models, 250–251
Equilibrium phase diagrams:
  computational strategies:
    extended sampling distributions, 51–57
    fast growth methhods, 41–43
    Gibbs Ensemble Monte Carlo, 39–40
    histogram reweighting, 57–59
    Monte Carlo techniques, 8–13
    NPT-TP method, 40–41
    path-based techniques, 13–18
      extended sampling, 26–29
      generic routes, 13–15
      meaning and specification, 13
      numerical integration and reference states, 19–23
      parallel tempering, 23–26
      sampling strategies, 13–18
      wormholes, phase switching, 29–38
    statistical mechanics and thermodynamics, 5–8
  finite-size effects, 46–48
  future research issues, 51
  imperfections, 49–50
  phase boundary extrapolation and tracking, 44–46
Ergodic blocks, equilibrium phase diagrams, hyper parallel tempering (HPT), 24–26
Exchange Monte Carlo, equilibrium phase diagrams, parallel sampling, 16
Expanded ensemble, equilibrium phase diagrams, 18
Exponential averaging, equilibrium phase diagrams, Fast Growth (FG) methods, 42–43
Exponential distribution, mean lifetime calculations, 75
Extended sampling (ES), equilibrium phase diagrams:
  barrier traverse, 26–29
  distribution building, 51–57
  path-based techniques, 16–18
  phase boundary tracking, 45–46
Extended Sampling Interface Traverse (ESIT), equilibrium phase diagrams, 26–29

Extended Sampling Phase Switch *vs.*, 38
Extended Sampling Phase Switch (ESPS), equilibrium phase diagrams, path-based transition to wormholes, 30–38

Fading oscillations, aqueous electrolyte solutions, dielectric response, frequency-dependent ionic conductivity, 279
Far infrared (FIR) spectra:
ice molecules, 82, 318–320
structural-dynamics (SD) models, isotopic effect, 308–310
Fast Growth (FG) methods, equilibrium phase diagrams, 41–43
Fcc lattices, equilibrium phase diagrams:
hard spheres, 35–38
Lennard-Jones fluid, 34–35
"Field" models, hat-curved models, polar fluids, 157
Finite-depth cosine-squared model, vibration lifetimes, 249
Finite-size effects, equilibrium phase diagrams, 46–48
phase boundary extrapolation and tracking, 46–48
Finite well depth:
hat-curved-elastic bond (HC-EB) model, 327
hat-flat potential, 77
"Fluctuation concept," wideband dielectric spectra, temperature dependence, 317–318
FORCE values, dielectric/far-infrared spectra, liquid water, average perturbation (AP) theorem, 91–92
Fourier amplitudes:
aqueous electrolyte solutions, dielectric response, frequency-dependent ionic conductivity, 276–279
dielectric/far-infrared spectra, liquid water, spectral function, local axisymmetric potential, 100–102
hat-curved models, librator dipoles, 192–193
ionic conductivity, aqueous electrolyte solutions, dielectric response, 273–274
Free charges, dielectric/FIR spectra of water, spectral function, complex susceptibility, 88–89
Free energy, equilibrium phase diagrams, 6–8
numerical integration, 20–23
path-based transition to wormholes, 30–38
Free parameters, hat-curved-harmonic oscillator (HC-HO) model, 228–232
Free rotors:
hat-flat models, statistical parameters, 130–132
specific interactions in liquid water, 203–206
Frequency dependence:
aqueous electrolyte solutions, dielectric response:
ionic conductivity, 274–279
ionic dispersion, 288–292
dielectric/FIR spectra, liquid water, 81
hat-curved-harmonic oscillator (HC-HO) model, 231–232
hat-curved-structural dynamical (HC-SD) model, wideband FIR absorption, 313–317
nonrigid oscillator composite models, 260–270
correlation factor, 268
Debye-like low frequency, 263
Gross collision model, 263
isothermal induced distribution, 260–263
Lorentz line integrated absorption, 264–265
resonance absorption, 265–266
static susceptibility, 263–264
Friction coefficient, dielectric/far-infrared spectra, liquid water, average perturbation (AP) theorem, 92

Gaussian curves:
equilibrium phase diagrams, finite-size effects, 48
specific interactions in liquid water, harmonically changing time-varying derivation, 219–221
Generalized coordinates, equilibrium phase diagrams, 29–38
Generic routes, equilibrium phase diagrams, path-based techniques, 13–15
Geometric parameters:
rectangular potential well protomodel, conical cavity dipole, 106–109
temperature dependence, wideband loss spectra, 317–318

Gibbs-Duhem integration (GDI), equilibrium phase diagrams:
  NPT-TP method, 41
  phase boundary tracking, 45–46
Gibbs Ensemble Monte Carlo (GEMC), equilibrium phase diagrams, 39–40
Global update, equilibrium phase diagrams, parallel sampling, 16
Gordon rules, absorption bandwidth, 105
Grand canonical ensemble, equilibrium phase diagrams, polydispersity, 49–50
Gross collision model:
  dielectric/far-infrared spectra, liquid water, spectral function (SF), complex susceptibility, 95–96
  frequency dependences, 263
    dielectric losses, 265–268
  hat-curved models, polar fluids, linear response theory, 160

Hard spheres, equilibrium phase diagrams:
  hcp-fcc lattices, 34–37
  liquid phases, 36–37
  numerical integration to reference macrostates (NIRM), 20–23
Harmonic change, spectral function (SF) derivation, 218–221
Harmonic oscillator (HO) model:
  liquid water, 221
  parabolic potential, nonrigid oscillator composite models, equation of motion, 250–251
  relaxation processes, 80–81
Harmonic radiation field, current density representation, 86–88
Hat-curved-cosine-squared potential composite model, liquid water, 241–250
  heavy water/liquid water spectral calculations, 241–246
  heuristic *vs.* nonheuristic methods, 249–250
  potential profile evolution, 246–248
Hat-curved-elastic bond (HC-EB) model, structure and properties, 322–327
Hat-curved-harmonic oscillator (HC-HO) model:
  dielectric/FIR spectra, liquid water, 80–82
  liquid water, 221–241
    complex permittivity components, 223–225
    dielectric relaxation mechanisms, 221–223
    free and statistical parameters, 228–232
    heavy water temperature-dependent spectral calculations, 232–235
    librational band narrowing, 236–237
    molecular motion, 239–240
    potential well profile, 237–239
    reorientation, 225–226
    vibration process, 226–228
    wideband spectra, temperature dependence, 235–236
  wideband dielectric spectra, 321–322
Hat-curved models:
  dielectric/FIR spectra of water, 79–80
  ice, 82
    FIR-dielectric response, 318–320
  polar fluids, 153–199
    librators, Fourier amplitudes, 192–193
    linear response theory, 158–160
    molecular subensembles, 160–173
    optical constants, 194–200
    precessors spectral function, integral transformation, 193–194
    spectral calculations, 174–181
      liquid water dielectric/FIR spectra, 174–177
      nonassociated liquids, dielectric/FIR spectra, 177
    spectral function derivation, 182–192
      analytical representation, 185
      Boltzmann distribution, norm *C*, 182–183
      librator dipoles, 186–188
      planar libration-regular precession (PL-RP) approximation, 185–186
      precessor ensembles, 188–190
      rotator dipoles, 190–191
      statistical averages, 191–192
      subensemble phase regions, 183–184
    statistical averages, 191–192
    unspecific interactions, 154–158
  specific interactions in liquid water:
    spectral calculations, 210–214
    spectral function modification, 207–210
  structural-dynamical model elastic interaction, 302–305
Hat-curved-structural dynamical (HC-SD) model:
  dielectric/FIR spectra, liquid water, 82–83
  evolution, 321–322

wideband FIR absorption frequency dependence, 313–317
Hat-flat (HF) models, dielectric/far-infrared spectra, liquid water, 120–153
hybrid model, 135–137
librator spectral function, 125–128
polar fluid applications, 140–153
complex permittivity, empirical formula, 144
Debye relaxation, 144–148
intermolecular cavity size, 148–150
spectral characteristics, 140–143
strongly-absorbing nonassociated liquid, 150–153
qualitative spectral dependencies, 137–140
rotator spectral function, 128–130
statistical parameters, 130–135
absorption-peak frequency estimation, 133–135
mean localization, 132–133
subensemble proportions, 130–132
trajectory classification, 120–125
Hat-flat potential, defined, 78–79
Hcp lattices, equilibrium phase diagrams:
hard spheres, 35–36
Lennard-Jones fluid, 34–35
Heavy water (D_2O):
double Debye approximation for permittivity in, 198–199
hat-curved-cosine-squared potential composite model, 241–246
hat-curved-harmonic oscillator (HC-HO) model:
statistical parameters, 232
temperature dependence, 232–235
hat-flat models:
failure of, 147–148
spectral characteristics, 142–143
spectral function calculations, 214
unspecific interactions, 216
Helmholtz inequality, equilibrium phase diagrams, Fast Growth (FG) methods, 42–43
Histogram reweighting, equilibrium phase diagrams, 57–59
Hopping length, hat-flat models, intermolecular cavity size, 149–150
HR techniques, equilibrium phase diagrams, phase boundary tracking, 46
Hybrid models:
aqueous electrolyte dielectric response, 271–274
hat-flat (HF) models:
dielectric/far-infrared spectra, liquid water, 135–137
free rotators, 156
qualitative spectral dependencies, 137–140
hat-flat potentials, 78
historical background, 84
Hydration sheath, dielectric response, 81–82
Hydrogen-bonded molecules:
aqueous electrolyte solutions, dielectric response, 290–292
dielectric FIR spectra of water, 73
"fluctuation concept," 317–318
hat-curved-harmonic oscillator (HC-HO) model:
complex permittivity, 224–225
dielectric relaxation, 222–223
molecular model evolution, 321–322
specific interactions in water, 203–206
stretching/bending vibrations, 84
structural-dynamical (SD) model, elastic interaction, 293–296
calculation protocol, 305–307
equation of motion, 296–298
FIR isotopic effect, 308–310
hydrogen-bond bending, 307–308
mean frequencies/amplitudes and density distributions, 300–301
numerical estimations, 302–305
transverse vibration dynamics, 298–300
Hyper parallel tempering (HPT), equilibrium phase diagrams:
Extended Sampling Interface Traverse (ESIT) *vs.*, 27–29
parallel sampling, 23–26

Ice, dielectric-FIR spectra, 318–320
Inaccuracy parameters, rectangular potential well protomodel, planar libration-regular precession (PL-RP) approximation, 117
Integral absorption, collision model frequency dependences, 264–265
Integral transformation:
hat-curved models, precessor spectral function, 193–194
nonrigid oscillator composite models, frequency dependence, 264–265

Intermolecular cavity:
hat-curved-cosine-squared potential composite model, heavy water (D_2O)/ liquid water spectral calculations, 242–246
hat-flat models, size measurements, 148–150
Intermolecular interactions:
dielectric response, liquid water, 81–82
string model, 82–83
Intermolecular potential, defined, 73
Internal-field correction:
hat-curved models, polar fluids, linear response theory, 160
hat-flat models, polar fluids, 141–143
Interpolation approximation, hat-flat models, rotator spectral function, 128–130
Ionic complex permittivity, dielectric response, liquid water, 81
Ionic conductivity, aqueous electrolyte solutions, dielectric response, 270–274
frequency dependence, 81, 274–279
linear motion, 273–274
spherical motion, 274
susceptibility parameters, 292
wideband spectra, 280–287
Ionic dispersion, aqueous electrolyte solutions, dielectric response, 287–292
Ion rattling phenomenon, aqueous electrolyte solutions, dielectric response, 290–292
Isothermal collisional model:
frequency dependences, 260–263
specific interactions in liquid water, 204–206
Isotopic effect, structural-dynamics (SD) models, FIR spectra, 308–310
Isotropic medium:
dielectric/far-infrared spectra, liquid water, local axisymmetric potential, dipoar reorientation, 97–102
dielectric response, 75

Kinetics, phase transformations, 3–4
Kirkwood correlation factor:
collision model, 75, 77
dielectric/far-infrared spectra, liquid water, spectral function (SF), complex susceptibility, 94–96
dielectric/FIR spectra of water, 75–76
hat-curved harmonic oscillator (HC-HO) composite model, 225–226
dissipation mechanism, 72
hat-flat models:
nonassociated fluids, 151–153
polar fluids, 141–143
spectral function modification in liquid water, 208–210
wideband spectra, ionic effect, 280–284
Kubo-like dipolar correlator, spectral function, 75
historical background, 84
ionic conductivity, 275

Lattice Switch Monte Carlo, equilibrium phase diagrams, path-based transition to wormholes, 29–38
Law of motion, dielectric/far-infrared spectra, liquid water:
current density representation, 86–87
dipole protomodel, conical cavity, 106–109
Lennard-Jones fluid, equilibrium phase diagrams:
parallel sampling, critical point tempering, 24–26
path-based transition to wormholes, 34–35
Librational band:
hat-curved-cosine-squared potential composite model, 244–248
hat-curved-elastic bond (HC-EB) model, 324–327
hat-curved-harmonic oscillator (HC-HO) model, narrowing, 236–237
structural-dynamical (SD) model, numerical calculations, 304–305
Librational phase volume, hat-flat (HF) models, mean number of reflections, absorption-peak frequency estimation, 133–135
Librational spectral function:
hat-curved models:
liquid water dielectric/FIR spectra, 176–177
polar fluids, phase volume, 169–173
rectangular potential well protomodel, 113
specific interactions in liquid water, 211–214
Librator dipoles:
hat-curved-harmonic oscillator (HC-HO) model, 229–232
hat-curved models:
Fourier amplitudes, 192–193
"partial" spectral functions, 166–168

spectral function:
polar fluids, 186–188
precessor boundaries, 183–184
representation, 185–188
unspecific interactions, 155–158
hat-flat (HF) models:
intermolecular cavity, 148–150
spectral function, 125–128
subensemble proportions, 130–132
trajectory classification, 121–125
periodic librations, structural-dynamical model elastic interaction, 297–298
specific interactions in liquid water, 204–206
Linear ionic motion, aqueous electrolyte solutions, dielectric response, 273–274
Linear response theory:
hat-curved-cosine-squared potential composite model, heavy water (D_2O)/ liquid water calculations, 241–248
hat-curved models, polar fluids, 158–160
parabolic potential, nonrigid oscillator composite models, 250–270
back-and-forth motion of charged particles, 251–260
equation of motion, 250–251
frequency dependences, 260–270
Linear transformations, equilibrium phase diagrams, path-based transition to wormholes, 30–38
Liquid-phase boundary, equilibrium phase diagrams, path-based techniques, 14–15
Liquid-state theory, dielectric/FIR spectra for water, 73, 75
Liquid-vapor boundary, equilibrium phase diagrams, parallel sampling, critical point tempering, 23–26
Liquid water:
composite models, 221–270
hat-curved-cosine-squared potential composite model, 241–250
hat-curved-harmonic oscillator (HC-HO) model, 221–241
complex permittivity components, 223–225
dielectric relaxation mechanisms, 221–223
free and statistical parameters, 228–232
heavy water temperature-dependent spectral calculations, 232–235
librational band narrowing, 236–237
molecular motion, 239–240
potential well profile, 237–239
reorientation, 225–226
vibration process, 226–228
wideband spectra, temperature dependence, 235–236
nonrigid oscillator, 250–270
back-and-forth motion of charged particles, 251–260
equation of motion, 250–251
frequency dependences, 260–270
specific interactions, 217
hat-curved-cosine-squared potential composite model, 241–250
heavy water/liquid water spectral calculations, 241–246
heuristic *vs.* nonheuristic methods, 249–250
potential profile evolution, 246–248
optical constants, 194–199
complex permittivity, double Debye approximation, 198–199
refractive index, 199–200
temperature dependence, 194–197
specific interactions, 75, 199, 201–221
composite model, relaxation times, 217
hat-curve model, 203–206
modified spectral function, 206–210
decaying terms, 220–221
harmonic changes, 218–220
nonrigidity implications, 215–216
spectral calculations, 210–214
submillimeter wavelength range (SWR), 199, 201–206
unspecific interactions, 216
Local axisymmetric potential, dielectric/far-infrared spectra, liquid water, dipolar reorienting spectral function, 96–102
ensemble averaging, 100
Fourier amplitudes, 100–102
Localization tecniques, molecular model correlation, dipolar ensemble, 102–103
Longitudinal dimensionless projections, structural-dynamical (SD) model, restricted rotator spectral function, 311–313
Lorentz representation:
hat-curved-elastic bond (HC-EB) model, 322–327

Lorentz representation (*Continued*)
hat-curved-harmonic oscillator (HC-HO) model:
dielectric/FIR spectra, liquid water, 80
vibrational susceptibility, 227–228
wideband spectra, 235–236
molecular model dielectric spectra correlation:
absorption bandwidthh, 105
absorption-peak frequency estimation, 103–104
nonrigid oscillator composite models:
frequency dependences:
isothermal models, 262–263
resonance-absorption, 265–266
spectral function analytic formula, 259–260
specific interactions in liquid water, harmonically changing time-varying derivation, 220
spectral function statistical parameters, 116–118
Low-frequency loss spectrum, hat-flat models, qualitative spectral dependencies, 138–140

Macrostate labels:
equilibrium phase diagrams, 7–8
extended sampling (ES), 16–18
Extended Sampling Interface Traverse (ESIT), 26–29
extended sampling (ES) distributions:
transition statistics, 56–57
visit statistics, 54–57
Macrovariables, extended sampling (ES) distributions, 52–57
Markov chain, equilibrium phase diagrams, Monte Carlo techniques, 8–13
Maximal probability, equilibrium phase diagram, 7–8
Maxwell-Boltzmann distribution:
dielectric/FIR spectra of water, 75
spectral function, complex susceptibility, 88–89
hat-curved models, polar fluids, "partial" spectral functions, 167–168
nonrigid oscillator composite models, perturbed trajectories, complex susceptibility, 253
steady-state trajectory, 75
Maxwell equations:
aqueous electrolyte solutions, dielectric response, 271–274
ionic conductivity susceptibility, 292
current density, 75
dielectric/far-infrared spectra, liquid water:
complex refraction index and current density, 85–86
dipole autocorrelator, 93
hat-curved models, polar fluids, linear response theory, 158–160
nonrigid oscillator composite models, perturbed trajectories, complex susceptibility, 251–253
spectral function (SF), 75
M-distribution, extended sampling (ES) distributions, 52–57
Mean frequencies/amplitudes, structural-dynamical model elastic interaction, 300–301
Mean localization, hat-flat (HF) models, 132–133
Metropolis acceptance function:
equilibrium phase diagrams, Monte Carlo techniques, 10–13
single-ensemble averages, 32–38
Metropolis Coupled Chain, equilibrium phase diagrams, parallel sampling, 16
Minimal free energy, equilibrium phase diagram, 7–8
Mirror symmetry, conical cavity, trajectory classification, 121–125
Molecular constants, hat-curved-structural dynamical (HC-SD) model, wideband FIR absorption, 315–317
Molecular dynamics (MD), phase transformations, 3–4
Molecular model:
dielectric/far-infrared spectra, autocorrelation, 102–106
absorption bandwidth, 104–105
absorption-peak frequency estimation, 103–104
dipolar ensemble localization, 102–103
relaxation spectrum, 105–106
hat-curved-harmonic oscillator (HC-HO) model:
liquid water, 239–240
potential well profile, 238–239
polar fluids, spectral characteristics, 143

Monte Carlo techniques, equilibrium phase diagrams, 4
basic tools, 8–13
Multicanonical ensemble, equilibrium phase diagrams, 18
Extended Sampling Interface Traverse (ESIT), 27–29
Extended Sampling Phase Switch, 37–38
Multicanonical weight function, extended sampling (ES) distributions, 53–57
Multihistogram methods, equilibrium phase diagrams, 58–59

Natural parameterizations, extended sampling (ES) distributions, 53–57
Near-critical systems, equilibrium phase diagrams, finite-size effects, 47–48
Nonassociated fluids:
hat-curved models, dielectric/FIR spectra, 177
hat-flat models:
spectral characteristics, 143
statistical parameters, 150–153
unspecific interactions, 216
Noncritical systems, equilibrium phase diagrams, finite-size effects, 46–47
Nonheuristic models:
hat-curved-cosine-squared potential composite model, 249–250
structural-dynamics model, 321
Nonperiodic function, current density representation, 86–88
Nonrigid dipole:
basic properties, 79–80
hat-curved-cosine-squared potential composite model, heavy water (D_2O)/ liquid water, 241–246
hat-curved-harmonic oscillator (HC-HO) model, complex permittivity, 224–225
hat-curved models, rectangular/parabolic well symbiosis, 158
spectral function modification in liquid water, 206–210
Nonrigid oscillator composite models, parabolic potential, linear response theory, 250–270
charged particle motion, 251–260
Boltzmann collision model, 256–260
complex susceptibility:
trajectory equilibrium, 253–256
trajectory perturbation, 251–253
equation of motion, 250–251
frequency dependences, 260–270
correlation factor, 268
Debye-like low frequency, 263
Gross collision model, 263
isothermal induced distribution, 260–263
Lorentz line integrated absorption, 264–265
resonance absorption, 265–266
static susceptibility, 263–264
Normalization constant, equilibrium phase diagrams, extended sampling (ES), 16–18
NPT-Test Particle method, equilibrium phase diagrams, 40–41
Numerical estimations, structural-dynamical model elastic interaction, 302–305
Numerical integration to reference macrostates (NIRM), equilibrium phase diagrams:
Extended Sampling Interface Traverse (ESIT) *vs.*, 28–29
Extended Sampling Phase Switch *vs.*, 38
path-based techniques, 20–23
NVT ensemble, equilibrium phase diagrams, finite-size effects, 47

Ohm's law:
aqueous electrolyte solutions, dielectric response, ionic conductivity susceptibility, 292
nonrigid oscillator composite models, frequency dependences, 262–263
One-dimensional rotation, dielectric/far-infrared spectra, liquid water, local axisymmetric potential, dipoar reorientation, 99–102
One-particle distribution function, dielectric/FIR spectra, liquid water, complex susceptibility expression through spectral function, 95–96
Optical constants, liquid water, 194–199
complex permittivity, double Debye approximation, 198–199
refractive index, 199–200
temperature dependence, 194–197
Order parameter, equilibrium phase diagrams, 6–8
Extended Sampling Interface Traverse (ESIT), 26–29
path-based transition to wormholes, 31–38

Parabolic well models:
hat-curved-cosine-squared potential composite model, heavy water (D_2O)/ liquid water spectral calculations, 242–246
hat-curved models:
polar fluids:
"partial" spectral functions, 168
subensemble proportions, 160–166
rectangular potential well symbiosis, 157–158
nonrigid oscillator composite models, linear response theory, 250–270
back-and-forth motion of charged particles, 251–260
equation of motion, 250–251
frequency dependences, 260–270
Parallel sampling, equilibrium phase diagrams, 15–16
critical point tempering, 23–26
Partition function, equilibrium phase diagram, 6
Path-based techniques:
basic principles, 4–5
equilibrium phase diagrams, 13–18
extended sampling, 16–18, 26–29
generic routes, 13–15
meaning and specification, 13
numerical integration and reference states, 19–23
parallel tempering, 23–26
sampling strategies, 13–18
extended sampling, 16–18
parallel sampling, 15–16
serial sampling, 15
wormholes, phase switching, 29–38
Periodic boundary conditions:
dielectric/far-infrared spectra, liquid water, local axisymmetric potential, dipoar reorientation, 97–102
equilibrium phase diagrams, finite-size effects, 47
Periodic librations, structural-dynamical (SD) model, elastic interaction, 297–298
Permittivity. *See* Complex permittivity
hat-curved-structural dynamical (HC-SD) model, wideband FIR absorption, 315–317
Perturbed oscillation amplitude, nonrigid oscillator models, 268–270
Phase behavior:
computational strategies, 2–3
defined, 2
equilibrium phase boundaries, 4
equilibrium phase diagrams, path-wormhole transition, 29–38
kinetics, 3
path-based techniques, 4–5
Phase boundary, equilibrium phase diagrams:
extrapolation, 44–45
tracking, 45–46
Phase space configuration, rectangular potential well protomodel, 109–110
Planar libration-regular precession (PL-RP) approximation:
defined, 77
hat-curved-harmonic oscillator (HC-HO) model, 229–232
hat-curved models, polar fluids:
spectral function, librator subensembles, 185–186
spectral function, norm *C* derivation, 182–183
subensemble proportions, 165–166
hat-flat models:
librator spectral function, 127–128
qualitative spectral dependencies, 137–140
simple water molecules, 146–148
hybrid hat-flat (HF) models, 135–137
rectangular potential well protomodel, 112–115
absorption frequency dependence, 115
statistical parameters, 116–118
Plasma phenomena, aqueous electrolyte solutions, dielectric response, 290–292
Polar angle, hat-curved models, subensemble proportions, 161–166
Polar fluids:
hat-curved models, 153–199
calculated spectra, 174–181
librators, Fourier amplitudes, 192–193
linear response theory, 158–160
molecular subensembles, 160–173
optical constants, 194–200
precessors spectral function, integral transformation, 193–194
spectral function derivation, 182–192
unspecific interactions, 154–158
hat-flat models, 140–153

complex permittivity, empirical formula, 144
Debye relaxation, 144–148
intermolecular cavity size, 148–150
spectral characteristics, 140–143
strongly-absorbing nonassociated liquid, 150–153
unspecific interactions, 216
Polar gases, resonance-absorption theory, 269–270
Polar molecules, hat-curved-cosine-squared potential composite model, 249
Polydispersity, equilibrium phase diagrams, 49–50
Potential wells:
absorption-peak frequency, finite depth, 104
dipolar spatial motion, 77
finite depth, trajectory classification, 121
hat-curved-cosine-squared potential composite model:
linear-response theory, 246–248
ordinary/heavy water comparisons, 242–246
hat-curved-elastic bond (HC-EB) model, 324–327
hat-curved-harmonic oscillator (HC-HO) model, 237–239
molecular model evolution, 321–322
structural-dynamical model elastic interaction, 302–305
Precessors:
dipole reorienting spectral function, 99–102
hat-curved-harmonic oscillator (HC-HO) model, 229–232
hat-curved models:
polar fluids:
"partial" spectral functions, 166–168
subensemble proportions, 164–166
spectral function:
integral transformation, 193–194
representation, 188–190
spectral function, subensemble phase region boundaries, 184–185
hat-curved models, polar fluids, spectral function calculation, 183–184
rectangular potential well protomodel, spectral function, 113–114
spectral function, calculation, 113–114
Probability density function (PDF), equilibrium phase diagrams, 6–8
hyper parallel tempering (HPT), 25–26
Monte Carlo techniques, 9–13
path-based transition to wormholes, 32–38
Protomodels:
aqueous electrolyte dielectric response, 271–274
dielectric/far-infrared spectra, liquid water, rectangular potential well protomodel, 106–120
conical cavity dipole, 106–110
planar librations-regular precession (PL-RP) approximation, 112–115
rigorous spectral function, 110–112
series expression, spectral function, 118–120
statistical parameters, 115–118

Qualitative spectral dependencies, hat-flat (HF) models, dielectric/far-infrared spectra, liquid water, 137–140
Quantitative analysis, dielectric/FIR spectra of water, 79–80
Quasi-independent ensembles, defined, 72

Radiation fields:
dielectric-FIR spectra, liquid water, average perturbation theorem, 90–92
nonrigid oscillator composite models, perturbed trajectories, complex susceptibility, 252–253
R-band:
dielectric/FIR spectra of water, 73
polar fluids, 146–148
elastic interactions, 321
hat-curved-elastic bond (HC-EB) model, 325–327
hat-curved-harmonic oscillator (HC-HO) model:
free and statistical parameters, 231–232
molecular motion in water, 239–240
ordinary/heavy water comparisons, 232–235
hat-curved models, polar fluids, 80
structural-dynamical (SD) model, restricted rotator dielectric function, 313
Rectangular potential well protomodel:
dielectric/far-infrared spectra, liquid water, 77, 106–120

Rectangular potential well protomodel (*Continued*)
  conical cavity dipole, 106–110
  planar librations-regular precession (PL-RP) approximation, 112–115
  rigorous spectral function, 110–112
  series expression, spectral function, 118–120
  statistical parameters, 115–118
 hat-curved model, polar fluids, spectral function calculations, 180–181
 hat-curved models:
  parabolic-well model symbiosis, 157–158
  unspecific interactions, 154–158
 hat-curved models, polar fluids, "partial" spectral functions, 168
Reference macrostate, equilibrium phase diagrams:
 numerical integration, 19–23
 path-based techniques, 13–15
Reflection mean numbers, hat-flat (HF) models, 133–135
Refractive index:
 aqueous electrolyte solutions, dielectric response, ionic conductivity susceptibility, 292
 liquid water optical constants, 194–200
Regular precession (RP) approximation, hat-curved models, spectral function, precessors, 189–190
Relaxation spectrum:
 ice, FIR-dielectric response, 318–320
 molecular model dielectric spectra correlation, 105–106
Relaxation times. *See also* Debye relaxation
 dielectric/FIR spectra, liquid water, 80–81
 specific interactions in liquid water, composite model, two relaxation times, 217
 specific interactions in liquid water and, 205–206
Renormalization group, equilibrium phase diagrams, finite-size effects, 47–48
Reorientation mechanism:
 aqueous electrolyte solutions, dielectric response, wideband spectra, ionic conductivity, 281–287
 hat-curved-harmonic oscillator (HC-HO) model:
  dielectric relaxation, 223
  potential well profile, 238–239
  spectral function (SF) calculations, 225–226
  vibration lifetime and, 240–241
 wideband dielectric spectra, temperature dependence, 317–318
Resonance-absorption, nonrigid oscillator composite models:
 frequency dependences, 265–266
 perturbed oscillation amplitude, 268–270
Restriction-rotation period, structural-dynamical (SD) model:
 dielectric function, 313
 elastic interaction, bending force constant, 306–307
 spectral function, 311–313
Rigid dipoles:
 aqueous electrolyte solutions, dielectric response, 287–292
 dielectric/FIR spectra of water, spectral function, complex susceptibility, 89
 hat-curved-cosine-squared potential composite model, 247–248
 specific interactions in liquid water, composite model, two relaxation times, 217
Rotational absorption band, structural-dynamical model elastic interaction, 300–301
Rotational deflection, structural-dynamical (SD) model, elastic interaction, 296
Rotational frequency, dielectric/FIR spectra, liquid water, 82
Rotational mobility:
 dielectric/FIR spectra, liquid water, 82–83
 ice, FIR-dielectric response, 319–320
Rotators:
 hat-curved-harmonic oscillator (HC-HO) model, 229–232
 hat-curved models:
  cosine-squared potentials, 244–246
  polar fluids:
   spectral function, 190–191
   subensemble proportions, 165–166
  unspecific interactions, 155–158
 hat-flat (HF) models:
  spectral function, 128–130
  subensemble proportions, 130–132

trajectory classification, 121–125
specific interactions in liquid water, 204–206
structural-dynamical (SD) model:
restricted rotator dielectric function, 313
restricted rotator spectral function, 311–313

Sampling strategies, equilibrium phase diagrams:
extended sampling, 16–18
extended sampling (ES), 16–18
parallel sampling, 15–16
series sampling, 15
Scalar quantity, dielectric/FIR spectra of water, spectral function, complex susceptibility, 89
Second Debye term, hat-curved-harmonic oscillator (HC-HO) model, dielectric relaxation, 223
Self-consistent wells:
dielectric/FIR spectra, liquid water, 82
hat-curved-cosine-squared potential composite model, 249–250
hat-curved model, polar fluids, spectral function calculations, 179–182
structural-dynamical model elastic interaction, 302–305
Semiphenomenological molecular model:
hat-curved model, polar fluids, spectral function calculations, 179–182
specific interactions in liquid water, 206
Serial sampling, equilibrium phase diagrams, 15
Series representation:
hat-flat models:
librator spectral function, 125–126
qualitative spectral dependencies, 137–140
spectral function, rectangular potential well protomodel, 110–111
summation, 118–120
Simulated tempering, equilibrium phase diagrams, 18
Simulation parameters, equilibrium phase diagrams, numerical integration, 22–23
Single-ensemble averages, equilibrium phase diagrams, path-based transition to wormholes, 32–38
Single Occupancy Cell Method (SOCM), equilibrium phase diagrams, numerical integration, 22–23
Single-phase simulation, equilibrium phase diagrams, Monte Carlo techniques, 11–13
parallel tempering, 24–26
Small β approximation:
hat-flat models, rotator spectral function, 128–129
hybrid hat-flat (HF) models, 135–137
rectangular potential well protomodel spectral function, 114–115
Solid-phase boundary, equilibrium phase diagrams, path-based techniques, 15
Space angles, hat-flat models, rotator spectral function, 129–130
Spatial motion:
aqueous electrolyte solutions, dielectric response, frequency-dependent ionic conductivity, 277–279
dipole representation, 75
structural-dynamical (SD) model, restricted rotator spectral function, 311–313
Specific interactions:
dielectric/FIR spectra for water, 73, 75, 199, 201–221
composite model, relaxation times, 217
hat-curve model, 203–206
modified spectral function, 206–210
decaying terms, 220–221
harmonic changes, 218–220
nonrigidity implications, 215–216
spectral calculations, 210–214
submillimeter wavelength range (SWR), 199, 201–206
unspecific interactions, 216
hat-curved-harmonic oscillator (HC-HO) model, dielectric relaxation, 222–223
nonrigid dipole, 79–80, 80
Spectral calculations:
hat-curved-cosine-squared potential composite model, heavy water (D_2O)/ liquid water, 241–246
hat-curved models, polar fluids, 174–181
liquid water dielectric/FIR spectra, 174–177
nonassociated liquids, dielectric/FIR spectra, 177
nonrigid oscillator composite models, frequency dependences, 263–265
specific interactions in liquid water, 210–214

Spectral function (SF):
complex dielectric permittivity, 83
dielectric/far-infrared spectra, liquid water:
average perturbation (AP) theorem, 91–92
basic principles, 75, 84
complex susceptibility, 85–96
average perturbation (AP) theorem, 90–92
complex refraction index and current density, 85–86
dipole autocorrelator, 92–93
expressions, 94–96
statistical distributions, 88–89
t_0 theorem, current density representation, 86–88
local axisymmetric potential, dipolar reorientation, 96–102
ensemble averaging, 100
Fourier amplitudes, 100–102
rectangular potential well protomodel
absorption frequency dependence, 115
analytical representation, 111–112
librational spectral function, 113
precessional spectral function, 113–114
series representation, 110–112, 118–120
small β approximation, 114–115
hat-curved-harmonic oscillator (HC-HO) model, reorientation mechanism, 225–226
hat-curved models, polar fluids:
derivation, 182–192
analytical representation, 185
Boltzmann distribution, norm *C*, 182–183
librator dipoles, 186–188
precessor boundaries, 183–184
precessor ensembles, 188–190
rotator dipoles, 190–191
statistical averages, 191–192
general derivation, 185–186
librator spectral function, 186–188
linear response theory, 159–160
subensemble proportions, 160–173
absorption spectrum analysis, 169–173
"partial" spectral functions, 166–168
particle potential form and classification, 160–166
hat-flat models:
librators, 125–128
polar fluids, 140–143
rotator dipoles, 128–130
trajectory classifications, 124–125
Kubo-like dipolar correlator, 75
Maxwell equations, 75
nonrigid oscillator composite models, charged particle motion, 251–260
Boltzmann collision model, 256–260
complex susceptibility:
trajectory equilibrium, 253–256
trajectory perturbation, 251–253
nonrigid water dipoles, 206–210
specific interactions in liquid water, 217–221
harmonically changing time-varying derivation, 218–221
structural-dynamical (SD) model, restricted rotators, 311–313
Spherical angle:
hat-curved-harmonic oscillator (HC-HO) model, 228–232
hat-flat models, rotator spectral function, 129–130
Spherical motion, aqueous electrolyte solutions, dielectric response, 274
Static susceptibility:
aqueous electrolyte solutions, dielectric response, wideband spectra, ionic conductivity, 280–287
nonrigid oscillator composite models, frequency dependence, 263–265
Statistical mechanics, equilibrium phase diagrams, 5–8
Statistical parameters:
dielectric/far-infrared spectra, liquid water, spectral function, complex susceptibility, 88–89
hat-curved-harmonic oscillator (HC-HO) model, 228–232
hat-curved models, polar fluids, spectral function, 191–192
hat-flat (HF) models, 130–135
absorption-peak frequency estimation, 133–135
mean localization, 132–133
nonassociated fluids, 150–153
subensemble proportions, 130–132
rectangular potential well protomodel, 115–118
Steady-state distribution, spectral function (SF) derivation, 217–221
Steric-restriction effect, rectangular potential well protomodel, spectral function, 112

Stochastic procedures:
equilibrium phase diagrams, Monte Carlo techniques, 9–13
hat-curved-harmonic oscillator (HC-HO) model, ordinary/heavy water (D_2O) comparisons, 232–235
Stratified approximation, specific interactions in liquid water, 204–206
Stretching vibrations, hat-curved-harmonic oscillator (HC-HO) model:
complex permittivity, 224–225
dielectric relaxation, 222–223
Strong-collision frequencdy, dielectric/far-infrared spectra, liquid water, average perturbation (AP) theorem, 92
Structural-dynamical (SD) model:
autocorrelation function application, 311–317
composite hat-curved (HC)-SD model frequency dependence, 313–317
restricted rotators:
dielectric response, 313
spectral function, 311–313
dielectric/FIR spectra, liquid water, 82
elastic interaction, 293–296
calculation protocol, 305–307
equation of motion, 296–298
FIR isotopic effect, 308–310
hydrogen-bond bending, 307–308
mean frequencies/amplitudes and density distributions, 300–301
numerical estimations, 302–305
transverse vibration dynamics, 298–300
future research issues, 327
Subensemble properties:
hat-curved models, polar fluids, spectral function, 160–173
absorption spectrum analysis, 169–173
"partial" spectral functions, 166–168
particle potential form and classification, 160–166
precessor boundaries, 183–184
hat-flat (HF) models, 130–132
Submillimiter wavelength range (SWR):
hat-curved-cosine-squared potential composite model, 247–248
hat-curved-harmonic oscillator (HC-HO) model:
dielectric relaxation, 222–223
heavy water (D_2O), 235–236
molecular motion in water, 239–240
ordinary/heavy water (D_2O) comparisons, 232–235
hat-curved models, polar fluids, R-band absorption, 80
hat-curved models, polar fluids, spectral function calculations, 181–182
hat-flat models, complex permittivity, 148
specific interactions in water, 199, 201–206
Susceptibility component, dielectric/far-infrared spectra, liquid water, spectral function (SF), complex susceptibility, 93–96
Symmetry axis, rectangular potential well protomodel, conical cavity dipole, 107–109

Temperature dependence:
hat-curved-harmonic oscillator (HC-HO) model:
heavy water (D_2O), 232–235
liquid water, 235–236
liquid water:
optical constants, 194–197
wideband spectra, 82
wideband dieledtric spectra, 317–318
Temperature scaling, equilibrium phase diagrams, 18
Test particle insertion method, equilibrium phase diagrams, 41
Thermodynamics, equilibrium phase diagrams, 5–8
Time-varying calculations, spectral function (SF) derivation, harmonic change, 218–221
Total effective weight, equilibrium phase diagrams, 12–13
Trajectory classification:
hat-flat model, dielectric/far-infrared spectra, liquid water, 120–125
nonrigid oscillator composite models:
complex susceptibility, perturbed trajectories, 251–253
equilibrium trajectory, 253–256
Transition probabilities, extended sampling (ES) distributions, 56–57
Translational band:
hat-curved-harmonic oscillator (HC-HO) model, heavy water (D_2O), 232–235
ice, FIR-dielectric response, 319–320
nonrigid oscillator composite models, perturbed trajectories, complex susceptibility, 253

Translational band (*Continued*)
structural-dynamical model elastic interaction, 303–305
Transverse deflection, structural-dynamical (SD) model:
elastic interaction, 295–296
numerical calculations, 303–305
restricted rotator spectral function, 311–313
Transverse vibrations, structural-dynamical model elastic interaction, 298–300
mean displacement formula, 301
T_0 theorem, dielectric/far-infrared spectra, liquid water, current density representation, 86–87

Umbrella sampling, equilibrium phase diagrams, 18
Undamped motion, dielectric/far-infrared spectra, liquid water, 75
Undamped reorientation, dielectric/far-infrared spectra, liquid water, average perturbation (AP) theorem, 92
Unspecific interactions:
dielectric/FIR spectra for water, 73, 75
hat-curved-harmonic oscillator (HC-HO) model, dielectric relaxation, 222–223
hat-curved models, polar fluids, 154–158
field model evolution, 157
rectangular-like models, 154–156
rectangular-well/parabolic-well models, 157–158
in liquid water, 203
dielectric-FIR spectra, 216
molecular model evolution, 321–322

Vibrational susceptibility:
hat-curved-cosine-squared potential composite model, heavy water (D_2O)/ liquid water spectral calculations, 242–246
hat-curved-elastic bond (HC-EB) model, 324–327
hat-curved-harmonic oscillator (HC-HO) model, 226–228
heavy water (D_2O), 232–235
reorientation lifetime, 240–241
specific interactions in liquid water, 215–216

Wang-Landau strategy, extended sampling (ES) distributions, macrostate visit statistics, 53–57
Well models, dielectric/far-infrared spectra, liquid water:
basic properties, 77–78
hat-curved models:
basic properties, 79–80
dielectric/FIR spectra of water, 79–80
ice, 82
polar fluids, 153–199
calculated spectra, 174–181
librators, Fourier amplitudes, 192–193
linear response theory, 158–160
molecular subensembles, 160–173
optical constants, 194–200
precessors spectral function, integral transformation, 193–194
spectral function derivation, 182–192
unspecific interactions, 154–158
hat-flat models, 120–150
hybrid model, 135–137
librator spectral function, 125–128
polar fluid applications, 140–153
qualitative spectral dependencies, 137–140
rotator spectral function, 128–130
statistical parameters, 130–135
trajectory classification, 120–125
Wideband spectra:
aqueous electrolyte solutions, dielectric response, ionic conductivity, 280–287
hat-curved-harmonic oscillator (HC-HO) model:
free and statistical parameters, 230–232
liquid water, temperature dependence, 235–236
ordinary/heavy water (D_2O) comparisons, 232–234
hat-curved-structural dynamical (HC-SD) model, 313–317
molecular model evolution, 321–322
temperature dependence, 317–318
Wormholes, equilibrium phase diagrams, path-based transition to, 29–38

Yukawa potential, equilibrium phase diagrams, Fast Growth (FG) methods, 43

Zwanzig formula, equilibrium phase diagrams:
Fast Growth (FG) methods, 42–43
NPT-Test Particle method, 41
path-based transition to wormholes, 32–38